New Concepts in Polymer Science

Stationary and Non-Stationary Kinetics of the Photoinitiated Polymerization

VSP
an imprint of Brill Academic Publishers
P.O. Box 346
3700 AH Zeist
The Netherlands

Tel: +31 30 692 5790
Fax: +31 30 693 2081
vsppub@brill.nl
www.vsppub.com
www.brill.nl

First published in 2004

ISBN 90-6764-415-3

Library of Congress Cataloging-in-Publication Data

A CIP record for this book is available from the Library of Congress

Printed in The Netherlands by Ridderprint bv, Ridderkerk.

New Concepts in Polymer Science

Stationary and Non-Stationary Kinetics of the Photoinitiated Polymerization

Yu.G. Medvedevskikh, A.R. Kytsya,
L.I. Bazylyak, A.A. Turovsky and G.E. Zaikov

UTRECHT • BOSTON - 2004

New Concepts in Polymer Science

Previous titles in this book series:

Thermal Stability of Engineering
Heterochain Thermoresistant Polymers
E.V. Kalugina, K.Z. Gumargalieva and G.E. Zaikov

The Concept of Micellar-Sponge
Nanophases in Chemical Physics of Polymers
Yu.A. Mikheev, G.E. Zaikov

Structure of the Polymer Amorphous State
G.V. Kozlov and G.E. Zaikov

The Statistical Nature of Strength and Lifetime in Polymer Films and Fibers
B. Tsoi, E.M. Kartashov and V.V. Shevelev

Cyclolinear Organosilicon Copolymers:
Synthesis, Properties, Application
O.V. Mukbaniani and G.E. Zaikov

Modern Polymer Flame Retardancy
S.M. Lomakin and G.E. Zaikov

Polymeric Biomaterials.
Part I. Polymer Implants
M.I. Shtilman

Biodegradation and Durability of
Materials under the Effect of Microorganisms
S.A. Semenov, K.Z Gumargalieva and G.E. Zaikov

Fire Resistant and Thermally Stable Materials
Derived from Chlorinated Polyethylene
A.A. Donskoi, M.A. Shashkina and G.E. Zaikov

Polymers and Polymeric Materials
for Fiber and Gradient Optics
N. Lekishvili, L. Nadareishvili, G. Zaikov and L. Khananashvili

Polymers Derived from Isobutylene.
Synthesis, Properties, Application
Yu.A. Sangalov, K.S. Minsker and G.E. Zaikov

Ecological Aspects of Polymer
Flame Retardancy
S.M. Lomakin and G.E. Zaikov

Molecular Dynamics of Additives in Polymers
A.L. Kovarski

Structure and Properties of
Conducting Polymer Composites
V.E. Gul'

Interaction of Polymers with
Bioactive and Corrosive Media
A.L. Iordanskii, T.E. Rudakova and G.E. Zaikov

Immobilization on Polymers
M.I. Shtilman

Radiation Chemistry of Polymers
V.S. Ivanov

Polymeric Composites
R.B. Seymour

Reactive Oligomers
S.G. Entelis, V.V. Evreinov and A.I. Kuzaev

Diffusion of Electrolytes in Polymers
G.E. Zaikov, A.L. Iordanskii and V.S. Markin

Chemical Physics of Polymer
Degradation and Stabilization
N.M Emanuel and A.L. Buchachenko

Announcement

The focus of this monograph (containing 8 chapters, 25 tables and 115 figures) is on kinetic models of the photoinitiated polymerization of mono- and bifunctional monomers up to the high conversion state, their derivation, analysis and comparison with experimental data. Indirectly, some general problems of chemical kinetics with regard to polymerization are also discussed in this monograph. An algorithm for construction of kinetic models for polymerization pathways, which can be applied to linear catalytic and unbranched radical-chain processes of polymerization in general, including the ones involving free radicals, is postulated.

CONTENTS

ABBREVIATIONS AND SYMBOLS

a, *b*, *c*	parameters of linear polymerization kinetics equations
BDMA	benzyl dimethyl amine
BDMC	1,4-butane diol dimethacrylate*

```
      CH3                   CH3
       |                     |
CH2=C–C–O–(CH2)4–O–C–C=CH2
          ||              ||
          O               O
```

bis-GMA	2,2-bis[4-(2-hydroxy-3-methacryloyloxypropoxy)phenyl]propane
c_0	starting concentration of the photoinitiator
Darocur 1173	2,2-dimethyl-2-hydroxyacetophenone (photoinitiator)
D_c	segmental diffusion coefficient
DCR	conception of the diffusible-controlled reactions
DDS	4,4′-diamino diphenyl sulfone
DEGDA	diethylenglycoldimethacrylate
d_f	fractal dimensionality of polymeric chains
d_L	fractal dimensionality of traps
d_t	dimensionality of fractal set of characteristic times
DMEG	ethylene glycol dimethacrylate*

```
      CH3                   CH3
       |                     |
CH2=C–C–O–(CH2)2–O–C–C=CH2
          ||              ||
          O               O
```

D_n	translational diffusion coefficient
DOPh	dioctyl phtalate
D_p	coefficient of reactive diffusion
d_s	fractal dimensionality of the solid polymeric phase
DSC	differential scanning calorimetry
d_v	fractal dimensionality of the liquid monomeric phase
E_0	UV-illumination intensity

$E_A{}^t$ activation energy of monomolecular chain termination

$E_A{}^\eta$ activation energy of viscous flow

f initiation efficiency

F free energy

F_N explosive force

F_v, F_s fractal characteristics of volume and surface of polymeric clusters

G a graph

GMC glycidyl methacrylate*

$CH_2{=}C(CH_3){-}C(O){-}O{-}CH_2{-}CH{-}CH_2$
```
                                  \ /
                                   O
```

$H(x_1, x_2, \ldots, x_N)$ Hamiltonian function

HDDA hexane-1,6-diol diacrylate

HEMA–2-hydroxyethyl methacrylate

i-BMC iso-butyl methacrylate*

```
CH2=C–C(O)–O–CH2–CH–CH3
    |            |
   CH3          CH3
```

IRGACURE 651 photoinitiator 2,2-dimethoxy-1,2-diphenyl ethane-1-on (photoinitiator)

I_0 light intensity falling on the surface of a photocomposition

J_0 light intensity of the UV-illumination

k rate constant

k_B Boltzmann constant

k_C chain termination constant rate limited by the segmental diffusion

k_p chain propagation rate constant

k_{pv} chain propagation rate constant in the liquid monomeric phase

k_R chain termination rate constant limited by the translational diffusion

k_t chain termination rate constant

$k_t{}^b$ bimolecular chain termination rate constant

$k_t{}^m$ monomolecular chain termination rate constant

k_{tv} bimolecular chain termination rate constant in the liquid monomeric phase

ℓ thickness of the photocomposition layer

L trap

M monomer molecule

[M] concentration of the functional groups

$[M_0]$ concentration of the monomer in bulk

$[M_s^0]$ concentration of the monomer in the polymer-monomeric phase

$[M_v^0]$ concentration of the monomer in the monomer-polymeric phase

MDF 2-α,ω-dimethacryloyl(oligodiethyleneoxideterephtalate)*

$$CH_2{=}C(CH_3){-}C(O){-}[O{-}(CH_2CH_2O)_2{-}C(O){-}C_6H_4{-}C(O)]_n{-}(OCH_2CH_2)_2{-}O{-}C(O){-}C(CH_3){=}CH_2$$

MGPh-9 α,ω-dimethacryloyl(tridiethyleneoxideterephtalate)*

$$CH_2{=}C(CH_3){-}C(O){-}O{-}(CH_2CH_2O)_3{-}C(O){-}C_6H_4{-}C(O){-}(OCH_2CH_2)_3{-}O{-}C(O){-}C(CH_3){=}CH_2$$

MMRP microheterogenous model of radical polymerization

MNA methyl nadic anhydride

MPPh monomer-polymeric phase

OCM-2 4,7,9,12,15,17,20-heptaoxa-2,22-dimethyl-1, 22-tricosadien-3,8,16,21-tetraon*

$$CH_2{=}C(CH_3){-}C(O){-}O{-}CH_2{-}CH_2{-}O{-}C(O){-}O{-}(CH_2{-}CH_2O)_2{-}C(O){-}O{-}CH_2{-}CH_2{-}O{-}C(O){-}C(CH_3){=}CH_2$$

OEDA 2,2′-oxybisethanol (diethylene glycol) diacrylate

OEDM 2,2′-oxybisethanol (diethylene glycol) dimethacrylate

OP oligoperoxide

p differentiated double-bond conversion

P — integrated double-bond conversion

P_0 — double-bond conversion starting in a dark period

PB — polybutadiene

P_s^0 — double-bond conversion at the moment of a polymer-monomeric phase separation

PS — polystyrene

P_v^0 — conversion of the monomer-polymeric solution saturated by a polymer

PEG — polyethylene glycol

PMPh — polymer-monomeric phase

R — cluster radius

R — active radical

R_z — “frozen” radical

[R] — concentration of active radicals

SBS — styrene-butadiene-styrene copolymer matrices

T — polymerization time

t_d, t_p — times characteristic for the move of the active center of the macroradical

T_g — glass transition temperature

TEGDM — triethylene glycol dimethacrylate

TEDA — 2,2'-thiobisethanol diacrylate

TEDM — 2,2'-thiobisethanol dimethacrylate

TGDDM — tetraglycidyl 4,4′-diamino diphenyl methane

TGM-3 — triethyleneglycol dimethacrylate*

```
        CH3                           CH3
        |                             |
CH2=C–C–O–(CH2–CH2–O)3–C–C=CH2
          ||                      ||
          O                       O
```

U — a route

V_0 — system volume

v_{im} — initiation rate

VMA — vinylmethacrylate

W — observed polymerization rate

w_m	specific rate of polymerization in the interface layer
w_p	propagation rate
W_0	postpolymerization starting rate
Z	active center of a catalyst
β	parameter determining the characteristic lifetime of a radical
$\overline{\delta}$	relative deformation
ε	molar extinction coefficient
ϕ	fraction of the non-glass share of PMPh
Γ	depth of the polymerization
Γ_0	final degree of double-bond conversion
γ	exponent in the stretched exponential law
γ_i	quantum yield for initiation
φ_m	volumetric part of the interface layer (3-D polymerization)
φ_s	volumetric part of the solid polymeric or polymer-monomeric phase
φ_v	volumetric part of the monomer-polymeric phase
φ_{vs}	volumetric part of the interface layer (linear polymerization)
λ	length of free run of particles
μ_j	chemical potential
v_f	Flory index
σ	shift strength
τ_p	time characteristic for chain propagation
τ_t	characteristic lifetime of the macroradicals
ξ	exponent in scaling dependence
$\psi(t)$	relaxation function

---

* The kinetics of the polymerization of the presented di- and monomethacrylates have been studied by the authors personally.

PREFACE

The main focus in this monograph is on models of the kinetics of photoinitiated polymerization of mono- and bifunctional monomers up to the high conversion state, their derivation, analysis and comparison with the experimental data.

Polymerization with up to the high conversion state is considered as a chemical process in a system changing its own phase state. Thus, the main accent is on the concentration at which the macro-phase separation of a polymerizing system forms new reactive zones; the contribution of every one of them to the total kinetics depends on the kinetic characteristics of the process in that particular zone and its volumetric part.

The kinetic models that are postulated include both bimolecular and monomolecular chain termination. It has been experimentally determined that the relaxation function of the monomolecular chain termination is follows the stretched exponential law; the rate constant has an activating nature and its scale depends on the concentration of the monomer in bulk. These characteristics are reflected in the proposed derivation of the stretched exponential law.

Furthermore, some general problems of chemical kinetics are also discussed. In particular, the dynamic characteristic number of the components and intermediate substances, elementary and final reactions that are needed for describing the dynamics of a system composition have been determined for every state of the evolution of a non-equilibrium chemical system.

An algorithm for the construction of kinetic models for routes, such as simple cycles of a graph assigned to a set of an elementary reaction and intermediate substances, is postulated. Such an algorithm can be applied to the linear catalytic and unbranched radical-chain processes of polymerization in general, including the free radical ones.

This monograph can be useful for scientists, engineers, post-graduate training students and students who are interested in problems both of the polymerization kinetics and general chemical kinetics.

YU. G. MEDVEDEVSKIKH

INTRODUCTION

Polymerization is an important method for obtaining multifunctional polymeric compositive materials and is the subject of interesting scientific investigations, first of all because of a wide range of kinetic characteristics. That is why the main focus of this monograph is on kinetic models of the photoinitiated polymerization of mono- and bifunctional monomers up to the high conversion state, the derivation of these models, their analysis and comparison with the experimental data (Chapters 5–8) obtained over the last 10 years.

Three states are highlighted in this monograph. Firstly, an analysis of the kinetics of the photoinitiated polymerization in the thin layers is given, taking into account the illumination gradient and, subsequently, the gradients of the photoinitiator concentration and polymerization depth on the layer of the polymerizing composition. The dynamics of the illumination intensity and photoinitiator concentration change on the coordinates of a layer from the irradiated surface is described. The influence of these dynamics on the kinetics of a layer polymerization is shown.

Secondly, polymerization with a high conversion rate is considered as a chemical process in a system changing by its own phase state. Therefore, assuming and taking into account the diffusible-controlled reactions conception, the main attention has been given to the microheterogeneous model, according to which the micro-phase separation of the polymerizing system in a defined state of its evolution creates a new reactive zone. Polymerizations in these zones are characterized by their own kinetic characteristics. The contribution of the reactive zones to the total kinetics of a process is determined by their volumetric part, changing with time, and by kinetic characteristics of the polymerization in these zones.

Thirdly, together with the bimolecular chain termination a monomolecular chain termination is also included in the kinetic models. The latter is controlled by the rate of chain propagation and represents the active center of a radical self-burial act. It has been determined, based on postpolymerization kinetics data (see Chapter 7), that the relaxation function of monomolecular chain termination follows the stretched exponential law and the rate constant has an activating nature with an activation energy typical for a constant of chain propagation and its scale depends on the molar-volumetric concentration of the monomer in bulk. All these characteristics are reflected in the proposed derivation of a stretched exponential law based on the statistics of self-avoiding random walk, a property of a fraction of the propagating polymeric chain, and on the basis of set of traps blocking the active center of a radical (see Chapter 8).

Some general problems of chemical kinetics are also discussed. They are not often discussed in monographs, maybe because their understanding has developed slower than the understanding of concrete chemical processes and tends to pass from the concrete to the abstract. However, clear formulation of the starting bases or principles of chemical kinetics accumulating all achievements in understanding the concrete chemical processes helps to contribute to the knowledge of the elements, bearing in mind the characteristics of concrete processes.

That is why, taking into account the hierarchy of the time scaling for partial processes and according to the principle of the short description of the non-equilibrium systems, the separate states of the evolution of a chemical system from the initial state to a final equilibrium have been analyzed (see Chapter 1). For the first time this analysis has been done on the basis of difference between the terms "stoichiometrically" and "dynamically", independent numbers of component and intermediate substances, an elementary and final reactions.

Another general problem, the development of an algorithm for the construction of kinetic models for the quasi-stationary state of the evolution of non-equilibrium chemical system, is solved by the method of linear routes as simple cycles of a graph assigned to sets of elementary reactions and intermediate substances (see Chapter 2). A general algorithm for construction of kinetic models for the linear catalytic and un-branched radical-chain processes, including a free radical polymerization, is proposed.

The modest size of this monograph did not allow us to consider the experimental reference material and to reflect on the different points of view and proposed approaches more elaborately (see Chapters 3 and 4). This might deprive the monograph of some needed elements of the discussing character. However, the available space is needed to present and discussion the problems described.

Therefore, the authors will be grateful for all remarks (including the critical ones) on the content of this monograph.

Chapter 1. Methodological principles of chemical kinetics

1.1. Introduction

A theory of kinetics is first of all a theory of the intermediate substances [**1**, **2**]. From this point of view any more or less complicated chemical reaction should proceed *via* the formation and decomposition of intermediate substances state-by-state, that is, *via* the total combination of elementary reactions. That is why a theory of an intermediate substances divides all reacting substances of a system into a component set $A = \{A_i\}$, $i = \overline{1,\ m}$ and a set of an intermediate substances $X = \{X_i\}$, $i = \overline{1,\ n}$, and all chemical transformations are divided into a final set $Q = \{Q_f\}$, $f = \overline{1,\ q}$ and a set $S = \{S_i\}$, $i = \overline{1,\ s}$ of an elementary chemical reaction representing the final one. In general the term "elementary reaction" (or state) means a chemical transformation between the components of a system and an intermediate substance proceeding *via* one act, that, is *via* the formation and the decomposition of one activated complex, assuming the kinetics follow the direct form of the mass action law. The term "final chemical reaction" means a reaction between the components of a system realized *via* a total combination of elementary reactions.

Assuming the stoichiometric coefficients of a component $\alpha_{ki}^{(1)}$ and intermediate substances $\alpha_{ki}^{(1)}$ and $\alpha_{kj}^{(2)} > 0$ for the products of elementary reaction k S_k, and $\alpha_{ki}^{(1)}$, and $\alpha_{kj}^{(2)} < 0$ for the starting substances, a stoichiometric equation of an elementary reaction can be written in algebraic form:

$$S_k \ : \ \sum_i \alpha_{ki}^{(1)} A_i + \sum_j \alpha_{kj}^{(2)} X_j = 0 \qquad (1.1)$$

Analogously, the stoichiometric note of the final reaction Q_f can be written as

$$Q_f \ : \ \sum_i \nu_{fi} A_i = 0 \qquad (1.2)$$

where again $\nu_{fi} > 0$ for the products and $\nu_{fi} < 0$ for the starting substances of the final reaction presented.

One of the main tasks of a formal kinetics is a description of the dynamics of chemical system composition *via* its transition from the initial indignant state into the final equilibrium one. Two factors have an influence on the dynamics of chemical system composition. Firstly, stoichiometric bonds caused by the conservation laws at chemical transformations, that is, the proportions of formation and consumption of component and intermediate substances that are assigned to equations of the final and an elementary reaction. Stoichiometric bonds have a constant influence and do not depend on the current state of a system. The second factor is correlative bonds (so-called interaction bonds); they are continuously formed in the process and represent a function of the system state and a function of its evolution step, respectively.

When discussing the characteristic numbers of a non-equilibrium chemical system, it is necessary to separate the ones determined by stoichiometric bonds and the ones determined on the basis of correlative bonds. Let us consider the first group [3–7].

1.2. Stoichiometric bonds

Stoichiometric notification of the final reaction according to its algebraic form is a linear combination of the component substances satisfying the conservation laws (namely, mass and number of atoms of an element) at chemical transformations. Thus, in a set $A = \{A_j\}$, $j = \overline{1, m}$ of component substances one always can separate such stoichiometrically independent components, by the linear combination of which (that is with the use of the final reaction) all remainder component substances of a system can be obtained. In turn, the linear combination of a final reaction is also the final reaction satisfying the stoichiometric notification (1.2). So, in the set $Q = \{Q_f\}$, $f = \overline{1, q}$ of the final reactions one can also separate stoichiometrically independent ones in the sense that all remaining components by themselves represent their linear combination. In accordance with equation (1.2) for a set Q we have

$$\nu A = 0 \qquad (1.3)$$

where $A = (A_j)$ is a column vector of component substances, $\nu = (\nu_{fi})$, $f = \overline{1,\ q}$, $i = \overline{1,\ m}$. A number of independent equations of the expression (1.3) equal to rank $(rk(\nu))$ of the stoichiometric matrix $\nu = (\nu_{fj})$ also determines a number $q^{(s)}$ of stoichiometrically independent final reactions, the so-called Juge's criterion [5, 6]:

$$q^{(s)} = rk(\nu) \qquad (1.4)$$

Relative to vector $A = (A_j)$ a system (1.3) has $m–rk(\nu)$ fundamental solutions, that is why in order to assign the vector in a similar manner it is necessary and sufficient to determine $m–rk(\nu)$ of its elements founding the remainder from solution of equation (1.3). This number of independent vector elements $A = (A_j)$ represents the number $m^{(s)}$ of stoichiometrically independent component substances of a chemical system

$$m^{(s)} = m-rk(\nu) \qquad (1.5)$$

For a set of an elementary reaction $S = \{S_k\}$, $k = \overline{1,\ s}$, by equation (1.1) assigned to a set of component and intermediate substances we have

$$\alpha^{(1)}A + \alpha^{(2)}X=0 \qquad (1.6)$$

where $\alpha^{(1)}=(\alpha_{kj}^{(1)})$ and $\alpha^{(2)}=(\alpha_{kj}^{(2)})$, $k = \overline{1,\ s}$ are the stoichiometric matrixes of component and intermediate substances in an elementary reaction, respectively; $X = (X_i)$ is a column vector of an intermediate substance.

Introducing the general stoichiometric matrix of an elementary reaction

$$\alpha = (\alpha^{(1)}, \alpha^{(2)}) \qquad (1.7)$$

bulks of which are $\alpha^{(1)}$ and $\alpha^{(2)}$, let us re-write equation (1.6) as follows

$$\alpha\begin{pmatrix} A \\ X \end{pmatrix} = 0 \qquad (1.8)$$

Analyzing equation (1.8) as equation (1.3) we found the number $s^{(s)}$ of stoichiometrically independent elementary reactions and the number $(m + n)^{(s)}$ of stoichiometrically independent component and intermediate substances:

$$s^{(s)} = rk(\alpha), \qquad (m + n)^{(s)} = m + n - rk(\alpha) \qquad (1.9)$$

A set $A = \{A_j\}$ can be assigned, regardless if a set $X = \{X_i\}$ in a view of a linear combination (that is final reaction (1.2)) of stoichiometrically independent component substances, the number of which is equal to $m-rk(\nu)$, is known or not. It allows to assign the set $X = \{X_i\}$, assuming that $A = \{A_j\}$ it is already known. Then, all intermediate substances X_i, $i = 1, n$, can be obtained by linear combination of component substances with the use of $n-1$ elementary reactions

$$A_i + X_i \Leftrightarrow A_j + X_j \qquad (1.10)$$

where A_i and A_j are the total of component substances. Thus, $rk(\nu)$ of the final reaction (1.2) and $n-1$ elementary reactions (1.10) cover all independent linear combinations between the component and intermediate substances, allowing to assign their sets $A = \{A_j\}$ and $X = X\ \{X_i\}$ if at least of $m-rk(\nu)$ elements of A and one element from X are known. Subsequently, $(m + n)^{(s)}$ in equation (1.9) can be divided into terms:

$$m^{(s)} = m - rk(\nu), \qquad n^{(s)} = 1 \qquad (1.11)$$

and to obtain an estimation of a rank of matrix α:

$$rk(\alpha) = rk(\nu) + n - 1 \qquad (1.12)$$

On the other hand, the rank of a matrix $\alpha^{(1)}$ should accurately coincide with the rank of a matrix ν, which determines the maximal number of independent combinations between component substances:

$$rk(\alpha^{(1)}) = rk(\nu) \qquad (1.13)$$

In turn, this means that

$$rk(\alpha^{(2)}) = n - 1 \qquad (1.14)$$

Summarizing the above, we obtain

$$q^{(s)} = rk(\nu) = rk(\alpha^{(1)}),$$

$$s^{(s)} = rk(\alpha) = rk(\nu) + n - 1 = rk(\alpha^{(1)}) + rk(\alpha^{(2)})$$

$$m^{(s)} = m - rk(\nu) = m - rk(\alpha^{(1)}),$$

$$n^{(s)} = 1$$

$$rk(\alpha^{(2)} = n - 1 \qquad (1.15)$$

There is another approach, which has been formulated by Brinkley [7] and has been discussed in detail a number of works, for instance Ref. [8]. In accordance with this approach, component and intermediate substances can be assigned as a linear combination of atomic composition L_ℓ, $l = \overline{1,\ p}$:

$$A_j = \sum_\ell \beta_{j\ell}^{(1)} L_\ell, \qquad X_i = \sum_\ell \beta_{i\ell}^{(2)} L_\ell \qquad (1.16)$$

where $\beta_{jl}^{(1)}$ and $\beta_{jl}^{(2)}$ are the ℓ numbers in component j component and intermediate substance i, respectively.

Ratios (1.16) can be re-written as

$$A = \beta^{(1)} L, \qquad X = \beta^{(2)} L \qquad (1.17)$$

where $\beta^{(1)}=(\beta_{jl}^{(1)})$ and $\beta^{(2)}=(\beta_{jl}^{(2)})$ are atomic matrices of component and intermediate substances having the dimensions $m\times p$ and $n \times p$, respectively; $L = (L_\ell)$ is a column vector of an atomic component. By introducing a general atomic matrix

$$\beta = \begin{pmatrix} \beta^{(1)} \\ \beta^{(2)} \end{pmatrix} \qquad (1.18)$$

let us re-write equation (1.17) as:

$$\begin{pmatrix} A \\ X \end{pmatrix} = \beta L \qquad (1.19)$$

Combining equation (1.3) and equation (1.17), and also equation (1.8) with equation (1.19) we obtain

$$\nu\beta^{(1)}L = 0, \qquad \alpha\beta L = 0 \tag{1.20}$$

It is necessary to note that equation (1.20) in any way does not limit the elements of L vector, since in accordance with the law of conservation of atom number under chemical transformations of the matrix $\nu\beta^{(1)}$ and $\alpha\beta$ they are equal to zero:

$$\nu\beta^{(1)}=0, \qquad \alpha\beta=0 \tag{1.21}$$

Introducing a column vector $\nu_f = (\nu_{fj})$, $f = \overline{1,\ q}$, $j = \overline{1,\ m}$ and $\alpha_k^T=(\alpha_{kj}^{(1)}, \alpha_{ki}^{(2)})^T$, $k = \overline{1,\ s}$, $i = \overline{1,\ n}$ instead of equation (1.21) we can write that

$$(\beta^{(1)})^T\nu_f = 0, \qquad \beta^T\alpha_k = 0 \tag{1.22}$$

First system of a linear homogeneous equations in an expression (1.22) has $m{-}rk(\beta^{(1)})$ fundamental solutions to which $m{-}rk(\beta^{(1)})$ of linear independent vectors ν_f correspond. Each of these vectors determines a stoichiometric note of the f of the final reaction. That is why, in accordance with the criterion of Brinkley, we have

$$q^{(s)} = m{-}rk((\beta^{(1)}), \qquad m^{(s)} = rk(\beta^{(1)}) \tag{1.23}$$

The second system of linear homogeneous equations in expression (1.22) has $m + n{-}rk(\beta)$ fundamental solutions determining a linearly independent vector α_k. Each of these vectors assign a stoichiometric note of k to the elementary reaction and that is why

$$s^{(s)} = m + n{-}rk(\beta), \qquad (m + n)^{(s)} = rk(\beta) \tag{1.24}$$

Taking into account equation (1.24) and equation (1.23) we have

$$n^{(s)} = rk(\beta){-}rk(\beta^{(1)}) \tag{1.25}$$

Comparing both approaches to an estimation of stoichiometrically independent final and elementary reactions, component and intermediate substances assigned by the ratios (1.15) and (1.23–1.25) we obtain the following expressions characterizing the stoichiometric bonds:

$$rk(\nu) = rk(\alpha^{(1)}) = m{-}rk(\beta^{(1)}) = m{-}rk(\beta) + 1 \tag{1.26}$$

$$rk(\alpha) = rk(\nu) + n-1 = m + n-rk(\beta) \quad (1.27)$$

$$rk(\beta) = rk(\beta^{(1)}) + 1 \quad (1.28)$$

Assigning the stoichiometrically independent final and an elementary reaction, components and intermediate substances completely determines sets Q, S, A and X, respectively. At the same time, knowledge of a set of final reactions realized in a system, a set of elementary reactions, resulting in realizing the final reaction, and also a set of components and intermediate substances is only the basis for the description of the dynamics of chemical system composition. That is why numbers $q^{(s)}$, $s^{(s)}$, $m^{(s)}$ and $n^{(s)}$ determined above are not characteristic numbers of a chemical system in the sense that we found in search of a synonymous assignment of the dynamics of its composition.

1.3. The principle of abridged description of non-equilibrium systems

Both stoichiometric and correlative bonds decrease the number of parameters of a chemical system needed and sufficient for the description of dynamics of a system's composition during its transition from the initial indignant state to the final equilibrium state. However, in order to understand their physical sense, let us consider briefly the characteristic numbers of the equilibrium chemical system. A minimal number of parameters, the assigning of which completely and synonymously determines the state of thermodynamical system, represents its fundamental characteristic, that is the number of degrees of freedom. In accordance with the phase Gibbs rule $c = k + 2-f$ (here k is the number of independent substances or system components, f is the number of phases, n is the number of external parameters influencing the equilibrium state of system). Usually, these are pressure and the temperature and that is why most often the phase Gibbs rule is denoted in the form $c = k + 2-f$ [9, 10].

Equations of a relation between the parameters of a system based on a phase Gibbs rule follows from the principle of detailed equilibrium in its different displays. Thus, for a multi-phase system the principle of a detailed equilibrium requires the equilibrium of any two phases with each other. This permits to separate them and to consider them separately from others. General conditions for an equilibrium in an isolated system are reduced to partial conditions of thermal (temperature of all phases is equal), mechanical (at plain boundaries of division the hydrostatic pressures in all phases are equal) and chemical equilibria in the form

$$\sum_j \mu_j \mathrm{d}N_j = 0 \tag{1.29}$$

where μ_j and N_j are chemical potential and a number of moles of substance j, respectively.

In turn, the general condition of equation (1.29) of chemical equilibrium is reduced to two partial ones; one of these partial conditions corresponds to the equilibrium of every possible chemical reaction of the system

$$\sum_j \nu_{ij}\mu_j = 0, \qquad i = \overline{1,\ q} \tag{1.30}$$

Here ν_{ij} is a stoichiometric coefficient of substance j in reaction i. For all sets of possible reactions in the system this can be written as

$$\nu\mu = 0 \tag{1.31}$$

where $\nu = (\nu_{ij})$ is a stoichiometric matrix of dimension $q \times m$ and $\mu = (\mu_j)$ is a column vector of chemical potentials.

The number of independent equations of relation (see 1.30), respectively, the number of independent chemical reactions is equal to a rank $rk(\nu)$ of the stoichiometric matrix; that is why a system of linear homogeneous equations (1.31) has $m-rk(\nu)$ fundamental solutions for un-known elements of the vector μ. So, in order for a vector μ to be determined synonymously, it is necessary and sufficient to assign only $m-rk(\nu)$, its elements giving the remainder from equation of relation (1.31).

Thus, in accordance with equation (1.31) a number of independent chemical reactions $(q^{(P)})$ and substances $(m^{(P)} = k)$ of the equilibrium system are determined *via* ratios:

$$q^{(p)} = rk(\nu), \qquad m^{(p)} = \kappa = m - rk(\nu) \tag{1.32}$$

If other limits are formed by composition of an equilibrium system and are not following from equation (1.31), then $m^{(p)} < m-rk(\nu)$. So, in a general case $m^{(p)} = k \leq m-rk(\nu)$.

The number of independent reactions and substances is an essential important characteristic, not only of an equilibrium, but also of non-equilibrium chemical systems. In the latter, however, conditions of chemical equilibrium (see equation 1.31) are infringed. Firstly, this means that the numbers of independent reactions and substances of non-equilibrium system cannot be determined on the basis of the same criteria as for equilibrium systems; successively, not only numbers but also their physical sense should be different. Secondly, during the evolution of chemical system from the initial state to a final equilibrium one, relations between their parameters are renewed continuously, but it is necessary to analyze them in such a way that characteristic numbers (that is numbers of independent reactions and substances of a non-equilibrium system) can be analyzed by a state function of a non-equilibrium system evolution. This question has not only a concrete scientific value, but also a methodological one and its solution is based on the principle of abridged description of non-equilibrium systems [11–13].

The main content of the above question can be split into two moments [12]. The first indicates the fact, that when two interacting processes (rapid and slow) proceeding in a system sharply differing by the relaxation times, then at times from the onset less than the relaxation time of the quick process, the latter can be studied independently from the slow one; at times longer than the relaxation time of the quick process the evolution of a system follows the characteristics of a slow process. Secondly, because of the evolution of an indignant system a continuous synchronization of a parameters and determination of correlative bonds between them takes place [13]; this decreases the number of degrees of freedom. However, taking into account the difference between the rates of partial processes and their relaxation times, some states of evolution can be found, every one of which is described by its own number of independent parameters changing with the transition from one state to another.

So, the principle of the abridged description in continuous synchronization of parameters and hierarchy of a time scale of partial processes allows to discriminate between continuous and discrete. At this, continuous synchronization of a parameter represents only a quantitative change of the number of correlative bonds. The last act differently on different parts of a chemical process: more strongly on the properties that are changed quickly, and less strongly on the ones that are changed slowly. That is why, in accordance with the hierarchy of a time scale of partial processes, the dynamics of a chemical system develop step-by-step. A typical measure of correlative bonds corresponds to each state. After achieving this measure a chemical system, by a sudden leap, transits to a new qualitative state. The last in a range of new measures again keeps correlative bonds between the parameters of a system storing this quality until a new jump takes place. Thus, every state of the dynamics of a chemical system can be characterized by its level of correlative bonds, its set of independent parameters and their characteristic numbers.

Mathematically, the most complete principle of abridged description is based and discovered in the statistical physics [11, 13], using as example a low-density single-component system. An analysis of this situation can be founded in Ref. [12]. According to Refs. [11, 12], for the determination of macroscopic characteristics of an indignant single-component system at the initial moment of time t_0, it is necessary to assign all its 2_f^N parameters (f is the number of degrees of freedom of the particle, N is the number of particles). This corresponds to the dependence:

$$\omega = \omega(x_1, x_2, ..., x_N) \tag{1.33}$$

in which ω is full distribution function or a density of probability and x_i is the total of variables determining the state of one particle.

In contrast, at equilibrium the single-component system is determined by the assignment of only two parameters, for example, free energy F and temperature T. Thus, the presented pair of variables completely determines the functional dependence ω for all x_i:

$$\omega = \omega(x_1, x_2, ..., x_N / F, T) \tag{1.34}$$

This has been established by Gibbs in the canonic distribution

$$\omega = exp\left\{ \frac{F - H(x_1, x_2, ..., x_N)}{\kappa_B T} \right\} \tag{1.35}$$

where $H(x_1, x_2, ..., x_N)$ is a Hamiltonian function.

Reduction of the number of independent parameters in the evolution of an indignant system proceeds continuously as a result of parameter synchronization and the determination of the correlative bonds between them. However, taking into account the difference between partial processes and their respective relaxation times, separate states of evolution can be noted, every one of which is described by its own number of independent parameters changing through transition from one state to another. For example, after the finishing time $t^{(k)} \sim r_0/V$ from the start of a single-component system, a so-called kinetic state of a process takes place [11, 12] (r_0 and V are characteristic size of a particle and heat rate of their moving, respectively; $t^{(k)}$ is an indication of particle interaction time under collision, in the order of 10^{-13}–10^{-12} *s*). In this state a state of a system is fully determined by a partial distribution function ω_1 that rules by the temporary evolution of a system. Multi-partial distribution function and as a result a full one represent a function of ω_1:

$$\omega = \omega(x_1, x_2,...x_N, / \omega_1) \tag{1.36}$$

For more times, namely at $t^{(r)} \sim \lambda/V$, where λ is the length of a free run of particles, a hydrodynamics state of the process takes place, when a system state is fully determined by the density of keeping values, which rule by a system developing with time. In the hydrodynamics state [11, 12]

$$\omega = \omega(x_1, x_2,..., x_N / \rho, \varepsilon, p) \tag{1.37}$$

where ω is a function of local values of density ρ, energy ε and impulse p. The value $t^{(r)}$, in the order of 10^{-9}–10^{-8} s, can be considered [11, 12] as relaxation time ω in the local equilibrium of Maxwell–Boltzmann's distribution function.

Finally, at times $t^{(p)} \sim L/V$, where L is a parameter characterizing the spatial inhomogeneity of a system, it transforms into a statistical equilibrium state, at which a number of independent parameters is reduced to 2, according to functional dependence (1.34) or (1.35). Since $r_0 << \lambda << L$, we obtain

$$t^{(\kappa)} << t^{(r)} << t^{(p)} \tag{1.38}$$

This equation represents a mathematical expression of existing hierarchy of the time scale of partial processes of single-component systems with little density evolution.

Depending on the character of the process and its concrete task, the above-mentioned evolution states of an indignant physical system can be described in detail starting from the relaxation time, for example, from the processes of transformation of translation energy into rotation energy, vibration energy, etc. As calculations show [14], relaxation times of these processes can are essentially different from the $t^{(k)} \sim r_0/V$, indicating average time of the interaction of particles under collisions, and $t^{(r)} \sim \lambda/V$, indicating average time between successive collisions of particles.

1.4. Relaxation times of activated complexes and a basic equation for a theory of absolute rates of reactions

A conception of abridged description of non-equilibrium systems is not principally a new theory of chemical kinetics, since the Bodenshtein–Semenov's principle has been used as a principle of quasi-stationary reactions for a long time. When applicable to a state, the Bodenshtein–Semenov's principle confirms that the quasi-stationary state of a chemical system is described essentially more simply than these processes, realization of which leads to the achievement of this state. Exactly from this point of view, Bodenshtein–Semenov's principle by itself represents a partial formulation of general conception of abridged description of non-equilibrium systems, according to which every state of the evolution can be and should be described by a different set of independent values. This, in turn, means that characteristic numbers of a chemical system need to be considered *via* an evolution waiting for their successive reduction when the chemical system approaches the equilibrium state.

Here, we will limit ourselves to an analysis [15] of the simplest chemical system assuming that it represents a catalytic system and all processes in it proceed with the participation of active centers Z of a catalyst of the same nature. We also assume that all intermediate substances have one center and elementary reactions are linear on intermediate substances. This means, that the stoichiometric coefficients $\alpha_{kj}^{(2)}$ in equation (1.1) can be equal only to values 0, +1 or –1, and for each elementary reaction $S_k : \sum_j \alpha_{kj}^{(2)} = 0$

In order to mark out the evolution states of an indignant chemical system, let us investigate the hierarchy of the time scales of partial processes starting from the elementary reaction. Let us stick to the main postulate of the absolute reaction rates theory [16], confirming the fact that any chemical act proceeds *via* the formation and decomposition of the activated complex, but reject [**17**] the other postulate according to which an activated complex is in equilibrium with the starting substances. Thus, the scheme of an elementary reaction proceeding *via* the formation and decomposition of an activated complex C should be written in such a way

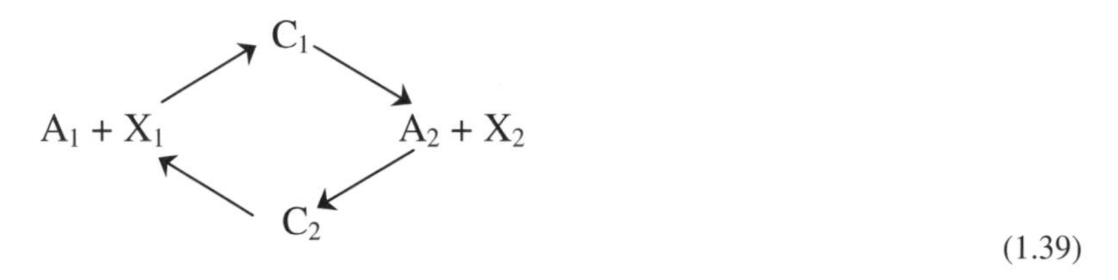

(1.39)

so that activated complexes C_1, formed from the starting substances, are decomposed only into the final ones and, in the contrast, activated complexes C_2, formed from the final products, are decomposed only into the initial ones. As we can see, scheme (1.39) predicts a difference of activated complexes depending on the method of their decomposition; this formally can satisfy, taking into account that C_1 and C_2 are activated complexes C that do not have different degrees of freedom connected with the decomposition. If we do not introduce this difference of activated complexes C_1 and C_2, then scheme (1.39) should be written as follows:

$$A_1 + X_1 \underset{\nu_1}{\overset{k_1}{\rightleftarrows}} C \underset{k_2}{\overset{\nu_2}{\rightleftarrows}} A_2 + X_2 \qquad (1.40)$$

Here k_1 and k_2 characterize the probabilities of the activated complexes formation per unit of time at single concentrations of starting and final reacting substances; ν_2 and ν_1, respectively, determine the probabilities of the decomposition of an activated complex in direct and indirect ways per unit of time at a single concentration of an activated complex.

Let us assume that the c, x and a are concentrations of an activated complex, intermediate substance and component, respectively, expressed in the same unit. Then, in accordance with the scheme (1.40) a total rate of the formation of an activated complex can be described by the equation

$$\frac{dc}{dt} = k_1 a_1 x_1 + k_2 a_2 x_2 - (\nu_1 + \nu_2)c \tag{1.41}$$

Let us assume in equation (1.41) that the values c, x and a have the characteristic relaxation times t_1^*, t_3^* and t_4^*, respectively. Here, t_1^* should be as same as particle interaction time $t^{(k)} \sim r_0/V$. If we will also assume that in the initial indignant state of a chemical system a non-equilibrium distribution of energy takes place, then values k_i and ν_i in equation (1.41) should be considered as variables with the characteristic relaxation time t_2^*, the same as the time $t^{(r)} \sim \lambda/V$ of relaxation of a single-partial function to an equilibrium Maxwell–Boltzmann distribution of energy.

Since $t_1^* << t_2^*$, t_3^*, t_4^*, in accordance with the main principles of abridged description in the range of time $0 < t < t_1$, the values k_i, ν_i, x_i and a_i by analysis of equation (1.41) can be considered as quasi-integrals of moving. Assuming that k_i, ν_i and a_i are constant after integration of equation (1.41) we will obtain

$$c = \frac{k_1 a_1 x_1 + k_2 a_2 x_2}{\nu_1 + \nu_2} [1 - exp\{-(\nu_1 + \nu_2)t\}] \tag{1.42}$$

It follows from this, that the next value is the relaxation time of an activated complex:

$$t_1^* \sim (\nu_1 + \nu_2)^{-1} \tag{1.43}$$

This value has been defined by us as the time of primary relaxation of an activated complex.

It follows from equation (1.42) that at $t >> t_1^*$ we can neglect the exponential components in comparison with 1. In this case c does not depend on time and represents the stationary concentration of an activated complex

$$c_s = \frac{k_1 a_1 x_1 + k_2 a_2 x_2}{\nu_1 + \nu_2} \tag{1.44}$$

The relation $c_s = c_s(a, x)$, according to equation (1.44), is the first level of a correlation in the indignant chemical system.

As was mentioned before, the time t_1^* of the first relaxation of an activated complex is less than the relaxation time $t^{(r)}$ of the single-partial function to the equilibrium of the Maxwell–Boltzmann distribution of energy. Thus, that in the range time $t_1^* << t \sim t^{(r)}$ from the start of an indignant system, the concentration of an activated complexes corresponds to a stationarity condition $dc/dt = 0$ and is determined by equation (1.44), but at this the energy distribution is not in equilibrium. That is why we assume a time relaxation of the system until equilibrium of the Maxwell–Boltzmann distribution of energy at time t_2^* of secondary relaxation of an activated complexes. Evidently, a time range $t >> t_2^*$ from the start of indignation is exactly a field of the formal kinetics to which the determination of stationary concentration of an activated complex and equilibrium distribution of energy correspond.

If the energy distribution between the particles proceeds only as a result of an interaction of translational degrees of the freedom (RR-transitions), then t_2^* coincides with the $t^{(r)} \sim \lambda/V$ and, therefore, will be in the order of 10^{-9} s. This, however, is true only for a system consisting of single-atom particles. In chemical systems consisting of multi-atom molecules, equilibrium distribution of energy not only between the particles but also in all freedom degrees of their motion suggests the interaction of the translational and oscillatiory (RT transitions), translational and rotary (RV-transitions), and oscillatory and rotary (TV-transitions) freedom degrees of motion. As shown in the calculations of Ref. [14], that have been done on the basis of model imaginations, relaxation times of these transitions can be thousands and more times longer than the relaxation time of the RR-transition equal to $t^{(r)}$. So, generally, $t_2^* \geq t^{(r)}$.

If the time from the indignant of the system starting is longer than the t_2^*, then in a chemical system another correlative relation is found between the parameters of a process. Formally, k_1 and ν_1 (see scheme (1.40)) are the constant rates of the direct and back process of the activated complex formation from the starting substances of an elementary reaction. In accordance with the mass action law and the principle of detailed balance the ratio k_1/ν_1 should be considered as the equilibrium constant (K_1) of the reaction leading to activated complex formation from the starting substances. Analogously, the ratio k_2/ν_2 is the equilibrium constant (K_2) of the reaction of activated complex formation from the products of an elementary reaction. That is why, under the equilibrium distribution of energy between k_1 and ν_1, k_2 and ν_2, we can determine the following relations:

$$k_1/\nu_1 = K_1, \qquad k_2/\nu_2 = K_2 \tag{1.45}$$

which we can call the second level of correlation in the indignant chemical system.

The rate v of an elementary reaction (1.40) can be determined; note that it has a physical sense only if $dc/dt = 0$.

In this case

$$v = k_1 a_1 x_1 - \nu_1 c_s = \nu_2 c_s - k_2 a_2 x_2 \tag{1.46}$$

Substitution of equation (1.44) into equation (1.46) gives

$$v = \frac{k_1 \nu_2}{\nu_1 + \nu_2} a_1 x_1 - \frac{k_2 \nu_1}{\nu_1 + \nu_2} a_2 x_2 \tag{1.47}$$

On the other hand, defining a scheme of an elementary reaction as

$$A_1 + X_1 \underset{\overleftarrow{k}}{\overset{\vec{k}}{\rightleftharpoons}} A_2 + X_2 \tag{1.48}$$

according to the mass action law we have

$$v = \vec{k} a_1 x_1 - \overleftarrow{k} a_2 x_2 \tag{1.49}$$

Comparing equations (1.47) and (1.49) we will obtain

$$\vec{k} = \frac{k_1 \nu_2}{\nu_1 + \nu_2}, \qquad \overleftarrow{k} = \frac{k_2 \nu_1}{\nu_1 + \nu_2} \tag{1.50}$$

A transition from the indignant chemical system to the equilibrium distribution of energy at $t >> t_2^*$ allows to use the relations of equation (1.45) in equation (1.50) below:

$$\vec{k} = \frac{\nu_1 \nu_2}{\nu_1 + \nu_2} K_1, \qquad \overleftarrow{k} = \frac{\nu_1 \nu_2}{\nu_1 + \nu_2} K_2 \tag{1.51}$$

By expressing the equilibrium constants K_1 and K_2 *via* the corresponding statistical sums F_i, let us re-write equation (1.51) as

$$\vec{k} = \frac{\nu_1\nu_2}{\nu_1+\nu_2}\frac{F_c}{F_{A_1}F_{X_1}}\exp\left\{-{E_1}\!\left/{RT}\right.\right\},$$
$$\overleftarrow{k} = \frac{\nu_1\nu_2}{\nu_1+\nu_2}\frac{F_c}{F_{A_2}F_{X_2}}\exp\left\{-{E_2}\!\left/{RT}\right.\right\} \tag{1.52}$$

Using the presented interpretation it is not taken to account that the activated complex is deprived of one degree of freedom belonging to the reaction's coordinate; that is why the statistical sum F_c of the activated complex is determined by all $3n$ degrees of motion freedom (here n is the number of atoms in the activated complex). Let us consider a partial case, in which the variable degrees of a motion freedom are responsible for the decomposition of activated complex. Let us detach a multiplier from the F_c characterizing the same degree of a freedom of variable motion, which determines, for example, the decomposition of activated complex directly. Under the classical approximation *($h\nu$ <<* *k_BT)* this multiplier is equal to $k_BT/h\nu_2$. That is why writing $F_c = F_c^* k_BT/h\nu_2$ we have

$$\vec{k} = \frac{\nu_1}{\nu_1+\nu_2}\frac{k_BT}{h}\frac{F_c^*}{F_{A_1}F_{X_1}}\exp\left\{-{E_1}\!\left/{RT}\right.\right\},$$
$$\overleftarrow{k} = \frac{\nu_1}{\nu_1+\nu_2}\frac{k_BT}{h}\frac{F_c^*}{F_{A_2}F_{X_2}}\exp\left\{-{E_2}\!\left/{RT}\right.\right\} \tag{1.53}$$

Using the comparison of the obtained ratios with the expressions in the absolute reaction rates theory [16] we can determine the transmission coefficient χ:

$$\chi_1 = \frac{\nu_1}{\nu_1+\nu_2} \tag{1.54}$$

Another option is to present the F_c by detaching from it a co-multiplier $k_BT/h\nu_1$ characterizing the contribution to the statistical sum of variable degree of freedom, which is responsible for the decomposition of the activated complex in the back direction: $F_c = F_c^{**} k_BT/h\nu_1$. Then,

$$\vec{k} = \frac{\nu_2}{\nu_1 + \nu_2} \frac{k_B T}{h} \frac{F_c^{**}}{F_{A_1} F_{X_1}} \exp\left\{-E_1/RT\right\},$$
$$\overleftarrow{k} = \frac{\nu_2}{\nu_1 + \nu_2} \frac{k_B T}{h} \frac{F_c^{**}}{F_{A_2} F_{X_2}} \exp\left\{-E_2/RT\right\} \quad (1.55)$$

As we can see, in this case the transmission coefficient has another value

$$\chi_1 = \frac{\nu_2}{\nu_1 + \nu_2} \quad (1.56)$$

due to the fact that F_c^* and F_c^{**} are different from one another, since they express the statistical sums of activated complexes without different degrees of freedom of variable motion.

Finally, combining equations (1.53) and (1.55) we obtain, for example

$$\vec{k} = \frac{\nu_1}{\nu_1 + \nu_2} \frac{k_B T}{h} \frac{F_c^{*}}{F_{A_1} F_{X_1}} \exp\left\{-E_1/RT\right\},$$
$$\overleftarrow{k} = \frac{\nu_2}{\nu_1 + \nu_2} \frac{k_B T}{h} \frac{F_c^{**}}{F_{A_2} F_{X_2}} \exp\left\{-E_2/RT\right\} \quad (1.57)$$

Such an expression of the constant rate corresponds the most accurately to the scheme presented above (1.39).

Thus, the approach based on taking into account the hierarchy of the time scales on the level of an elementary reaction usually leads to equations of the absolute rates of reaction theory, but, at this, allows to escape the contrasting postulate about an equilibrium of activated complexes with the initial substances, does not request the artificial introduction into the expression for the constant rate of the transmission coefficient and gives its analytical determination of the type (1.54) and (1.56).

The approach presented here does not pay attention to which degree of freedom of an activated complex leads to its decomposition in either direction. In particular, not only oscillatory, but also translational, rotary and electronic degrees of a freedom are responsible for the decomposition of an activated complex. That is why one of the tasks of the absolute rates of reactions theory is determination of the degree of freedom for the motion of the activated complex responsible for its decomposition and the calculation of the corresponding frequencies.

The next expressions for the decomposition frequencies can be proposed as semi-empirical ones determined by translational and rotary motions. The statistical sum f_R of the translational motion of a molecule per degree of a freedom is equal to

$$f_R = \frac{(2\pi m k_B T)^{1/2} e^{1/3} \ell}{h} \tag{1.58}$$

Here $\ell = \upsilon^{1/3}$, and υ is the volume of one molecule. Assuming that $e^{1/3} \approx 1$, let us re-write equation (1.58) in the form

$$f_R = \left(\frac{2\pi m}{k_B T}\right)^{1/2} \ell \frac{k_B T}{h} \tag{1.59}$$

In accordance with the multiplier division in equation (1.59) we can see [16] that $(2\pi m/k_B T)^{1/2}\ell$ represents the characteristic time of the translational motion. Thus, we can write

$$\nu_R = \left(\frac{k_B T}{2\pi m}\right)^{1/2} \frac{1}{\ell} \tag{1.60}$$

The expression (1.60) with the accuracy of the constant multiplier can also be obtained from the following simple reflections. Energy per degree of freedom of translational motion is equal to $k_B T/2$. Then, writing that $k_B T/2 = mv^2/2$ (v is an average rate of the translational motion) we will obtain $v = (k_B T/m)^{1/2}$. Dividing the rate by the characteristic size ℓ we have the expression

$$\nu_R = \left(\frac{k_B T}{m}\right)^{1/2} \frac{1}{\ell} \tag{1.61}$$

which differs from equation (1.60) only by the constant multiplier $(2\pi)^{-1/2}$.

Analogously, let us consider the rotary motion, the statistical sum of which is equal to

$$F_v = \frac{\pi^{1/2}}{\sigma}\left(\frac{8\pi^2 k_B T}{h^2}\right)^{3/2} J_x^{1/2} J_y^{1/2} J_z^{1/2} \tag{1.62}$$

Here σ is a symmetry number of a molecule, I_i is the inertia moment respective to a rotation axis. Denoting I as $I_x I_y I_z$, we will obtain for the degree of a freedom (assuming that $\pi^{1/6} \approx 1$)

$$f_v = \sigma^{1/3}\left(\frac{8\pi^2 k_B T}{h^2}\right)^{1/2} J^{1/2} \tag{1.63}$$

As we can see, the next value represents the decomposition frequency connected with the rotary motion

$$\nu_v = \sigma^{1/3}\frac{(k_B T / 2J)^{1/2}}{2\pi} \tag{1.64}$$

This value can also be estimated starting from the ratio for energy of the rotary motion $k_B T/2 = I\omega^2/2$, in which ω is an angle rate of the rotation. From this we have $\omega = (k_B T/I)^{1/2}$, and assuming the value 2π as a characteristic "size" of the rotary motion we will obtain $\nu_v = (k_B T/I)^{1/2}/2\pi$. This, together with the accuracy of the constant multiplier $\sigma^{1/3}/2^{1/2}$ coincidences with the previous expression (1.64) and is well-known [18].

Calculations show that the ν_R and ν_v are close in value and can assume values in the range 10^{11}–10^{13} s^{-1}. Starting from the scattering values ν_R, ν_v and ν_r, it follows that the transmission coefficient at its lower limit has a value of 10^{-2} if additional limitations on the chemical act do not take place.

However, let us note that the mean free path length $\ell = \lambda$ can be characteristic of the size ℓ of the translational motion in the expression (1.60). Then,

$$\nu'_R = \frac{(k_B T / 2\pi m)^{1/2}}{\lambda} \tag{1.65}$$

As we can see, this formula describes the situation when the decomposition of the activated complex needs to collide with a particle. As λ is 3–4 orders of magnitude longer than the effective diameter of a molecule, v_R' is 3–4 orders of magnitude smaller than v_R. As a result, in this case the transmission coefficient can be equal to 10^{-6} and depends on the concentration of other particles or general pressure.

1.5. Characteristic numbers of a pre-quasistationary system

In accordance with the above, the field of formal kinetics begins at $t >> t_2^*$. On this state of evolution of the indignant chemical system the stationary concentration of an activated complex and equilibrium distribution of energy on all degrees of a molecule's motion are determined. A correlation by the view $c_s = c_s(a, x)$ (see equation (1.44)) *via* the relaxation time t_1^* determines the functional dependence of the activated complexes concentration from the concentration of the composing and the intermediate substances and, furthermore, decreases the number of independent chemical transformations to the number of an elementary reaction. The second level of the correlation *via* the relaxation time t_2^* determines the connection (see equation 1.45) allowing to express the constants of the elementary reactions rates in straight and back directions *via* the frequencies of the activated complexes decomposition and the reaction equilibrium constants of their formation from the starting and final substances (see equation (1.51)). Both levels of the correlation determine the transition of an elementary reaction to the stationary conditions allowing the determination of its rate in accordance with the mass action law and the absolute rates of the reactions theory.

In order to detach the other states of an indignant chemical system evolution, let us estimate the relaxation times of intermediate and composing substances. Since the relaxation time is a time from the onset of chemical system indignation which always can be considered as being far from the equilibrium state, and putting in equation (1.49) that $K_1a_1x_1 >> K_2a_2x_2$, we will consider the reaction (1.48) as the non-reversible one. Then it can be written

$$\frac{\mathrm{d}a}{\mathrm{d}t} = \frac{\mathrm{d}x}{\mathrm{d}t} = \nu_0 Kax \qquad (1.66)$$

in which we introduce the notification

$$\nu_0 = \frac{\nu_1 \nu_2}{\nu_1 + \nu_2} \tag{1.67}$$

Determining the relaxation times of the intermediate (t_3^*) and composing (t_4^*) substances by the expressions $t_3^*(\mathrm{d}x/\mathrm{d}t) \sim x$, $t_4^*(\mathrm{d}a/\mathrm{d}t) \sim a$ we will obtain

$$t_3^* \sim (\nu_0 K a)^{-1}, \qquad t_4^* \sim (\nu_0 K x)^{-1} \tag{1.68}$$

As we can see, the difference between the relaxation times of the intermediate and component substances taking part in the one elementary reaction depends on the ratio of their concentrations. Thus, the relaxation time of the intermediate substances will be less than the relaxation time of the composing ones only in the cases that $x \ll a$. In such a way the above-mentioned reflection is confirmed in Ref. [19], according to which a parameter dividing the relaxation times of composing and intermediate substances is their concentration but not a constant of separate states. This conclusion is also in accordance with the well-known theory of solution principle following from the analysis of the Gibbs–Durham equation, according to which properties of a component change more when its concentration decreases.

In the following analysis we will consider that the inequality is performed, so $t_3^* \ll t_4^*$. Comparing with the previous results we will obtain the succession $t_1^* \ll t_2^* \ll t_3^* \ll t_4^*$, characterizing the hierarchy of the time scale in the chemical kinetics. In accordance with this kinetics the time interval $t_2^* \ll t \sim t_3^*$ corresponds to an indignant chemical system with the equilibrium distribution of energy and stationary concentration of an activated complex does not depend on time: $c_s = c_s(a, x)$. But at this $x = x(t)$, $a = a(t)$. This state of evolution of the chemical system can be called pre-quasistationary.

Assigning in advance a dynamics of the chemical system composition to the pre-quasistationary state of its evolution can be done in two ways. First is the indirect or experimental way due to determination of the concentrations of reacting substances in time. The second is the direct or theoretical way *via* determination of the rate of an elementary reaction. Both methods assign in advance a phase trajectory $a = a(t)$ and $x = x(t)$ only at well-known starting concentrations of the reacting substances.

Under condition of the stationarity of a system upon the concentrations of activated complexes $(\mathrm{d}c/\mathrm{d}t = 0)$ for each elementary reaction the next ratio can be obtained:

$$\frac{dN_{kj}}{\alpha_{kj}^{(1)}dt} = \frac{dN_{kj}}{\alpha_{ki}^{(2)}dt} = \frac{d\chi}{dt} \qquad (1.69)$$

in which χ_k is the run of an elementary reaction. The value $v_k = d\chi_k/dt$ determines of the rate of elementary reaction S_k. Given the rates of formation of *j*-composing (ω_j) and *i*-intermediate (ω_i) substances in the whole system as $\omega_j^{(1)} = \sum_k \alpha_{kj}^{(1)} v_k$, $\omega_i^{(2)} = \sum_k \alpha_{ki}^{(2)} v_k$ for sets $A = \{A_j\}$ and $X = \{X_i\}$ respectively, we have

$$\omega^{(1)} = \left(\alpha^{(1)}\right)^T v, \qquad \omega^{(2)} = \left(\alpha^{(2)}\right)^T v \qquad (1.70)$$

where $\omega^{(1)} = (\omega_j^{(1)})$, $j = \overline{1,\ m}$, $\omega^{(2)} = (\omega_i^{(2)})$, $i = \overline{1,\ n}$ are vector columns of the formation rates of composing and intermediate substances; $v = (v_k)$, $k = \overline{1,\ s}$ is a column vector of the rates of elementary reactions.

Combining the expressions into equation (1.70) we can write

$$\omega = \begin{pmatrix} \omega^{(1)} \\ \omega^{(2)} \end{pmatrix} = \alpha^{(T)} v \qquad (1.71)$$

Let us assume that *via* the chemical system evolution the possibility arises, for example, an experimental one, to determine the rates of composing and intermediate substances formation independently on fact, in which elementary reactions they are formed. The question is: which minimal number of composing and intermediate substances concentrations do we need to take into account in order to have the whole image of the dynamics of composition of all substances? This question implies that the number $(m + n)^*$ of independent compositions and intermediate substances is equal to the number of independent elements of vector ω. Let us use the approach proposed in Ref. [20].

In the closed system a law of conservation of atom number for any ℓ chemical element can be written as

$$\sum_j \beta_{j\ell}^{(1)} N_j + \sum_i \beta_{i\ell}^{(2)} N_i = N_\ell \tag{1.72}$$

where N_ℓ is a number of atoms of the ℓ sort in a system. Differentiating equation (1.72) by time and taking into account that $dN_\ell/dt = 0$ we will obtain

$$\sum_j \beta_{j\ell}^{(1)} \omega_j^{(1)} + \sum_i \beta_{i\ell}^{(2)} \omega_i^{(2)} = 0 \tag{1.73}$$

Combining equation (1.73) upon all indexes we turn to the matrix form

$$\beta^T \omega = 0 \tag{1.74}$$

A system of linear homogeneous equations (1.74), the relatively un-known vector ω has $m + n - rk(\beta)$ independent solutions, and the rest can be found by linear combination of independent ones. So, for assigning in advance the dynamics of a composition of pre-quasistationary system to vector ω it is necessary and sufficient to determine the concentrations $m + n - rk(\beta)$ of compositions and intermediate substances:

$$(m+n)^* = m+n-rg(\beta) \tag{1.75}$$

In accordance with the determined stoichiometric bonds we also have

$$(m+n)^* = rg(\alpha) = rg(\nu) + n - 1 = m + n - rg(\beta) \tag{1.76}$$

$$m^* = rg(\nu) = m - rg(\beta) + 1 = m - rg\left(\beta^{(1)}\right) \tag{1.77}$$

$$n^* = n - 1 \tag{1.78}$$

At this, the number $m^* = rk(\nu)$ of independent compositions of substances was defined as the number of key substances [21].

Let us consider the second method of assigning in advance a dynamics of composition, namely *via* the vector of elementary reaction rates. Since every element v_k of the vector v represents by itself one of the possible elementary reactions of a system, the number of independent elements of vector v and a number s^* of independent elementary reactions are the same.

It follows from the law of conservation of atoms of every sort in the form (1.74) and determination of vector ω *via* the vector v according to equation (1.77) that

$$(\alpha\beta)^T v = 0 \qquad (1.79)$$

However, taking into account that $\alpha\beta = 0$ we conclude that the law of conservation in the form (1.79) is valid for any vector v and is not limited in any way by its composition.

So, all compositions of the vector v in the pre-quasistationary system are lineally independent, also causing the independency of all elementary reactions assigned by the total kinetic scheme of a process:

$$s^* = s \qquad (1.80)$$

The statement proven above can be considered a principle of independency of an elementary reaction in pre-quasistationary systems, the essence of it beiung that the rate of any possible elementary reaction in the pre-quasistationary systems cannot be found as a linear combination of the rates of the rest of the elementary reactions.

1.6. Characteristic numbers of a quasistationary closed (stationary opened) system

Let us assume that in the closed system the condition $t_3^* << t_4^*$ is valid, and a time from onset of indignation is in the range $t_3^* << t \sim t_4^*$. At this state of evolution a closed system, according to the Bodenshtein–Semenov principle, becomes quasistationary: $\omega^{(1)} = \mathrm{d}a/\mathrm{d}t \neq 0$, $\omega^{(2)} = \mathrm{d}x/\mathrm{d}t = 0$. The null vector $\omega^{(2)}$ does not contain independent elements; thus,

$$n^* = 0 \qquad (1.81)$$

At the quasistationary state of evolution the concentration of intermediate substances is completely determined by the concentration of the composing substances, $x = x(a)$, and does not depend on time. The connection $x = x(a)$ is the third level of correlation in the indignant chemical system.

Under condition $\omega^{(2)}$, the conservation law (1.74) in the quasistationary system becomes $(\beta^{(1)})\omega^{(1)} = 0$. It follows from this that the number of independent composition substances is the same as earlier:

$$m^* = m - rk\left(\beta^{(1)}\right) = rk(\nu) \tag{1.82}$$

From the condition $\omega^{(2)}$ also a system of equations for a connection between the compositions of a vector v of elementary reactions rates follows.

$$\left(\alpha^{(2)}\right)^T v = 0 \tag{1.83}$$

Note that not all vectors in the quasistationary system are lineally independent. Just as the number of independent equations of connection in equation (1.83) is equal to $rk(\alpha^{(2)})$, the number of independent elements of the vector $v = (v_k)$ is equal to $s - rk(\alpha^{(2)})$. It exactly determines the number of independent elementary reactions in the quasistationary system s^*

$$s^* = s - rg\left(\alpha^{(2)}\right) = s - (n-1) \tag{1.84}$$

In the stationary opened system not only dx/dt = 0, but also da/dt = 0. However, the latter means only that the vector $\omega^{(1)}$ of the rates of composition substances formation in the chemical process is equal to vector W of the rates of their input into a system and output from it:

$$\omega^{(1)} = W, \qquad \left(\alpha^{(1)}\right)^T v = W \tag{1.85}$$

That is why da/dt = $\omega^{(1)} - W = (\alpha^{(1)})^T v - W = 0$ is performed, but it does not mean that $\omega^{(1)}=0$ and $v = 0$. Thus, the stationarity of the opened system does not additionally limit the elements of vectors $\omega^{(1)}$, $\omega^{(2)}$ and v in comparison with the quasistationary closed system and both by analysis of characteristic numbers and the deduction of kinetic models these systems can be considered conjointly on the basis of the general condition of the stationarity on concentrations of intermediate substances.

In the quasistationary system, under condition dx/dt = 0 for the f final reaction, the following ratio is valid:

$$\frac{dN_{fj}}{\nu_{fj}dt} = \frac{d\chi_f}{dt}, \qquad j = \overline{1,\ m} \tag{1.86}$$

in which χ_f is a run of the f of the final reaction. Expression (1.86) determines the rate of a final reaction R_f = $dN_{fj}/\nu_{fj}dt$ and shows that the rate of a final reaction has a physical sense only in the quasistationary systems. Just as at this $x = x(a)$, the dynamics of a system composition can be assigned in advance by another method, namely *via* the vector $R = (R_f)$, $f = \overline{1,\ q}$, of the final reactions rates. From the determination of R_f it follows that $\omega_j^{(1)} = \sum_f \nu_{fj} R_f$. Conversion to a matrix form, we obtain

$$\omega^{(1)} = \nu^T R \tag{1.87}$$

Assigning in advance the dynamics of the composition of quasistationary system *via* the vector of the final reaction rates, one needs to take into account the number of its independent elements. Since every element of the vector R represents one of the possible final reactions, we again conclude that the number q^* of independent final reactions exactly represents the number of independent elements of the vector R. In order to determine this number it is necessary to determine the equation of the connection between R_f similar, for example, to equation of connection (1.74) between the elements of the vector ω. By combining equation (1.87) with the expression $(\beta^{(1)})^T\omega^{(1)}$ = 0, which is true for the quasistationary system we obtain

$$\left(\nu\beta^{(1)}\right)^T R = 0 \tag{1.88}$$

However, independent from equation (1.88), from the condition of the material balance follows, that $\nu\beta^{(1)}$ = 0. This means, that in the ratio (1.88) on the left of the vector R there is a null matrix, leading to the conclusion that equation (1.88) will be trustful at any vector R. So, equation (1.88) does not limit the elements of the vector R. All of them are lineally independent and this means that the rate of any possible final reaction in the quasistationary system cannot be found by linear combination of the rates of the rest final reactions. Exactly this postulate is necessary to consider a formulation of the main principle of chemical kinetics, that is, a principle of an independency of the final reactions in the non-equilibrium systems.

At this point we would like to underline once more that the principle of independent proceeding of the final reactions cannot mean that the rate of one final reaction absolutely does not depend on the rates of other reactions; naturally, the rates of the different final reactions are interconnected *via* the composition of a system's intermediate and compositions substances; this interconnection is close, so that the rate of a final reaction can be expressed as a function of the rates of others. Exactly this condition establishes the principle of an independency of the final chemical reactions in the non-equilibrium systems. In accordance with this condition the number of independent final reactions in a non-equilibrium system is equal to the number of possible ones:

$$q^* = q \tag{1.89}$$

The principle of independency of the final reactions in non-equilibrium systems complicates the chemical system and it becomes indefinite, since from this point of view it is necessary to consider all thermodynamically allowed directions of the chemical process described by the different final equations as independent ones. However, in practice, this indefinition always can be minimal, for example, as a result of neglecting the final reactions of which the rate is small and does not essentially influence the change of composition of a system. A similar kinetic choice cannot be applied to the theoretical estimation but allows, on the basis of the straight lines or indirect experimental data, to simplify the considered complicated system leading to the most important reactions from the kinetic point of view.

Finally, let us note the following [15]: two factors have an influence on the dynamics of chemical system composition. The first is stoichiometric bonds caused by the conservation laws under chemical transformations. They are not a function of the system statement. The second factor is correlative bonds of the interaction; they are created in the process and represent a function of the system state. In accordance with the principle of simplified description, as a result of the difference in the relaxation times of the partial processes, proceeding of the parameters synchronization can also be considered as a discrete process, detaching the evolution states of a chemical system, every one of which is characterized by its own level of correlative bonds. This level is changed at the transition from one evolution state to another. This allows in every state to detach its own set of independent variables needed and sufficient for assigning in advance the dynamics of the non-equilibrium chemical system composition.

There are two main methods for assigning in advance the dynamics of the chemical system composition, *via* the determination of a concentration of reacting substances in time and *via* the rates of the elementary and the final reaction. The first method leads to the number of independent composites and intermediate substances as a number of independent elements of the vectors $\omega^{(1)}$ and $\omega^{(2)}$. The second method leads to the number of independent elementary and final reactions as a number of the independent composites vectors v and R. Both methods allow the assignment in advance of the dynamics of the chemical system composition, only at known starting composition. The characteristic numbers determined in this way can be called the dynamic ones. As shown, they do not coincide with the numbers of the stoichiometric independent composites and intermediate substances, and elementary and final reactions. In the equilibrium system (at $t >> t_4$) the dynamic characteristic numbers are equal to zero: its decomposition at the well-known start is determined by the external parameters. So, the dynamic characteristic numbers of the equilibrium system do not coincide with the characteristic numbers determined in accordance with the Gibbs phase rule. This is connected with the fact that the Gibbs characteristic numbers of the equilibrium system determine its state (composition) in an unknown starting state (composition).

Thus, the successful application of the abridged description principle allows to look at the methodological content of the characteristic numbers of the equilibrium and non-equilibrium systems.

REFERENCES

[1] Sabat'je P. *Kataliz v organicheskoj khimiji*, Leningrad: Goskhimtiekhizdat, **1932**, 416 p.

[2] Christiansen J. A. *The elucidation of reaction mechanisms by the method of intermediate in quasi-stationary concentrations*, Advan. Catalysis. N. Y.: Acad. Press Inc., **1953** (5), p. 311.

[3] Aris R. *The fundamental arbitrariness in stoichiometry // Chem. Ing. Sci.*, **1953**, 18 (8), p. 554.

[4] Sellers P. *Algebraic complexes applied to chemistry // Proc. Nat. Acad. Sci.*, **1966**, 55 (4), p. 693.

[5] Jouguet E. *Observations sur les principles et les theorems generaux de la statique chimique // J. de l'Ecole Poly., II Ser.*, **1921**, 21, s. 61.

[6] Jouguet E. *Sur les lois de la dynamique relatives chimique aux sens des reactions irreversibles // J. de l'Ecole Poly., II Ser.*, **1921**, 21, s. 181.

[7] Brinkley S. *Note on the conditions of equilibrium for systems of maxy constituents // J. Chem. Phys.*, **1946**, 14 (2), p. 563.

[8] Stepanov N., Yerlykina M., Philippov G. *Metody liniejnoy algebry v phizicheskoj khimiji*, Moskva: MGU, **1976**, 360 p.

[9] Gibbs J. *Thermodinamicheskije raboty*, Moskva: Gostiehteoretizdat, **1950**, 492 p.

[10] Storonkin A., Zharov V., Marinichev A. *Pravilo phaz Gibbsa i ego razvitije // Zhurnal phys. khimiji*, **1976**, 50 (12), p. 3048.

[11] Bogoliubov N. *Probliemy dinamicheskoy tieoriji v statisticheskoj phisikie*, Kyjiv: Naukova Dumka, **1970**, 522 p.

[12] Kuni F. *Pravilo phaz Gibbsa i ideya o sokrashchionnom opisanii v statisticheskoy phisikie // Zhurnal phys. khimiji*, **1976**, 50 (12), p. 3031.

[13] Akhijezier A., Peletminskiy S. *Metody statisticheskoy phisiki*, Moskva: Nauka, **1977**, 368 p.

[14] Nikitin E. *Tieoriya elemientarnykh atomno-molekuliarnykh processov v gazakh*, Moskva: Khimija, **1970**, 455 p.

[15] Medvedevskikh Yu., Kucher R. *Princyp sokrashchionnogo opisaniya i kharaktieristicheskije chisla nieravnoviesnoj khimicheskoj sistiemy*, Kyjiv: Naukova Dumka, **1986**, p. 22.

[16] Glesston S., Lejder K., Earring G *Theoria absolutnykh skorostiej reakcij*, Moskva: IL, **1948**, 583 p.

[17] Tiomkin M. Appendix to the N. N. Semenov's monograph *"O niekotorykh probliemakh khimicheskoj kinetiki i reakcionnoj sposobnosti"*, Moskva: AS USSR, **1958**, 686 p.

[18] Buchachenko A. *Dinamika elemientarnykh processov v zhydkosti // Uspiekhi khimiji*, **1979**, 48 (10), p. 1713.

[19] Yablonskiy G., Bykov V., and Gorban' A. *Kineticheskije modeli kataliticheskikh reakcij*, Novosibirsk: Nauka, **1983**, 253 p.

[20] Snagovsiy Yu., Ostrovskiy G. *Modelirovanije kinetiki heterogennykh kataliticheskikh processov*, Moskva: Khimija, **1976**, 248 p.

[21] Tiomkin M. *Kinetika stacionarnykh slozhnykh reakcij* (in book *"Mekhanizm i kinetika slozhnykh kataliticheskikh reakcij"*), Moskva: Nauka, **1970**, p. 57.

Chapter 2. The method of routes and kinetic models

2.1. Topology of the kinetic scheme and the routes

The theory of kinetics as a theory of an intermediate substance serves first of all to solve two main problems. The first problem is the quantum-chemical description of elementary acts, the total of which makes up a chemical process. This problem includes questions about the determination of the nature of intermediate substances and their thermodynamic characteristics, composition and structure of activated complexes of an elementary reaction for the calculation of the activation energy and entropy. Thus, we talk about a full or partial description of a topographic map of the potential energy of a system, which provides the information about potential holes, corresponding to the states of an intermediate substance, and an activated barrier separating one hole from another. The presence of such a topographic map of the potential energy of a system exactly determines the so-called chemical mechanism of a reaction.

The second problem is the formulation of general methods of description of the evolution of chemical system composition from the indignant starting state into a final equilibrium one and also the construction of formalized kinetic models as mathematical manners of separated states of a chemical process. This problem can be considered as phenomenological to the first problem, taking into account that the constants of a kinetic equation are assigned in advance as a function of an external parameter, the nature of composites and an intermediate substance.

Since the first problem is far from its solution, a postulated (not real) chemical mechanism of a reaction, that is, grounded and probable only in one or another degree, is taken as the basis of a kinetic model construction. Therefore, the task is not to obtain the kinetic model of a concrete catalytic or radical-chain process, but to develop the algorithm of its construction based only on the fact that the sets of composites $A=\{A_i\}$, $i=\overline{1,\ m}$, an intermediate substance $X=\{X_j\}$, $j=\overline{1,\ n}$, final $Q=\{Q_f\}$, $f=\overline{1,\ q}$ and an elementary reaction $S=\{S_k\}$, $k=\overline{1,\ s}$ are known.

More often, taking into account the time scale, we are interested in the quasi-stationary state of chemical system evolution, on which, as was shown in the Chapter 1, both the terms "elementary reaction rate" and "final reaction rate" have a physical sense and the concentrations of intermediate substances are not a function of time, but represent the functions of composite substances concentrations. This principle of the stationary reactions theory, also called the Bodenshtein–Semenov's principle, is the base of the algorithm for the construction of the kinetic models of a linear catalytic and radical-chain processes on the quasi-stationary state of a chemical system evolution.

There is a number of references [1–13] in which an algorithm of the kinetic models construction is described, but mainly two widely used methods are applied, namely linear algebra [1, 2, 7, 10–13] and the theory of graphs [5, 6, 8, 9]. In the most of the proposed algorithms the main attention is paid into obtaining the expression for the rate of an elementary reaction. Principally, it suffices to use the vector of a rate of an elementary reaction to determine the vector of the rate of a composite substance's formation and in such a way to describe the evolution of a chemical system's composition. In particular cases, however, the expressions for the final reactions rates are retained, since in complicated systems with a set of final reactions the knowledge of an elementary reaction rate does not mean knowledge of the final reactions rates.

The route method allows to overcome the above-mentioned lack and to determine the relation between a rate of a composite substance's formation, elementary and final reactions and the rates on the routes. In this sense the presented method more fully reflects the kinetics of a chemical process than other methods, is more visible, convenient and needs only an elementary knowledge of the matrix theory and the theory of graphs.

The term "route" was introduced by Horiuti [14–16]. From his point of view, any linear combination of an elementary, linear and non-linear factor with an interface substance's reactions previously multiplied on some stoichiometric numbers satisfying the condition of equality to zero of a stoichiometric coefficient of an intermediate substance represents the route realizing the final chemical transformation. According to Horiuti a stoichiometric number can take any whole-number value in order to satisfy the above-mentioned condition, implying that the number of the runs of the presented elementary reaction *via* route can be more than 1.

Using the interpretation proposed below we consider the term "route" in a more specific way; first, let us only refer to linear on intermediate substance mechanisms; secondly, the number of runs of an elementary reaction should not exceed 1.

So, let us assign sets *Q, A, X* and *S*. Likewise, the stoichiometric equations of the final and elementary reactions are assigned:

$$Q_f : \sum_i \nu_{fi} A_i = 0 \tag{2.1}$$

$$S_k : \sum_i \alpha_{ki}^{(1)} A_i + \sum_j \alpha_{kj}^{(2)} X_j = 0 \tag{2.2}$$

As mentioned earlier, let us assume the stoichiometric coefficients ν_{fi}, $\alpha_{ki}^{(1)}$, $\alpha_{kj}^{(2)} >$ 0 for a product and ν_{fi}, $\alpha_{ki}^{(1)}$, $\alpha_{kj}^{(2)} < 0$ for a starting material under assigned stoichiometric note of the final and an elementary reaction.

All elementary reactions are monomolecular or linear ones upon an intermediate substances, so that $\alpha_{kj}^{(2)}$=0, +1 or –1.

Next, let us consider a linear combination of elementary reactions previously multiplying their stoichiometric equations by stoichiometric numbers γ_{uk}, $u = \overline{1,\ p}$, $k = \overline{1,\ s}$ in order to obtain the linear combination $\sum_k \gamma_{uk} S_k$, the stoichiometric coefficient of an intermediate substance will be equal to zero. In such a way a stoichiometric coefficients will be ordered to the condition

$$\sum_k \gamma_{uk} \alpha_{kj}^{(2)} = 0 \tag{2.3}$$

for all $u = \overline{1,\ p}$ and $j = \overline{1,\ n}$.

The second requirement for the stoichiometric numbers γ_{ik} is that in a linear combination $\sum_k \gamma_{uk} S_k$, according to equation (2.3) reflecting final chemical transformation, an elementary reaction can be represented only one time. For the back elementary reaction this means that γ_{uk} is ± 1; if γ_{uk} cannot be presented, then γ_{uk}=0.

Since the linear combination $\sum_k \gamma_{uk} S_k$ also realizes the final chemical transformation, it can be written as a linear combination of composite substances, otherwise

$$\sum_k \gamma_{uk} S_k = \sum_i \lambda_{ui} A_i \tag{2.4}$$

In turn, a linear combination of the composite substances can be obtained by the linear combination of the final reactions. So,

$$\sum_k \gamma_{uk} S_k = \sum_i \lambda_{ui} A_i = \sum_f \delta_{uf} Q_f \tag{2.5}$$

The topology of the kinetic scheme assigned in advance to a set of linear elementary reactions $S=\{S_k\}$, $k=\overline{1,\ s}$ and on a set of an intermediate substances $X=\{X_j\}$, $j=\overline{1,\ n}$ is visibly reflected by a graph $G=\{S, X\}$, the top of which is formed by intermediate substances X_j and ribs are elementary reactions connecting these tops. Ribs can be directed if an elementary reaction is not reversible and in-directed, in other words if they are reversible.

Under condition $\gamma_{uk}=0, \pm 1$, the ratio (2.3) will be satisfied then and only at that time, when an intermediate substance X_j in the linear combination $\sum_k \gamma_{uk} S_k$ will appear twice: one time with $\alpha_{kj}^{(2)}=1$, and the second time with $\alpha_{kj}^{(2)}=-1$, that is, the first time as a "product" of an elementary reaction, the second time as a "starting substance". Since this corresponds to all intermediate substances for which $\alpha_{kj}^{(2)} \neq 0$, in the linear combination presented here $\sum_k \gamma_{uk} S_k$ in the latter case is a cyclic sequence of an elementary reaction on a graph $G=\{S, X\}$. Since $\gamma_{uk}=0$, 1 or -1 in the assigned cyclic sequence, any elementary reaction can participate (and cannot participate if $\gamma_{ik}=0$) only once. Thus, from the requirements $\gamma_{ik}=0 \pm 1$, $\alpha_{kj}^{(2)}=0 \pm 1$ follows that a simple cycle without repeating ribs and tops corresponds to a linear combination $\sum_k \gamma_{uk} S_k$ on a graph $G=\{S, X\}$.

Note that the transition from X_j into X_i *via* the direction of an elementary reaction and after that from X_i into X_j *via* a back direction will not be considered as a simple cycle, since it realizes a null final chemical transformation.

A simple cycle of a graph $G=\{S, X\}$ we will call a route U_u, realizing a final chemical transformation. In accordance with equation (2.5) we have

$$U_u = \sum_k \gamma_{uk} S_k = \sum_i \lambda_{ui} A_i = \sum_f \delta_{uf} Q_f \tag{2.6}$$

So, in accordance with equation (2.6), route U_u yields a final reaction in the stoichiometric form $\delta_{uf}Q_f$ in such a way, that δ_{uf} can be named by the order of the route U_u on the presented final reaction Q_f; λ_{ui} is a stoichiometric coefficient of the composite substance A_i *via* the presented route. At this, taking into account equations (2.1), (2.2) and (2.6), the following ratios hold:

$$\lambda_{ui} = \sum_k \gamma_{uk} \alpha_{ki}^{(1)} \tag{2.7}$$

$$\lambda_{ui} = \sum_f \delta_{uf} \nu_{fi} \tag{2.8}$$

Let us determine the term "rate" r_u of the route U_u in addition to the terms of the final reaction rate R_f and an elementary reaction rate v_k:

$$R_f = \frac{\mathrm{d}n_{fi}}{\nu_{fi}\mathrm{d}t} = \frac{\mathrm{d}\chi_f}{\mathrm{d}t} \tag{2.9}$$

$$v_k = \frac{\mathrm{d}n_{ki}}{\alpha_{ki}^{(1)}\mathrm{d}t} = \frac{\mathrm{d}\chi_k}{\mathrm{d}t} \tag{2.10}$$

$$r_u = \frac{\mathrm{d}n_{ui}}{\lambda_{ui}\mathrm{d}t} = \frac{\mathrm{d}\chi_u}{\mathrm{d}t} \tag{2.11}$$

Here χ_f, χ_k and χ_u are the number of runs of the final reaction, elementary reaction and *u*-route, respectively.

It follows from these determinations that the rate of the *i*-composite of substance formation can be found from the expressions

$$\omega_i^{(1)} = \sum_f \nu_{fi} R_f \tag{2.12}$$

$$\omega_i^{(1)} = \sum_k \alpha_{ki}^{(1)} v_k \tag{2.13}$$

$$\omega_i^{(1)} = \sum_u \lambda_{ui} r_u \tag{2.14}$$

At this, taking into account relations (2.7) and (2.8) we have

$$\omega_i^{(1)} = \sum_u \lambda_{ui} r_u = \sum_u \sum_k \gamma_{uk} \alpha_{ki}^{(1)} r_u = \sum_k \alpha_{ki}^{(1)} \sum_u \gamma_{uk} r_u \tag{2.15}$$

$$\omega_i^{(1)} = \sum_u \lambda_{ui} r_u = \sum_u \sum_f \delta_{uf} v_{fi} r_u = \sum_f v_{fi} \sum_u \delta_{uf} r_u \tag{2.16}$$

By comparing equation (2.15) with equation (2.13), and equation (2.16) with equation (2.12), we find

$$v_k = \sum_u \gamma_{uk} r_u \tag{2.17}$$

$$R_f = \sum_u \delta_{uf} r_u \tag{2.18}$$

Thus, applying the vector of rates on the routes $r=(r_u)$, $u = \overline{1,\ p}$ we can express all main characteristics of a chemical process: the vector of the rate of a composite substance's formation $\omega^{(1)}=(\omega_I^{(1)})$, $i = \overline{1,\ m}$

$$\omega^{(1)} = \lambda^T r \tag{2.19}$$

the vector of the rates of the final reaction $R = (R_f)$, $f = \overline{1,\ q}$

$$R = \delta^T r \tag{2.20}$$

and the vector of the rates of an elementary reaction $v = (v_k)$, $k = \overline{1,\ s}$

$$v = \gamma^T r \tag{2.21}$$

Here $\lambda=(\lambda_{ui})$, $u=\overline{1,\ p}$, $i=\overline{1,\ m}$ is a matrix with the dimension $p \times m$; $\delta=(\delta_{uf})$, $u=\overline{1,\ p}$, $f=\overline{1,\ q}$ is a matrix with the dimension $p \times q$; $\gamma=(\gamma_{uk})$, $u=\overline{1,\ p}$, $k=\overline{1,\ s}$ is a matrix with the dimension $p \times s$, where p is the number of routes of graph $G=\{S, X\}$.

Excluding the possible trivial cases, matrices λ, δ and γ do not have back directions; respectively, in general the back direction to equations (2.19)-(2.21) transformations is absent. This means that the vector r is more informative than the vectors $\omega^{(1)}$, R_f and v.

Next, let us consider the question about the number of independent routes, as before dividing them into stoichiometrically and dynamically independent ones. By introducing the column vector γ_u, $\gamma_u^T=(\gamma_{u1}, .., \gamma_{un})$, the elements of which are stoichiometric numbers of k elementary reactions in the u-route, let us re-write equation (2.3) in the matrix form

$$\left(\alpha^{(2)}\right)^T \gamma_u = 0 \tag{2.22}$$

It follows from this that the number $p^{(c)}$ of a linearly independent vector γ_u or the number of a linearly independent elements of a vector γ_u is equal to $s\text{–}rk(\alpha^{(2)})$. Since a rank $\alpha^{(2)}$ is equal to n–1, we have $p^{(c)}=s-n+1$. Since every vector γ_u assigns the stoichiometric equations of the u-route, the number of stoichiometric independent routes is equal to the number of independent vectors γ_u:

$$p^{(c)} = s - n + 1 \tag{2.23}$$

This equation represents the Horiuti rule [13], but it is specified by us as a number of a stoichiometric independent routes. Expression (2.23) means that, knowing the basis from the $p^{(c)}$ linearly independent vectors γ_u the rest can be founded from the solution of equation (2.22) as a linear combination of the base ones.

The condition for the quasi-stationarity of a system on the basis of an intermediate substance can be written (see Chapter 1) as

$$\omega^{(2)} = \left(\alpha^{(2)}\right)^T v \tag{2.24}$$

This equation, as it was shown in Chapter 1, limits the number of linearly independent elements of the vector V and in the quasi-stationarity system the number s^* of dynamically independent elementary reactions is equal to $s^*=s-n+1$. By comparing this ratio with equation (2.23) we have

$$s^* = p^{(c)} \tag{2.25}$$

However, by substituting in expression (2.24) a vector v according to equation (2.21) we will obtain

$$\omega^{(2)} = \left(\gamma\alpha^{(2)}\right)^T r = 0 \tag{2.26}$$

However, at the same time, in accordance with equation (2.22) all elements of the matrix $\gamma\alpha^{(2)}$ are equal to zero; so on the left from the vector r in equation (2.26) is a null matrix. In such a way, the condition of the quasi-stationarity of a system upon the intermediate substances in a form of equation (2.26) does not limit the elements of the vector r.

Next, let us consider the conservation law in the form $(\beta^{(1)})^T\omega^{(1)}=0$. It limits the vector $\omega^{(1)}$ of the rates formation of the composites substances but not upon the vector r. Using equation (2.19) in the conservation law we have

$$\left(\beta^{(1)}\right)^T \omega^{(1)} = \left(\lambda\beta^{(1)}\right)^T r = 0 \tag{2.27}$$

where $\lambda\beta^{(1)}=0$.

So, neither the condition of quasi-stationarity in the form (2.26) nor the conservation law in the form (2.27) limit in any way the elements of the rates of route r vector; all of them are linearly independent in the sense that the rate upon the any route can not be founded as a linear combination of the rates upon the rest routes. This conception can be considered as a principle of the route independency in a non-equilibrium quasi-stationary system according to which

$$p^* = p \tag{2.28}$$

To illustrate this let us consider the example of route formation of a catalytic and radical-chain process.

2.1.1. Ili-Ridil's chemical mechanisms of the catalytic reaction

$$2CO + O_2 \longleftrightarrow 2CO_2 \tag{2.29}$$

In the simplest variant it is proposed that the presented reaction is realized *via* the totality of an elementary reaction

$$\begin{aligned} &1.\ O_2 + Z \longrightarrow ZO_2 \\ &2.\ CO + ZO_2 \longrightarrow CO_2 + ZO \\ &3.\ CO + ZO \longrightarrow CO_2 + Z \end{aligned} \tag{2.30}$$

Here Z is an active center of a catalyst, X={Z, ZO_2, ZO}. A graph G={S, X} reflecting the topology (2.30) on a set X is represented in Figure 2.1. All ribs of a graph are directed and the number of routes is equal to 1; this graph is formed by a sequence of an elementary reaction (2.30), a vector γ^T=(1, 1, 1). The route realizes a final reaction in the stoichiometric form (2.29).

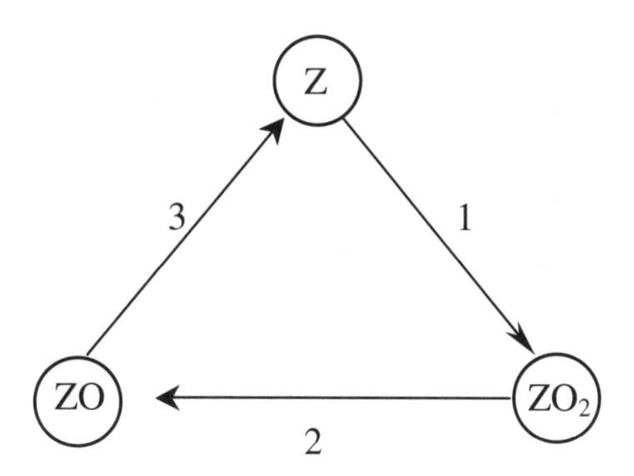

Figure 2.1. A graph G={S, X} reflecting the topology (2.30).

Let us complicate scheme (2.30) by introducing a possible elementary reaction, but by own stoichiometry.

$$\begin{aligned} &4.\ 2CO + ZO_2 \longrightarrow 2CO_2 + Z \\ &5.\ CO + O_2 + ZO \longrightarrow CO_2 + ZO_2 \\ &6.\ CO + O_2 + Z \longrightarrow CO_2 + ZO \end{aligned} \tag{2.31}$$

An elementary reaction ((1)–(3) in scheme (2.30)) and a reaction (4) in scheme (2.31) represent a so-called sequential redox mechanism, that is, oxidization–reduction of a catalyst by one of the components of reactive mixture, whereas elementary reactions (5) and (6) in equation (2.31) represent a so-called associated mechanism in which two components, namely oxidizing and reducing agents, take part in an elementary act simultaneously, "helping" one another.

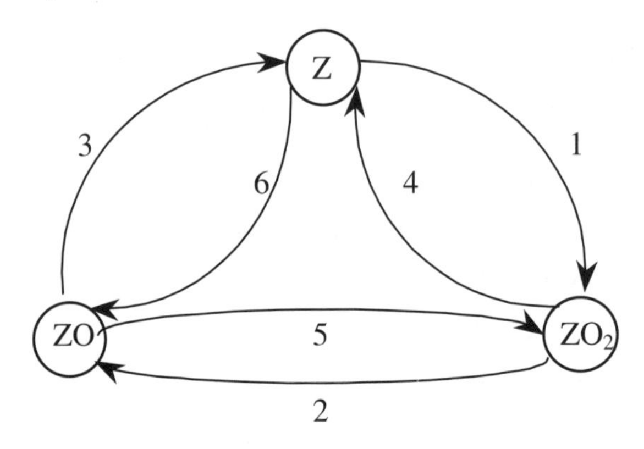

Figure 2.2. A graph $G=\{S, X\}$ reflecting the topology (2.30) and (2.31).

A graph $G=\{S, X\}$ reflecting the topology (2.30) and (2.31) is represented in Figure 2.2. As we can see, we have three simple cycles or routes between every pair of intermediate substances:

$$
\begin{aligned}
&U_1 = S_1 + S_4, && \gamma_1^T = (1\ 0\ 0\ 1\ 0\ 0), && \delta_1 = 1 \\
&U_2 = S_2 + S_5, && \gamma_2^T = (0\ 1\ 0\ 0\ 1\ 0), && \delta_2 = 1 \\
&U_3 = S_3 + S_6, && \gamma_{31}^T = (0\ 0\ 1\ 0\ 0\ 1), && \delta_3 = 1
\end{aligned}
\tag{2.32}
$$

All of these realize a final reaction in the stoichiometric form (2.29), that is why δ_l=1. Furthermore, we have another two routes formed by a cyclic sequence of the three elementary reactions (the first of them is already known):

$$U_4 = S_1 + S_2 + S_3, \qquad \gamma_4^T = (1\ 1\ 1\ 0\ 0\ 0), \qquad \delta_4 = 1$$
$$U_5 = S_6 + S_5 + S_4, \qquad \gamma_5^T = (0\ 0\ 0\ 1\ 1\ 1), \qquad \delta_5 = 2 \tag{2.33}$$

Route U_5 yields a final reaction in the stoichiometric form $4CO + 2O_2 \rightarrow 4CO_2$, that is why the stoichiometric number of the final reaction (2.29) is equal to 2. Subsequently, the rate of the reaction (2.29) is equal to

$$R = r_1 + r_2 + r_3 + r_4 + 2r_5$$

Let us once more complicate the kinetic schemes (2.30) and (2.31), assuming that all elementary reactions involved are reversible. Then a graph $G=\{S, X\}$ (see Figure 2.3) will be undirected.

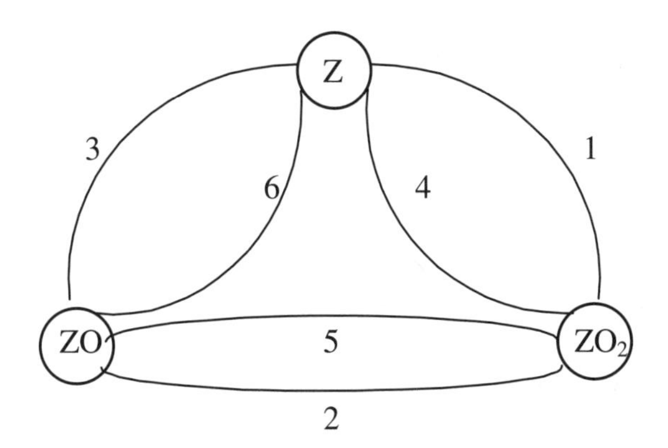

Figure 2.3. A graph $G=\{S, X\}$ reflecting the topology of kinetic schemes (2.30) and (2.31) under the reversibility of the all elementary reactions.

Besides the above calculated five routes U_1–U_5 the undirected graph in Figure 2.3 also has a new, sixth, one, namely:

$$\begin{aligned}
&U_6 = S_1 - S_5 + S_3, && \gamma_6^T = (1\ 0\ 1\ 0\ -1\ 0), && \delta_6 = 0 \\
&U_7 = S_6 + S_5 - S_1, && \gamma_7^T = (-1\ 0\ 0\ 0\ 1\ 1), && \delta_7 = 1 \\
&U_8 = S_1 + S_2 - S_6, && \gamma_8^T = (1\ 1\ 0\ 0\ 0\ -1), && \delta_8 = 0 \\
&U_9 = -S_3 - S_2 + S_4, && \gamma_9^T = (0\ -1\ -1\ 1\ 0\ 0), && \delta_9 = 0 \\
&U_{10} = S_6 - S_2 + S_4, && \gamma_{10}^T = (0\ -1\ 0\ 1\ 0\ 1), && \delta_{10} = 1 \\
&U_{11} = -S_3 + S_5 + S_4, && \gamma_{11}^T = (0\ 0\ -1\ 1\ 1\ 0), && \delta_{11} = 1
\end{aligned} \tag{2.34}$$

As we can see, the routes U_6, U_8 and U_9 are empty, that is the stoichiometric numbers of the final reaction (2.29) in these routes are equal to zero. So, from the 11 routes of the undirected graph $G=\{S, X\}$ in Figure 2.3 seven of them realize the chemical transformations in the stoichiometric form (2.29) and for them $\delta_i=1$, one route (U_5) satisfies (2.29) in the twin stoichiometric form, for $\delta_5=2$, and three are empty. That is why the rate of the final reaction (2.29) is equal to

$$R=r_1 + r_2 + r_3 + r_4 + 2r_5 + r_7 + r_{10} + r_{11}$$

Taking into account the reversibility of all elementary reactions, all routes are also reversible, and every one can be written as follows

$$\begin{aligned}
&U_1 = S_1 + S_4 = -(-S_1 - S_4) \\
&U_1 = (S_1 - S_5 + S_3) = -(-S_1 + S_5 - S_3)
\end{aligned} \tag{2.35}$$

It is obvious that the combination of the $S_1 + S_4$ and $-S_1-S_4$, or $S_1-S_5 + S_3$ and $-S_1 + S_5-S_3$ type are not independent, that is, they represent the same reversible route, but written in direct and indirect ways of the final chemical transformation.

2.1.2. Radical-chain process: co-oxidization of two hydrocarbons

Let us consider the situation, when more than one final reaction proceeds in the system, for example, liquid-phase co-oxidization of two hydrocarbons R_1H and R_2H:

$$Q_1 : R_1H + O_2 \longrightarrow R_1OOH, \tag{2.36}$$

$$Q_2 : R_2H + O_2 \longrightarrow R_2OOH \tag{2.37}$$

Under the simplest variant the following elementary reactions of chain propagation are considered:

$$
\begin{aligned}
&1.\ O_2 + R_1^{\bullet} \longrightarrow R_1OO^{\bullet}\\
&2.\ O_2 + R_2^{\bullet} \longrightarrow R_2OO^{\bullet}\\
&3.\ R_1H + R_1OO^{\bullet} \longrightarrow R_1OOH + R_1^{\bullet}\\
&4.\ R_1H + R_2OO^{\bullet} \longrightarrow R_2OOH + R_1^{\bullet}\\
&5.\ R_2H + R_1OO^{\bullet} \longrightarrow R_1OOH + R_2^{\bullet}\\
&6.\ R_2H + R_2OO^{\bullet} \longrightarrow R_2OOH + R_2^{\bullet}
\end{aligned}
\qquad (2.38)
$$

Thus, $X=\{R_1^{\bullet}, R_2^{\bullet}, R_1OO^{\bullet}, R_2OO^{\bullet}\}$; a graph $G=\{S, X\}$ reflecting the topology (2.38) is represented on a Figure 2.4. This graph is directed and that is why only three routes are easy visible:

$$
\begin{aligned}
&U_1 = S_1 + S_3, && \gamma_1^T = (1\ 0\ 1\ 0\ 0\ 0), && \delta_{11} = 1,\ \delta_{12} = 0\\
&U_2 = S_2 + S_6, && \gamma_2^T = (0\ 1\ 0\ 0\ 0\ 1), && \delta_{21} = 0,\ \delta_{22} = 1\\
&U_3 = S_1 + S_5 + S_2 + S_6, && \gamma_3^T = (1\ 1\ 0\ 0\ 1\ 1), && \delta_{31} = 1,\ \delta_{32} = 1
\end{aligned}
\qquad (2.39)
$$

Thus, the route U_1 realizes only the final reaction Q_1, U_2 realizes only the final reaction Q_2 and the route U_3 realizes both final reactions in the stoichiometric form ((2.36) and (2.37)).

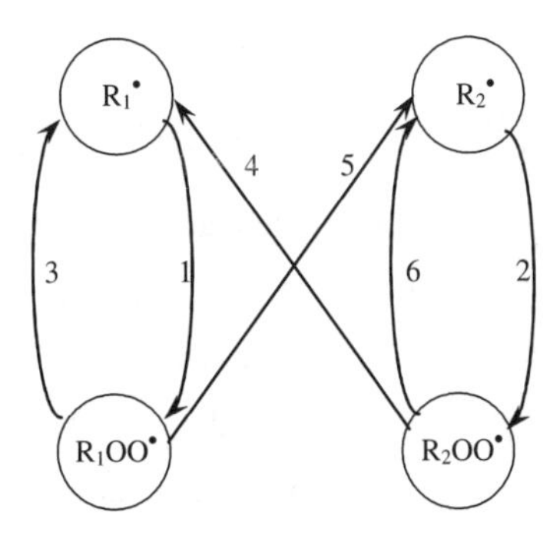

Figure 2.4. A graph $G=\{S, X\}$ reflecting the topology of a kinetic scheme (2.38).

So, the rates of final reactions R_1 and R_2 are:

$R_1=r_1 + r_3$, $R_2=r_2 + r_3$

For an illustration let us complicate the kinetic scheme (2.38) by adding the reversible elementary reaction of a change between the radicals $R_1^{\bullet}$ and $R_2^{\bullet}$:

$$7.\ R_1H + R_2^{\bullet} \longleftrightarrow R_2H + R_2^{\bullet} \quad (2.40)$$

As a result, two another routes are obtained

$$\begin{aligned} U_4 &= S_1 + S_5 + S_7, \quad & \delta_{41} = 1,\ \delta_{42} = 0 \\ U_5 &= S_2 + S_3 - S_7, \quad & \delta_{51} = 0,\ \delta_{52} = 0 \end{aligned} \quad (2.41)$$

Every one of these routes realizes only Q_1 or Q_2. Under this variant their rates are:

$R_1=r_1 + r_3 + r_4$, $R_2=r_2 + r_3 + r_5$

A total rate of the oxygen consumption R_{O2} is equal to a sum $R_1 + R_2=r_1 + r_2 + 2r_3+ r_4 + r_5$.

The characteristics of the formation of the free-radical polymerization routes will be considered below.

2.2. A kinetic model of a route of linear catalytic reactions

Let us assume again, that the whole kinetic scheme of the final chemical transformations $Q=\{Q_f\}$, $f=\overline{1,\ q}$ is determined as a set $S=\{S_k\}$, $k=\overline{1,\ s}$ or $S=\{S_{ij}^{m(ij)}\}$, $m(ij)=\overline{1,\ s(ij)}$ of an elementary reaction assigned on a set $X=\{X_i\}$, $i=\overline{1,\ n}$ of an intermediate substance. For the linear catalytic reactions $X=\{X_i\}$ this is a set of states of an active center of a catalyst. The topology of a kinetic scheme is reflected by a graph $G=\{S, X\}$, tops of which are formed by intermediate substances X_i, ribs usually by reversible elementary reactions $S_{ij}^{m(ij)}=-S_{ji}^{m(ji)}$, connecting the i-top with the j-top. To every top X_i corresponds its weight x_i, that is, relative concentration of an intermediate substance X_i:

$$x_i = L_i / L \qquad (2.42)$$

where L_i is a number of active centers in a state X_i and L is total number of active centers in a system. For the catalytic processes L=constant.

Weights $b_{ij}^{(m(ij))}$ and $b_{ji}^{(m(ji))}$ correspond to a rib $S_{ij}^{m(ij)}=-S_{ji}^{m(ji)}$; they express the frequencies of a transition from one top to another and are determined in accordance with the active mass law and by the stoichiometry of the elementary reaction, $S_{ij}^{m(ij)}$. If

$$S_{ij}^{(m(ij))} : A_j^{(m(ij))} + X_j \longleftrightarrow A_i^{(m(ji))} + X_i : S_{ji}^{(m(ji))} \qquad (2.43)$$

then

$$b_{ij}^{(m(ij))} = k_{ij}^{(m(ij))}\left[A_j^{(m(ij))}\right], \quad b_{ji}^{(m(ji))} = k_{ji}^{(m(ji))} A_i^{(m(ji))} \qquad (2.44)$$

Taking into account all $s(ij)$ methods of transition from X_j into X_i and back, we have

$$b_{ij} = \sum_{m(ij)} b_{ij}^{(m(ij))}, \qquad b_{ji} = \sum_{m(ji)} b_{ji}^{(m(ji))} \qquad (2.45)$$

Under this assumption the rate of transition from X_j to X_i is equal to $b_{ij}x_j$, and that is why the condition of stationarity of a system of the intermediate substances can be written as follows:

$$x_i \sum_{j \neq i} b_{ij} = \sum_{j \neq i} b_{ij} x_j \qquad (2.46)$$

or

$$\sum_{j \neq i} (b_{ij} x_j - b_{ji} x_i) = 0 \qquad (2.47)$$

In the matrix form

$$Bx = 0 \qquad (2.48)$$

where x is a column vector of the concentrations of intermediate substances; $B=(b_{ij})$ is a kinetic square matrix of n-order, the elements of which, b_{ij}, $i \neq j$, are determined in accordance with the ratios (2.44) and (2.45); diagonal elements b_{ii} are equal with the back sign to a sum of the elements of the corresponding column:

$$b_{ii} = -\sum_{j \neq i} b_{ij} \tag{2.49}$$

Next, let us introduce the auxiliary terms [17].

A sub-set $V(v, l) \subseteq X$, containing l tops of a graph $G=\{X, S\}$ will be called a cardinal number group l. A vector $v=(v_1, \ldots, v_n)$, the elements of which can take a value 1 or 0, depending on the presence in this group of a top X_i, can be a characteristic of the above-said group. For example, at $n=5$ the vector $v=(0, 1, 0, 1, 1)$ represents the group $V(X_2, X_4, X_5)$ or, more simply, V(2, 4, 5); the cardinal number group $l=3$. The number $N(v, l)$ of a group is the number of sub-sets of the cardinal number group l of the set $X=\{X_i\}$, $i=\overline{1,\ n}$:

$$N(v, \ell) = C_n^{\ell} \tag{2.50}$$

A cardinal number group l with the sequence of regional tops assigned to it, for example, clockwise or counter-clockwise, will be called a basis(set) $W(v, w, l)$ of the length l. In the one of the l tops of a group we have l–1 variants of a motion into the rest of the tops; in the next top the numbers of motion variants into the remanding ones is equal to l–2, etc. A total number of different sequences of the round of l tops is equal to $(l-1)!$. Just as the opposite sequence of a counter-clockwise round corresponds to every sequence of a clockwise round, the total number of different sequences of the l tops is equal to $(l-1)!\,/\,2$. However, in this ratio a correction should be made for the case when $l=2$. So, the number $N(v, w, l)$ of the basis(sets) $W(v, w, l)$ in the groups $V(v, l)$ is determined by the expression

$$N(v, w, \ell) = {}^{(\ell - \delta_k)!}\!/_{2}, \qquad \ell \geq 2 \tag{2.51}$$

where

$$\delta_k = \begin{cases} 0 & if \quad \ell = 2 \\ 1 & if \quad \ell > 2 \end{cases} \tag{2.52}$$

In accordance with (2.51) and (2.52) every cardinal number groups l=2 and l=3 have a one basis(set), the cardinal number group l=4 has three basis(sets), represented in Figure 2.5. The cardinal number group l=5 has 12 basis(sets).

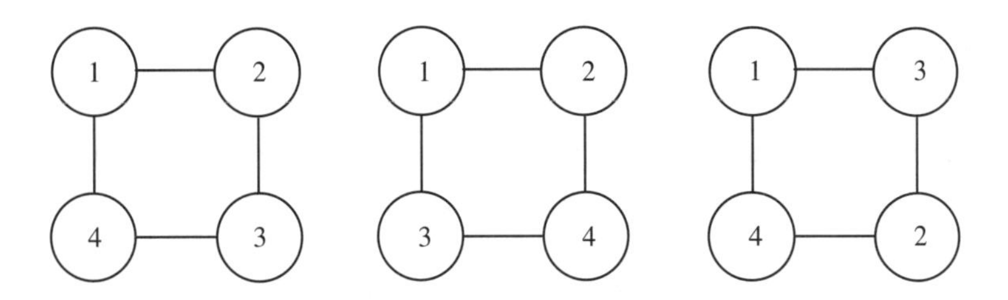

Figure 2.5. A basis(set) $W(v, w, l=4)$ of a group $V(v, l=4)$.

Let us reiterate that among all possible cycles of a graph $G=\{X, S\}$ we investigate only the simple ones, since they do not contain other cycles, but realize the final chemical transformations in which the stoichiometric coefficients of the intermediate substances are equal to zero. A simple cycle is a closed route which does not contain the repeating ribs and a top. It follows from this, that the stoichiometric number of an elementary reaction in the simple cycle is equal only to 0, –1 or +1, depending on if the presented elementary reaction is included in a simple cycle and, if so, in which direction? Thus, the term "route" will be used for a simple cycle, that is a closed route which does not contain repeating ribs and a top of a graph $G=\{X, S\}$ representing the topology of a kinetic scheme.

In a simple cycle the number of a ribs and tops coincide; let us call it the length l of a route. A minimal length of a route is 2. Let $U(v, w, l=2)=\{(s_{ij} + s_{ji})\}$ be the assignment of a set of such routes in the basis(set) $W(X_i, X_j)$. Taking into account the reversibility of an elementary reaction we have

$$U(v,w,\ell=2)=\{(S_{ij}^{(m(ij))}+S_{ji}^{(m(ji))})\}=-\{(S_{ji}^{(m(ij))}+S_{ij}^{(m(ji))})\},\quad m(ij)\neq m(ji) \qquad (2.53)$$

From (2.53) we obtain the number $N(v, w, u, l=2)$ of a route in the basis(set) $W(X_i, X_j)$:

$$N(v,w,u,\ell=2)=s(ij)(s(ij)-1)/2 \qquad (2.54)$$

For example, at $s(ij)$ = 3 a set $U(v, w, l = 2)$ contains three routes:

$$U(v,w,u=1)=\left(S_{ij}^{(1)}+S_{ji}^{(2)}\right)=-\left(S_{ji}^{(1)}+S_{ij}^{(2)}\right),$$

$$U(v,w,u=2)=\left(S_{ij}^{(1)}+S_{ji}^{(3)}\right)=-\left(S_{ji}^{(1)}+S_{ij}^{(3)}\right), \qquad (2.55)$$

$$U(v,w,u=3)=\left(S_{ij}^{(2)}+S_{ji}^{(3)}\right)=-\left(S_{ji}^{(2)}+S_{ij}^{(3)}\right),$$

Let us assign, in accordance with a set of routes (2.53), the set of delta-factors or cyclic characteristics [5, 6, 8] $\Delta(v, w, l=2)=\{\Delta(v, w, u, l=2)\}$:

$$\Delta(v,w,\ell=2)=\left\{\left(b_{ij}^{(m(ij))}b_{ji}^{(m(ji))}-b_{ji}^{(m(ij))}b_{ij}^{(m(ji))}\right)\right\},\quad m(ij)\neq m(ji) \qquad (2.56)$$

For example, for route (2.55) we have

$$\Delta(v,w,u=1)=b_{ij}^{(1)}b_{ji}^{(2)}-b_{ji}^{(1)}b_{ij}^{(2)},$$

$$\Delta(v,w,u=2)=b_{ij}^{(1)}b_{ji}^{(3)}-b_{ji}^{(1)}b_{ij}^{(3)}, \qquad (2.57)$$

$$\Delta(v,w,u=3)=b_{ij}^{(2)}b_{ji}^{(3)}-b_{ji}^{(2)}b_{ij}^{(3)},$$

For a basis(set) $W(v, w, l = 3) = W(X_i, X_j, X_k)$ a set of routes is $U(v, w, l = 3) = \{U(v, w, u, l = 3)\} = \{(S_{ij} + S_{jk} + S_{ki})\}$. Taking into account the reversibility of an elementary reactions we find

$$U(v,w,\ell=3)=\left\{\left(S_{ij}^{(m(ij))}+S_{jk}^{(m(jk))}+S_{ki}^{(m(ki))}\right)\right\}=-\left\{\left(S_{ik}^{(m(ki))}+S_{kj}^{(m(jk))}+S_{ji}^{(m(ij))}\right)\right\} \qquad (2.58)$$

A set of a delta-factors corresponds to a set of routes (2.58)

$$\Delta(v,w,\ell=3)=\left\{\left(b_{ij}^{(m(ij))}b_{jk}^{(m(jk))}b_{ki}^{(m(ki))}-b_{ik}^{(m(ki))}b_{kj}^{(m(jk))}b_{ji}^{(m(ij))}\right)\right\} \qquad (2.59)$$

It follows from this, that the number $N(v, w, u, l=3)$ of a route $U(v, w, u, l=3)$ in the basis(set) $W(X_i, X_j, X_k)$ is equal to $s(ij)\ s(jk)\ s(ki)$.

Analogous to (2.58) and (2.59), sets of the routes and delta-factors in the basis(sets) of more lengths are determined. The number $N(v, w, u, l \geq 3)$ of routes in the basis(sets) $W(v, w, l \geq 3)$ we find from the expression

$$N(v,w,u,\ell \geq 3) = \prod_{W(X_i,X_j) \in W(v,w,\ell \geq 3)} s(ij) \qquad (2.60)$$

where $W(X_i, X_j)$ is a component of the basis(sets) with a length $l=2$ of a basis(set) $W(v, w, l \geq 3)$.

In accordance with connection (2.49), the rank of the kinetic matrix in (2.48) is less than n and its determinant is equal to zero. On the other hand, a rank of the kinetic matrix B should be exactly equal to $n–1$ and a system of the linear homogeneous equations $Bx=0$ should only have a single solution which can be written as

$$x_i = cA_{ii} \qquad (2.61)$$

where c is a arbitrary constant. Taking into account the condition of the normalization $\sum_i x_i = 1$ we finally find

$$x_i = A_{ii} \Big/ \sum_i A_{ii} \qquad (2.62)$$

where A_{ii} are an algebraic additives of an elements b_{ii} of the matrix B.

It follows from the properties of the kinetic matrix B that the algebraic additives A_{ii} are positive at even-numbered $n–1$ and are negative in the opposite case. That is why we can conveniently use the values

$$M_i = A_{ii}(-1)^{n-1} = [B_i](-1)^{n-1} \qquad (2.63)$$

which are always positive. Here $[B_i]$ is a determinant of the matrix B_i obtained from the starting B by the deleting the i-line and a column. Let us re-write the solution of (2.62) as follows

$$x_i = M_i \Big/ \sum_i M_i \qquad (2.64)$$

Using (2.64) in the expression of an elementary reaction rate $v_{ij}^{(m(ij))}=b_{ij}^{(m(ij))}x_j– b_{ji}^{(m(ij))}x_i$, we will obtain the components, every one of which represents the rate of one of the routes proceeding *via* the presented elementary reaction. The studies of these components allow to describe the rate $r(v, w, u)$ of a route $U(v, w, u)$ in the basis(set) $W(v, w)$ of a group $V(v)$ by the following equation [17]:

$$r(v,w,u) = \left(\sum_i M_i \right)^{-1} M(v) \Delta(v,w,u) \tag{2.65}$$

Here

$$M(v) = [B(v)](-1)^{n-l} \tag{2.66}$$

can be called an additional minor of the group $V(v)$ of the cardinal number group l, since $[B(v)]$ is a determinant of the matrix $B(v)$ obtained from the starting one by deleting those lines and columns, of which the indexes coincide with the indexes of a graph $G=\{X, S\}$ represented in this $V(v)$ group. For example, if $V(v)=V(X_i, X_j X_k)$ of the cardinal number group $l=3$, then $B(v)$ will be obtained from the starting one B by deleting the i-, j- and k-lines and columns. For a group $V(v, l=n)$ including all tops of a graph $G=\{X, S\}$ we have $M(v)=1$. For a single group $V(X_i)$: $M(v)=M_i$.

In turn, from the rates of a route according to relations (2.14), (2.17) and (2.18) we can find the rates of composite formation and the rates of the elementary and final reactions:

$$\begin{aligned}
\omega_j^{(1)} &= \left(\sum_i M_i \right)^{-1} \sum_v M(v) \sum_w \sum_u \lambda(v,w,u,j) \Delta(v,w,u) , \\
v_k &= \left(\sum_i M_i \right)^{-1} \sum_v M(v) \sum_w \sum_u \gamma(v,w,u,k) \Delta(v,w,u) , \\
R_f &= \left(\sum_i M_i \right)^{-1} \sum_v M(v) \sum_w \sum_u \delta(v,w,u,f) \Delta(v,w,u) ,
\end{aligned} \tag{2.67}$$

As a example, let us consider the final reaction (2.29) and its kinetic schemes (2.30) and (2.31), assuming that all elementary reactions are reversible.

By denoting

$$\mathrm{Z} = X_3 \qquad \mathrm{ZO_2} = X_2 \qquad \mathrm{ZO} = X_1 \quad . \tag{2.68}$$

Le us re-write the elementary reactions S_1-S_6 in the (2.30) and (2.31) in a new form:

$$\begin{aligned}
&S_1 = S_{23}^{(1)} : \mathrm{O_2 + Z \longleftrightarrow ZO_2} \ : && S_{32}^{(1)} = -S_1 \\
&S_4 = S_{32}^{(2)} : \mathrm{2CO + ZO_2 \longleftrightarrow 2CO_2 + Z} \ : && S_{23}^{(2)} = -S_4 \\
&S_2 = S_{12}^{(1)} : \mathrm{CO + ZO_2 \longleftrightarrow CO_2 + ZO} \ : && S_{21}^{(1)} = -S_2 \\
&S_5 = S_{21}^{(2)} : \mathrm{CO + O_2 + ZO \longleftrightarrow CO_2 + ZO_2} \ : && S_{12}^{(2)} = -S_5 \\
&S_3 = S_{31}^{(1)} : \mathrm{CO + ZO \longleftrightarrow CO_2 + Z} \ : && S_{13}^{(1)} = -S_3 \\
&S_6 = S_{13}^{(2)} : \mathrm{CO + O_2 + Z \longleftrightarrow CO_2 + ZO} \ : && S_{31}^{(2)} = -S_6
\end{aligned} \tag{2.69}$$

The frequencies of a transition can be written as follows:

$$\begin{aligned}
&b_{23}^{(1)} = k_{23}^{(1)}[\mathrm{O_2}] \ , && b_{32}^{(1)} = k_{32}^{(1)} \ , \\
&b_{32}^{(2)} = k_{32}^{(2)}[\mathrm{CO}]^2 \ , && b_{23}^{(2)} = k_{23}^{(2)}[\mathrm{CO_2}]^2 \ . \\
&b_{12}^{(1)} = k_{12}^{(1)}[\mathrm{CO}] \ , && b_{21}^{(1)} = k_{21}^{(1)}[\mathrm{CO_2}] \ , \\
&b_{21}^{(2)} = k_{21}^{(2)}[\mathrm{CO}][\mathrm{O_2}] \ , && b_{12}^{(2)} = k_{12}^{(2)}[\mathrm{CO_2}] \ , \\
&b_{31}^{(1)} = k_{31}^{(1)}[\mathrm{CO}] \ , && b_{13}^{(1)} = k_{13}^{(1)}[\mathrm{CO_2}] \ , \\
&b_{13}^{(2)} = k_{13}^{(2)}[\mathrm{CO}][\mathrm{O_2}] \ , && b_{31}^{(2)} = k_{31}^{(2)}[\mathrm{CO_2}]
\end{aligned} \tag{2.70}$$

The elements of the third-order matrix B are equal to:

$$\begin{aligned}
&b_{12} = b_{12}^{(1)} + b_{12}^{(2)}, && b_{21} = b_{21}^{(1)} + b_{21}^{(2)}, \\
&b_{13} = b_{13}^{(1)} + b_{13}^{(2)}, && b_{31} = b_{31}^{(1)} + b_{31}^{(2)}, \\
&b_{23} = b_{23}^{(1)} + b_{23}^{(2)}, && b_{32} = b_{32}^{(1)} + b_{32}^{(2)}, \\
&b_{11} = -(b_{21}^{(1)} + b_{31}^{(2)}), && b_{22} = -(b_{12}^{(1)} + b_{32}^{(2)}), \\
&b_{33} = -(b_{13}^{(1)} + b_{23}^{(2)})
\end{aligned} \tag{2.71}$$

Writing a route with the same number system as in (2.32), (2.33) and (2.34) in a new form gives:

$$U_1 = S_{23}^{(1)} + S_{32}^{(2)} = -(S_{32}^{(1)} + S_{23}^{(2)})$$

$$U_2 = S_{12}^{(1)} + S_{21}^{(2)} = -(S_{21}^{(1)} + S_{12}^{(2)})$$

$$U_3 = S_{31}^{(1)} + S_{13}^{(2)} = -(S_{13}^{(1)} + S_{31}^{(2)})$$

$$U_4 = S_{23}^{(1)} + S_{12}^{(1)} + S_{31}^{(1)} = -(S_{32}^{(1)} + S_{21}^{(1)} + S_{13}^{(1)})$$

$$U_5 = S_{13}^{(2)} + S_{21}^{(2)} + S_{32}^{(2)} = -(S_{31}^{(2)} + S_{12}^{(2)} + S_{23}^{(2)})$$

$$U_6 = S_{23}^{(1)} + S_{12}^{(2)} + S_{31}^{(1)} = -(S_{32}^{(1)} + S_{21}^{(2)} + S_{13}^{(1)})$$

$$U_7 = S_{13}^{(2)} + S_{21}^{(2)} + S_{32}^{(1)} = -(S_{31}^{(2)} + S_{12}^{(2)} + S_{23}^{(1)})$$

$$U_8 = S_{23}^{(1)} + S_{12}^{(1)} + S_{31}^{(2)} = -(S_{32}^{(1)} + S_{21}^{(1)} + S_{13}^{(2)})$$

$$U_9 = S_{13}^{(1)} + S_{21}^{(1)} + S_{32}^{(2)} = -(S_{31}^{(1)} + S_{12}^{(1)} + S_{23}^{(2)})$$

$$U_{10} = S_{23}^{(2)} + S_{12}^{(1)} + S_{31}^{(2)} = -(S_{32}^{(2)} + S_{21}^{(1)} + S_{13}^{(2)})$$

$$U_{11} = S_{13}^{(1)} + S_{21}^{(2)} + S_{32}^{(2)} = -(S_{31}^{(1)} + S_{12}^{(2)} + S_{23}^{(2)}) \tag{2.72}$$

Having this image of routes, it is easy to write their delta-factors, for example:

$$\begin{aligned}
\Delta_1 &= b_{23}^{(1)} b_{32}^{(2)} - b_{32}^{(1)} b_{23}^{(2)} \\
\Delta_4 &= b_{23}^{(1)} b_{31}^{(1)} b_{12}^{(1)} - b_{32}^{(1)} b_{21}^{(1)} b_{13}^{(1)} \\
\Delta_6 &= b_{23}^{(1)} b_{31}^{(1)} b_{12}^{(2)} - b_{32}^{(1)} b_{21}^{(2)} b_{13}^{(1)} \\
\Delta_7 &= b_{13}^{(2)} b_{32}^{(1)} b_{21}^{(2)} - b_{31}^{(2)} b_{12}^{(2)} b_{13}^{(1)} \\
\Delta_8 &= b_{23}^{(1)} b_{31}^{(2)} b_{12}^{(1)} - b_{32}^{(1)} b_{21}^{(1)} b_{13}^{(2)}
\end{aligned} \tag{2.73}$$

etc.

A solution for relative concentrations of an intermediate substance in the presented case is written as follows:

$$x_i = \frac{M_i}{M_1 + M_2 + M_3} \tag{2.74}$$

where

$$M_1 = (-1)^2 \begin{vmatrix} b_{22} & b_{23} \\ b_{32} & b_{33} \end{vmatrix},\ M_2 = (-1)^2 \begin{vmatrix} b_{11} & b_{13} \\ b_{31} & b_{33} \end{vmatrix},\ M_3 = (-1)^2 \begin{vmatrix} b_{11} & b_{12} \\ b_{21} & b_{22} \end{vmatrix} \tag{2.75}$$

By substituting this solution into the expression of the rate of an elementary reaction, for example, S_1 from (2.69), we will obtain

$$\begin{aligned} v_1 - b_{23}^{(1)} x_3 - b_{32}^{(1)} x_2 = &\left(-b_{11}\left(b_{23}^{(1)} b_{32}^{(2)} - b_{32}^{(1)} b_{23}^{(2)}\right) + \left(b_{23}^{(1)} b_{31}^{(1)} b_{12}^{(1)} - b_{32}^{(1)} b_{21}^{(1)} b_{13}^{(1)}\right) + \right. \\ &+ \left(b_{23}^{(1)} b_{31}^{(1)} b_{12}^{(2)} - b_{32}^{(1)} b_{21}^{(2)} b_{13}^{(1)}\right) + \left(b_{13}^{(2)} b_{21}^{(2)} b_{32}^{(1)} - b_{31}^{(2)} b_{12}^{(2)} b_{23}^{(1)}\right) + \\ &+ \left.\left(b_{23}^{(1)} b_{31}^{(2)} b_{12}^{(1)} - b_{32}^{(1)} b_{21}^{(1)} b_{13}^{(2)}\right)\right) / \left(M_1 + M_2 + M_3\right) \end{aligned} \tag{2.76}$$

Here the number of components is equal to the number of a route proceeding *via* the presented elementary reaction S_1: that is U_1, U_4, U_6, U_7 and U_8, with delta-factors as given in (2.73). That is why, in accordance with the general form (2.65) every component in (2.76) is identified as a rate of its own route, for example

$$\begin{aligned} r_1 &= \frac{\left(b_{21} + b_{31}\right)\left(b_{23}^{(1)} b_{32}^{(2)} - b_{32}^{(1)} b_{23}^{(2)}\right)}{M_1 + M_2 + M_3} \\ r_4 &= \frac{b_{23}^{(1)} b_{31}^{(1)} b_{12}^{(1)} - b_{32}^{(1)} b_{21}^{(1)} b_{13}^{(1)}}{M_1 + M_2 + M_3} \end{aligned} \tag{2.77}$$

2.3. A kinetic model of a route of un-branched radical-chain processes

As stated before (see section 2.1), a set of elementary reactions included in chain propagation assigned to a set of radicals is the same as a set of an elementary reaction of the catalytic process assigned to a set of states of the active center. The difference is that the graph $G' = \{S, X\}$ reflects only the topology of chain propagation, which is only a part of the elementary reactions of un-branched radical-chain process, whereas the full kinetic scheme includes all elementary reactions [18]:

(i) initiation, including first and the second ones,

$$I_{ij} \xrightarrow{k_{diff}} X_i + X_j \tag{2.78}$$

ii) chain propagation

$$A_j^{(k)} + X_j \underset{k_{pji}^{(k)}}{\overset{k_{pij}^{(k)}}{\longleftrightarrow}} A_i^{(k)} + X_i \tag{2.79}$$

(iii) monomolecular

$$C_j + X_j \xrightarrow{k_{sji}} \text{nonactive product} \tag{2.80}$$

and bimolecular chain termination

$$X_i + X_j \xrightarrow{k_{tij}} \text{nonactive product} \tag{2.81}$$

Let us re-write the rate of the elementary reaction (2.79) of the chain propagation as follows:

$$v_{pij}^{(k)} = k_{pij}^{(k)}[A_j^{(k)}]L_j - k_{pji}^{(k)}[A_i^{(k)}]L_i = L\left(b_{ij}^{(k)}x_j - b_{ji}^{(k)}x_i\right) \tag{2.82}$$

where $L = \sum_i L_i$ is molar concentration and $x_i = L_i/L$ is relative concentration of radicals X_i. In contrast to a catalytic process, the general concentration L of the intermediate substances or radicals is not constant but depends on the ratio of the rates of initiation and radical termination processes.

The values

$$b_{ij}^{(k)} = k_{pij}^{(k)}[A_j^{(k)}], \qquad b_{ij} = \sum_k b_{ij}^{(k)} \tag{2.83}$$

represent the intensity of a probability of transition from X_j into X_i *in* k- and all k-elementary reactions of chain propagation $S_{ij}^{(k)}$ connecting X_i and X_j, respectively

Next, in accordance with the mass action law determining the rates of initiation $v_{dij}=k_{dij}[I_{ij}]$, monomolecular $v_{sij}=Lk_{sij}c_jx_i$ and bimolecular $v_{tij}=L^2k_{tij}x_jx_i$ chain termination, we obtain a system of usual differential equations describing the dynamics of intermediate substances X_i in the reactive system:

$$\frac{\mathrm{d}L_i}{\mathrm{d}t} = L\sum_j (b_{ij}x_j - b_{ji}x_i) + \sum_j v_{dij} - L^2\sum_j k_{tij}x_i x_j - L\sum_j k_{sij}c_j x_i$$

(2.84)

By summing (2.84) upon i and taking into account that $\sum_{ij}(b_{ij}x_j - b_{ji}x_i) = 0$ we find

$$\frac{\mathrm{d}L}{\mathrm{d}t} = \sum_{ij} v_{dij} - L^2\sum_{ij} k_{tij}x_i x_j - L\sum_{ij} k_{ji}c_j x_i \tag{2.85}$$

Under the approximation of the method of quasi-stationary concentration $\frac{\mathrm{d}L_i}{\mathrm{d}t} = 0$, $\frac{\mathrm{d}L}{\mathrm{d}t} = 0$ and in line with the condition of a long chain $L\sum_j(b_{ij}x_j - b_{ji}x_i) >> \sum_j v_{dij} - L^2\sum_j k_{tij}x_i x_j - L\sum_j k_{sij}c_j x_i$, we have the following system of algebraic equations instead of (2.84) and (2.85):

$$\sum_j(b_{ij}x_j - b_{ji}x_i) = 0 \tag{2.86}$$

$$\sum_{ij} v_{dij} - L^2\sum_{ij} k_{tij}x_i x_j - L\sum_{ij} k_{ji}c_j x_i = 0 \tag{2.87}$$

Denoting

$$b_{jj} = \sum_{i \neq j} b_{ij} \tag{2.88}$$

let us re-write (2.86) in the matrix form

$$Bx = 0 \tag{2.89}$$

where $B=(b_{ij})$ is a kinetic square matrix of the order n and $x=(x_j)$ is a column vector of the relative concentrations of radicals X_i.

A system of linear homogeneous equations has a single solution which, taking into account $\sum_i x_i = 1$, can be written as follows:

$$x_i = \frac{M_i}{\sum_i M_i} \tag{2.90}$$

where $M_i = (-1)^{n-1}|B_i|$ is additional minor of a single group $V(X_i)$ and $|B_i|$ is a determinant of the matrix B_i obtained from the initial matrix B by deleting the i-line and column.

By substituting (2.90) into expression (2.82) of the rate of an elementary reaction of the chain propagation we find the terms which determine the rates $r(v, w, u)$ of routes $U(v, w, u)$ in the basis(set) $W(v, w)$ of a group $V(v)$ [18]:

$$r(v,w,u) = L\left(\sum_i M_i\right)^{-1} M(v)\Delta(v,w,u) \tag{2.91}$$

Here $M(v)=(-1)^{n-l}|B(v)|$ is additional minor of a group $V(v)$ with intensity l, consisting of the tops of a graph $G=\{X, S\}$, since $|B(v)|$ is a determinant of the matrix $B(v)$ obtained from the starting matrix B by deleting those lines and columns of which the indexes coincide with the indexes of tops represented in the $V(v)$; $\Delta(v, w, U)$ is a delta-factor or cyclic characteristic of a route.

Next, solving (2.87) relative to L and taking into account (2.90) we find [18]

$$L = \left(\sum_i M_i\right)^{-1} \left\{ \left[\left(\frac{\sum_{ij} k_{sji} c_j M_i}{2\sum_{ij} k_{tij} M_i M_j} \right)^2 + \frac{\sum_{ij} v_{dij}}{\sum_{ij} k_{tij} M_i M_j} \right]^{1/2} - \frac{\sum_{ij} k_{sji} c_j M_i}{2\sum_{ij} k_{tij} M_i M_j} \right\} \tag{2.92}$$

Expressions (2.91) and (2.92) determine a general kinetic model of a route of a radical-chain process with mixed chain termination. Here, let us write only its partial forms, which are true at bimolecular

$$r(v,w,u)=\left(\frac{\sum_{ij} v_{dij}}{\sum_{ij} k_{tij} M_i M_j}\right)^{1/2} M(v)\Delta(v,w,u) \tag{2.93}$$

and monomolecular chain termination

$$r(v,w,u)=\frac{\sum_{ij} v_{dij}}{\sum_{ij} k_{tij} M_i M_j} M(v)\Delta(v,w,u) \tag{2.94}$$

Let us consider as an example the process of a liquid-phase co-oxidization of the hydrocarbons upon the final reactions (2.36) and (2.37); a set of the elementary reactions of a chain propagation is represented in (2.38) and a graph $G = \{S, X\}$ is shown in Figure 2.4. A route of this directed graph is given in (2.39). Let us introduce the notifications

$$X_1 = \mathrm{R_1^\bullet},\ X_2 = \mathrm{R_1OO^\bullet},\ X_3 = \mathrm{R_2^\bullet},\ X_4 = \mathrm{R_2OO^\bullet} \tag{2.95}$$

Taking into account (2.95) and also the initiation (input of the radicals into a system) and chain termination reactions (output of the radicals from a system), a graph $G' = \{S, X\}$ isobtained and shown in *Figure* 2.6.

At this, the elementary reaction in (2.38) can take the forms:

$$\begin{aligned}
S_1 = S_{21} &: \mathrm{O_2 + R_1^\bullet \longrightarrow R_1OO^\bullet} \\
S_2 = S_{43} &: \mathrm{O_2 + R_2^\bullet \longrightarrow R_2OO^\bullet} \\
S_3 = S_{12} &: \mathrm{R_1H + R_1OO^\bullet \longrightarrow R_1OOH + R_1^\bullet} \\
S_4 = S_{14} &: \mathrm{R_1H + R_2OO^\bullet \longrightarrow R_2OOH + R_1^\bullet} \\
S_5 = S_{32} &: \mathrm{R_2H + R_1OO^\bullet \longrightarrow R_1OOH + R_2^\bullet} \\
S_6 = S_{34} &: \mathrm{R_2H + R_2OO^\bullet \longrightarrow R_2OOH + R_2^\bullet}
\end{aligned} \tag{2.96}$$

We have two routes of length $l = 2$ in the groups $V(X_1X_2)$ and $V(X_3X_4)$:

$$\begin{aligned}
U_1 &= S_{21} + S_{12} \\
U_2 &= S_{43} + S_{34}
\end{aligned} \tag{2.97}$$

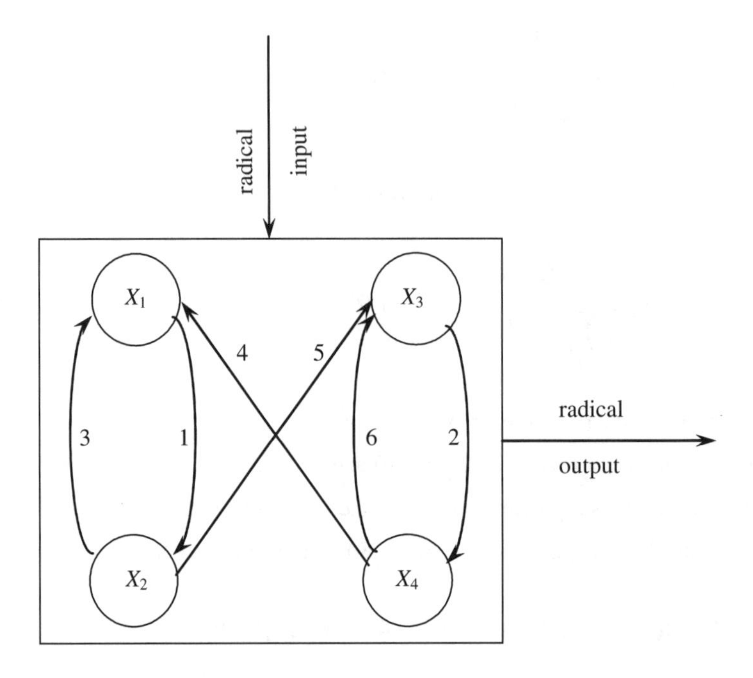

Figure 2.6. A graph reflecting the topology of kinetic scheme (2.96) taking into account the initiation and a chain termination.

Their delta-factors are equal to

$$\Delta_1 = b_{21}b_{12}$$
$$\Delta_2 = b_{43}b_{34} \qquad (2.98)$$

We also have a route of length $l = 4$ in the group (basis(set)) $V(X_1, X_2, X_3, X_4)$:

$$U_3 = S_{21}^{(1)} + S_{32}^{(1)} + S_{43}^{(1)} + S_{14}^{(1)} \qquad (2.99)$$

for which

$$\Delta_3 = b_{21}b_{32}b_{43}b_{14} \qquad (2.100)$$

A kinetic fourth-order matrix B is as follows:

$$B = \begin{pmatrix} -b_{21} & b_{12} & 0 & b_{14} \\ b_{21} & -(b_{12}+b_{32}) & 0 & 0 \\ 0 & b_{32} & -b_{43} & b_{34} \\ 0 & 0 & b_{43} & -(b_{14}+b_{34}) \end{pmatrix} \tag{2.101}$$

with the elements

$$b_{12} = k_{12}[R_1H],\ b_{21} = k_{21}[O_2],\ b_{14} = k_{14}[R_1H]$$
$$b_{43} = k_{43}[O_2],\ b_{34} = k_{34}[R_2H],\ b_{32} = k_{32}[R_2H] \tag{2.102}$$

Additional minors of the single groups M_i can be founded according to form (2.63) by deleting from the kinetic matrix B the i-line and column:

$$M_1 = (-1)^3 \begin{vmatrix} -(b_{12}+b_{32}) & 0 & 0 \\ b_{32} & -b_{43} & b_{34} \\ 0 & b_{43} & -(b_{14}+b_{34}) \end{vmatrix} = b_{11}b_{43}b_{14} + b_{32}b_{43}b_{14} \tag{2.103}$$

$$M_2 = (-1)^3 \begin{vmatrix} -b_{12} & 0 & b_{14} \\ 0 & -b_{43} & b_{34} \\ 0 & b_{43} & -(b_{14}+b_{34}) \end{vmatrix} = b_{12}b_{14}b_{43} \tag{2.104}$$

$$M_3 = (-1)^3 \begin{vmatrix} -b_{21} & b_{12} & 0 \\ b_{21} & -(b_{12}+b_{32}) & 0 \\ 0 & 0 & -(b_{14}+b_{34}) \end{vmatrix} = b_{14}b_{21}b_{32} + b_{34}b_{21}b_{32} \tag{2.105}$$

$$M_4 = (-1)^3 \begin{vmatrix} -b_{21} & b_{12} & 0 \\ b_{21} & -(b_{12}+b_{32}) & 0 \\ 0 & 0 & -b_{43} \end{vmatrix} = b_{21}b_{32}b_{43} \tag{2.106}$$

For the routes U_1 and U_2 in the basis(sets) $W(X_1, X_2)$ and $W(X_3, X_4)$ additional minors of the respecting groups $V(X_1, X_2)$ and $V(X_3, X_4)$ are equal to

$$M(X_1, X_2) = \begin{vmatrix} -b_{43} & b_{34} \\ b_{43} & -(b_{14}+b_{34}) \end{vmatrix} = b_{14}b_{43} \tag{2.107}$$

$$M(X_3, X_4) = \begin{vmatrix} -b_{12} & b_{12} \\ b_{21} & -(b_{12}+b_{32}) \end{vmatrix} = b_{32}b_{21} \tag{2.108}$$

For the route U_3 in the basis(set) $W(X_1, X_2, X_3, X_4)$ an additional minor of a group is equal to 1.

Thus, assuming that only a bimolecular chain termination takes place, we can write the following expressions for the rates of a route:

$$r_1 = \left(\frac{v_{in}}{\sum_{ij} k_{tij} M_i M_j} \right)^{1/2} b_{14}b_{43}b_{21}b_{12} \tag{2.109}$$

$$r_2 = \left(\frac{v_{in}}{\sum_{ij} k_{tij} M_i M_j} \right)^{1/2} b_{32}b_{21}b_{43}b_{34} \tag{2.110}$$

$$r_3 = \left(\frac{v_{in}}{\sum_{ij} k_{tij} M_i M_j} \right)^{1/2} b_{21}b_{14}b_{43}b_{32} \tag{2.111}$$

Here $v_{in} = \sum_{ij} v_{dij}$ is a total rate of an initiation,

$$\sum_{ij} k_{tij} M_i M_j = k_{t11}M_1^2 + 2k_{t12}M_1M_2 + 2k_{t13}M_1M_3 + 2k_{t14}M_1M_4 + k_{t22}M_2^2 + \\ + 2k_{t23}M_2M_3 + 2k_{t24}M_2M_4 + k_{t33}M_3^2 + 2k_{t34}M_3M_4 + k_{t44}M_4^2 \tag{2.112}$$

Rates of the oxidation of the hydrocarbons R_1H and R_2H, respectively, are equal to

$$R_1 = r_1 + r_2, \qquad R_2 = r_2 + r_3 \tag{2.113}$$

A rate of the oxygen absorption at the expense of both reactions is equal to

$$\omega_{O_2} = r_1 + r_2 + 2r_3 \tag{2.114}$$

So, in such a way a kinetic model of a process constructed.

If the concentration of alkyl radicals is considerably less than the peroxy ones, that is $[R_i] << [R_iOO]$ and M_1, M_3 << M_2, M_4 then we can neglect by the participation of alkyl radicals via chain termination keeping in the expression (2.112) only terms $k_{t22}M_2^2$ + $2k_{t24}M_2M_4$ + $k_{t44}M_4^2$. In this simpler variant we will obtain a kinetic model earlier proposed in a work [19] on the basis of the Mayo–Walling equation [20] for the co-polymerization.

2.4. A kinetic models of routes of free radical polymerization

Although free radical polymerization is an example of an un-branched radical-chain process, it has characteristics which should be taken into account when interpreting a route and an algorithm of the construction of their kinetic models.

The main peculiarity of a free radical polymerization is that the final chemical transformation of a monomer into a polymer is realized *via* elementary chain termination reactions and does not reflect a stoichiometric equation of a route assigned through a sequence of elementary chain propagation reactions of the type

$$A_i + R_i \longrightarrow R_i \qquad (2.115)$$

or

$$A_j + R_i \longrightarrow R_j , \qquad (2.116)$$

but reflects only the monomer consumption *via* elementary chain propagation reactions. An exception is the chain transfer reaction, for example on a monomer with the formation of active radicals and a macromolecule of a polymer *П*:

$$A_i + R_i \longrightarrow \Pi + R_j \qquad (2.117)$$

Thus, in general routes as cyclic sequences of elementary reactions of the chain propagation (2.115) and (2.116) realize the final chemical transformation into a stoichiometric form in which the product of the reaction is not denoted:

$$A_i \longrightarrow \qquad (2.118)$$

or

$$\sum_i A_i \longrightarrow \tag{2.119}$$

Since elementary reactions of the chain propagation are assumed as being far from the equilibrium state and this allows to ignore their reversible directions, the stoichiometric notations of elementary (2.115) and (2.116) and also the final reactions (2.118) and (2.119) do not limit in any way the use of an algorithm of construction of the kinetic models of the routes. Furthermore, formally it can always be assumed that elementary (2.115) and (2.116) and the final reactions (2.118) and (2.119) have an "empty" composite substance, the concentration of which is equal to zero; for example

$$A_i \longrightarrow 0 \tag{2.120}$$

$$\sum_i A_i \longrightarrow 0 \tag{2.121}$$

A formalisation of the final algorithm of the kinetic models for route construction is wholly stored under this variant of the stoichiometric notations of the non-reversible elementary and final reactions.

However, the stoichiometric notation (2.115) of the reaction of chain propagation foresees or postulates the same reactive ability of the macroradicals with different degree of the polymerization. The identity of the radicals on the left and on the right in (2.115) implies the appearance of a cycle or a route consisting of only a single elementary reaction, a cycle forming a hinge closing the top of graph $G = \{S, R\}$. For these routes an additional minor of group $M(v)$ and an additional minor of top M_i coincide. This is the second peculiarity of the free radical polymerization which should be taken into account in the construction of a kinetic model of a process according to the proposed algorithm.

Next, let us illustrate its application in the analysis of some well-known variants of free radical polymerization, underlining its simplicity.

2.4.1. Homopolymerization

Homopolymerization is represented only by one elementary reaction of the chain propagation

$$A + R \longrightarrow R \tag{2.122}$$

This reaction forms a single cycle, a hinge on the single top of graph *G' = {S, R}* as shown in Figure 2.7.

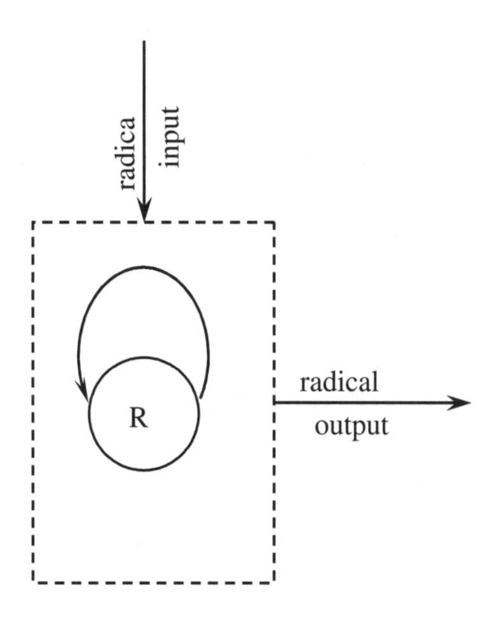

Figure 2.7. Graph *G' = {S, R}* of the topology of a kinetic scheme of the chain propagation in a homopolymerization process.

Subsequently, a kinetic matrix *B* contains only one element reflecting the intensity of a single-top transition into oneself:

$$B = \left(k_p[A]\right) \tag{2.123}$$

It follows from this that the additional minors of the group and tops are equal to 1. That is why in the equations (2.93) and (2.94) under bimolecular chain termination

$$\left(\sum_{ij} k_{tij} M_i M_j\right)^{1/2} = k_t^{1/2} \tag{2.124}$$

and under monomolecular chain termination

$$\left(\sum_{ij} k_{sij} c_j M_i \right)^{1/2} = k_s c \qquad (2.125)$$

Taking into account that the delta-factor of a route-hinge is equal to

$$\Delta = k_p [A] \qquad (2.126)$$

we obtain from the general models (2.93) and (2.94) a partial model characterizing the rate of the homopolymerization at the bimolecular

$$r = \left(\frac{v_{\text{in}}}{k_t} \right)^{1/2} k_p [A] \qquad (2.127)$$

and monomolecular chain termination

$$r = \frac{v_{\text{in}}}{k_s c} k_p [A] \qquad (2.128)$$

where v_{in} is the initiation rate.

In the case that the monomer is a substance leading the monomolecular chain termination, that is, when chain termination is the last act of its propagation, we have $c = [A]$ and

$$r = v_{\text{in}} \frac{k_p}{k_s} \qquad (2.129)$$

From this we conclude that the homopolymerization rate does not depend on the concentration of the monomer. Exactly this mechanism will be proposed by us below for the polymerization process of bifunctional monomers in the interface layer on the boundary of a liquid monomer–solid polymer.

2.4.2 Co-polymerization

For co-polymerization of two monomers the kinetic scheme of the chain propagation in the simplest variant consists of only elementary reactions (2.115) and (2.116). If we introduce for them a single numeration, we have

$$\begin{aligned} S_1 &: A_1 + R_1 \xrightarrow{k_{p11}} R_1 \\ S_2 &: A_2 + R_2 \xrightarrow{k_{p22}} R_2 \\ S_3 &: A_1 + R_2 \xrightarrow{k_{p12}} R_1 \\ S_4 &: A_2 + R_1 \xrightarrow{k_{p21}} R_2 \end{aligned} \tag{2.130}$$

A graph *G* = *{S, R}* presenting the topology of a kinetic scheme (2.130) is shown in Figure 2.8.

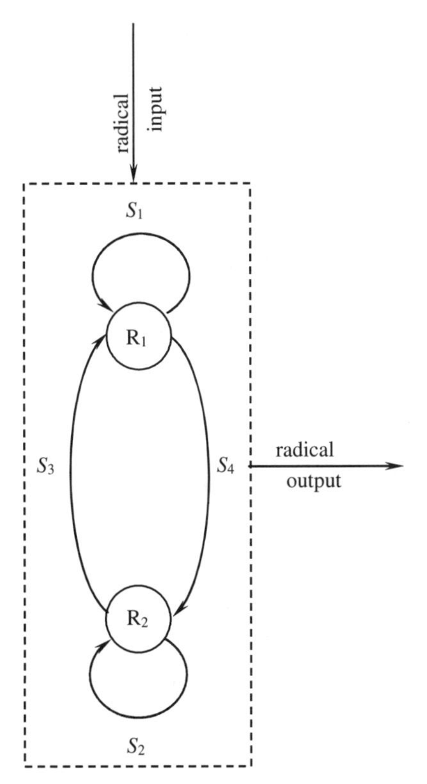

Figure 2.8. Graph $G' = \{S, R\}$ presenting the topology of a kinetic scheme of a chain propagation (2.130) for the co-polymerization process.

This graph contains three simple cycles or routes. Two of them are hinges from S_1 and S_2 and they can be called the routes of the monomers A_1 and A_2, since they only consume one of the monomers: $A_1 \rightarrow$ and $A_2 \rightarrow$. The third route, *via* a sequence of elementary reactions S_3 and S_4, yields the final transformation $A_1 + A_2$ and can be called a mixed one.

In accordance with the (2.130) the kinetic matrix has a second order with the elements

$$B = \begin{pmatrix} b_{11} & b_{12} \\ b_{21} & b_{22} \end{pmatrix} = \begin{pmatrix} -k_{p21}[A_2] & k_{p12}[A_1] \\ k_{p21}[A_2] & -k_{p12}[A_1] \end{pmatrix} \tag{2.131}$$

In this matrix there are no intensities of the transitions of the elementary reactions S_1 and S_2, since they do not change the ratio between the concentrations of radicals R_1 and R_2. These elementary reactions are taken into account only as an own-route hinge for which

$$U_1 \ : \ \Delta_1 = k_{p11}[A_1], \ \ M(v) = M_1 = k_{p12}[A_1] \tag{2.132}$$

$$U_2 \ : \ \Delta_2 = k_{p22}[A_2], \ \ M(v) = M_2 = k_{p21}[A_2] \tag{2.133}$$

For the third route

$$U_3 \ : \ \Delta_3 = k_{p21}k_{p12}[A_1][A_2], \ \ M(v) = 1 \tag{2.134}$$

In accordance with the expressions for M_1 and M_2 we find

$$\left(\sum_{ij} k_{tij} M_i M_j\right)^{1/2} = \left[k_{t11}\left(k_{p12}[A_1]\right)^2 + 2k_{t12}k_{p21}k_{p12}[A_1][A_2] + k_{t22}\left(k_{p21}[A_2]\right)^2\right]^{1/2} \tag{2.135}$$

Expressions (2.132)–(2.135) are in accordance with the kinetic models of the routes of co-polymerization. At this, the rate of monomer A_1 consumption is equal to $r_{A1} = r_1 + r_3$, monomer A_2 consumption is equal to $r_{A2} = r_2 + r_3$ and the ratio between them

$r_{A1} / r_{A2} = (r_1 + r_3) / (r_2 + r_3)$ determines a composition of the forming polymer. The total rate of the co-polymerization is equal to $r = r_{A1} + r_{A2} = r_1 + r_2 + 2\ r_3$, that is

$$r = \frac{k_{p11}k_{p12}[A_1]^2 + 2k_{p21}k_{p12}[A_1][A_2] + k_{p22}k_{p21}[A_2]}{\left[k_{t11}\left(k_{p12}[A_1]\right)^2 + 2k_{t12}k_{p21}k_{p12}[A_1][A_2] + k_{t22}\left(k_{p21}[A_2]\right)^2\right]^{1/2}}, \quad (2.136)$$

which is exactly the well-known Mayo–Walling equation [20].

2.4.3. Terpolymerization

Most clearly the simplicity of the derivation of an algorithm of a kinetic model of a route can be illustrated by the co-polymerization process of three monomers. In the simplest variant the kinetic scheme of chain propagation consists of only elementary reactions of the (2.115) and (2.116) type:

$$\begin{aligned}
S_1 &: A_1 + R_1 \xrightarrow{k_{p11}} R_1 \\
S_2 &: A_2 + R_2 \xrightarrow{k_{p22}} R_2 \\
S_3 &: A_3 + R_3 \xrightarrow{k_{p33}} R_3 \\
S_4 &: A_2 + R_1 \xrightarrow{k_{p21}} R_2 \\
S_5 &: A_1 + R_2 \xrightarrow{k_{p12}} R_1 \\
S_6 &: A_3 + R_2 \xrightarrow{k_{p32}} R_3 \\
S_7 &: A_2 + R_3 \xrightarrow{k_{p23}} R_2 \\
S_8 &: A_3 + R_1 \xrightarrow{k_{p31}} R_3 \\
S_9 &: A_1 + R_3 \xrightarrow{k_{p13}} R_1
\end{aligned} \quad (2.137)$$

A graph $G = \{S, R\}$ representing the topology of a kinetic scheme of chain propagation (2.137) is shown in Figure 2.9.

We can see three own routes U_1, U_2 and U_3 in the basis(sets) $W(R_1)$, $W(R_2)$ and $W(R_3)$ represented by hinges from elementary reactions S_1, S_2 and S_3, yielding the final chemical transformation $A_i \rightarrow$ $i = \overline{1,\ 3}$. *Via* the sequence of elementary reactions 4 and 5 in the basis(set) $W(R_1, R_2)$, 6 and 7 in the basis(set) $W(R_2, R_3)$ and 8 and 9 in the basis(set) $W(R_3, R_1)$ mixed routes U_4, U_5 and U_6, yielding the final chemical transformations

$A_i + A_j \rightarrow i, j = \overline{1,\ 3}$ are formed. Another two routes, U_7 and U_8 in the basis(set) $W(R_1, R_2, R_3)$, are formed *via* the sequence of elementary reactions 4, 6, 9 and 8, 7, 5, respectively; they yield the final chemical transformation $A_1 + A_2 + A_3 \rightarrow$. A kinetic matrix of the third order is as follows:

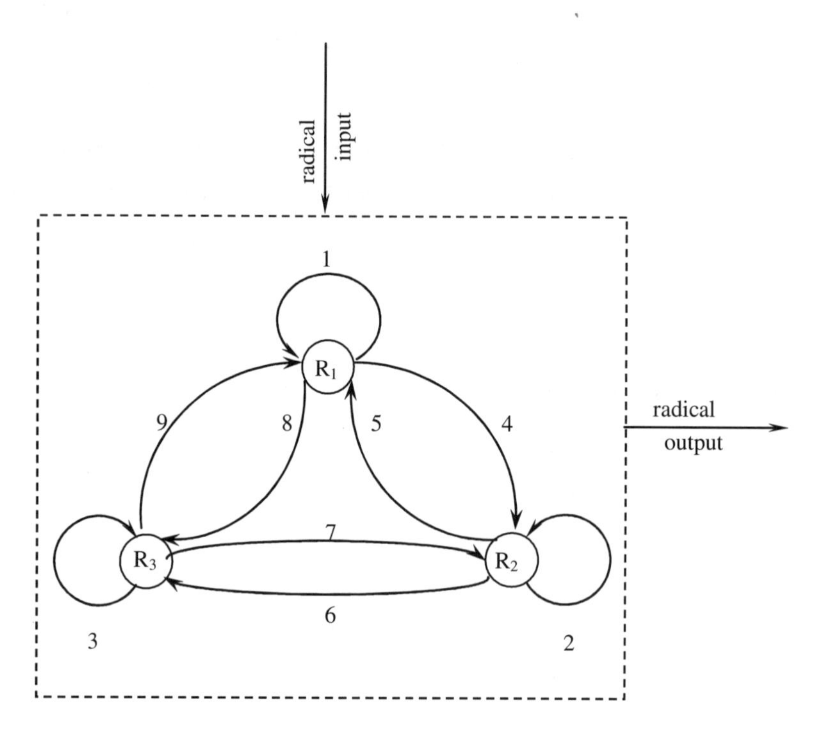

Figure 2.9. Graph $G = \{S, R\}$ representing the topology of a kinetic scheme of chain propagation (2.137) in the terpolymerization process.

$$B = (b_{ij}) = \\ = \begin{pmatrix} -(k_{p21}[A_2]+k_{p31}[A_3]) & k_{p12}[A_1] & k_{p13}[A_1] \\ k_{p21}[A_2] & -(k_{p12}[A_1]+k_{p32}[A_3]) & k_{p23}[A_2] \\ k_{p31}[A_3] & k_{p32}[A_{32}] & -(k_{p13}[A_1]+k_{p23}[A_2]) \end{pmatrix} \quad (2.138)$$

As we can see, in the kinetic matrix *B'* again there are no intensities of the transitions on the elementary reactions S_1-S_3, since they do not change the ratio between relative concentrations of radicals. These elementary reactions are taken into account only as own routes for which:

$$\Delta_1 = k_{p11}[A_1], \quad M(v) = M(1) = M_1$$
$$\Delta_2 = k_{p22}[A_2], \quad M(v) = M(2) = M_2 \qquad (2.139)$$
$$\Delta_3 = k_{p33}[A_3], \quad M(v) = M(3) = M_3$$

and in accordance with (2.138) additional minors of the single groups are determined as follows:

$$M_1 = k_{p12}k_{p13}[A_1]^2 + k_{p12}k_{p23}[A_1][A_2] + k_{p13}k_{p32}[A_1][A_3]$$
$$M_2 = k_{p21}k_{p23}[A_2]^2 + k_{p21}k_{p13}[A_2][A_1] + k_{p23}k_{p31}[A_2][A_3] \qquad (2.140)$$
$$M_3 = k_{p31}k_{p32}[A_3]^2 + k_{p13}k_{p12}[A_3][A_1] + k_{p32}k_{p21}[A_3][A_2]$$

We have for the routes U_4, U_5 and U_6 in the basis(sets) $W(R_1, R_2)$, $W(R_2, R_3)$ and $W(R_3, R_1)$

$$\Delta_4 = k_{p12}k_{p21}[A_1][A_2], \quad M(v) = M(1,2) = k_{p13}[A_1] + k_{p23}[A_2]$$
$$\Delta_5 = k_{p23}k_{p32}[A_2][A_3], \quad M(v) = M(2,3) = k_{p21}[A_2] + k_{p31}[A_3] \qquad (2.141)$$
$$\Delta_6 = k_{p13}k_{p21}[A_1][A_3], \quad M(v) = M(1,3) = k_{p12}[A_1] + k_{p32}[A_3]$$

For the routes U_7 and U_8 in the basis(set) $W(R_1, R_2, R_3)$ we have

$$\Delta_7 = k_{p13}k_{p32}k_{p21}[A_1][A_2][A_3], \quad M(v) = M(1,2,3) = 1$$
$$\Delta_8 = k_{p12}k_{p23}k_{p31}[A_1][A_2][A_3], \quad M(v) = M(1,2,3) = 1 \qquad (2.142)$$

A factor of the bimolecular chain termination can be denoted as

$$Z = \sum_{ij} k_{tij}M_iM_j = k_{t11}M_1^2 + k_{t22}M_2^2 + k_{t33}M_3^2 +$$
$$+ 2k_{t12}M_1M_2 + 2k_{t13}M_1M_3 + 2k_{t23}M_2M_3 \qquad (2.143)$$

Then the rate of a route can be written in the shorter form:

$$r_1=\left(\frac{v_{in}}{Z}\right)^{1/2}M_1\Delta_1,\ r_2=\left(\frac{v_{in}}{Z}\right)^{1/2}M_2\Delta_2,\ r_3=\left(\frac{v_{in}}{Z}\right)^{1/2}M_3\Delta_3$$

$$r_4=\left(\frac{v_{in}}{Z}\right)^{1/2}M(1,2)\Delta_4,\ r_5=\left(\frac{v_{in}}{Z}\right)^{1/2}M(2,3)\Delta_5,$$

$$r_6=\left(\frac{v_{in}}{Z}\right)^{1/2}M(1,3)\Delta_6 \tag{2.144}$$

$$r_7=\left(\frac{v_{in}}{Z}\right)^{1/2}\Delta_7,\ r_8=\left(\frac{v_{in}}{Z}\right)^{1/2}\Delta_8$$

in which the all values are as determined in (2.139)–(2.143).

At this, the rates of monomer A_1, A_2 and A_3 consumption are determined in accordance with their stoichiometric participation in the routes:

$$\omega_{A_1}=r_1+r_4+r_6+r_7+r_8$$

$$\omega_{A_2}=r_2+r_4+r_5+r_7+r_8 \tag{2.145}$$

$$\omega_{A_3}=r_3+r_5+r_6+r_7+r_8$$

The rate of a polymerization R is equal to $\omega_{A1}+\omega_{A2}+\omega_{A3}$:

$$R=r_1+r_2+r_3+2r_4+2r_5+2r_6+3r_7+3r_8 \tag{2.146}$$

If we substitute in this expression all the values determined above we obtain the well-known equation of the terpolymerization rate.

Taking into account the ω_{Ai} values in (2.145) we can easily determine the ratio between the concentrations of monomers in the polymer:

$$[A_1]:[A_3]:[A_2]=\omega_{A_1}:\omega_{A_2}:\omega_{A_3}$$

So, the proposed algorithm of the kinetic models of route construction for a graph $G = \{S, R\}$ assigned to the set of radicals and elementary reactions in the chain propagation is clear and simple.

REFERENCES

[1] Christiansen J. *The elucidation of reaction mechanisms by the method of intermediates in quasi-stationary concentrations* // In: *Advances in Catalysis*, 5, New York, Academic Press, **1953**, p. 311.

[2] King E., Altman C. *A schematic method of deriving the rate laws for enzyme-catalyzed reaction* // *J. Phys. Chem.*, **1956**, 60 (10), p. 1375.

[3] Walkenshtein M., Goldstein B. *Primienienije tieoriji* Graphov *k raschiots slozhnykh reakcij* // *Dokl. AS* USSR, **1966**, 170 (4), p. 963.

[4] Walkenshtein M., Goldstein B. *Novyj mietod reshenija zadach stacionarnoj kinetiki phermentativnykh reakcij* // *Biokhimiya*, **1966**, 31 (3), p. 541.

[5] Yevstignieyev V., Yablonskiy G., Bykov V. *Obshchaya forma stacionarnogo kineticheskogo uravnieniya slozhnoy kataliticheskoj reakcji (mnogomarshrutnyj liniejnyj mechanism)* // *Dokl. AS* USSR, **1979**, 245 (4), p. 871.

[6] *Analiz obshchej formy stacionarnogo kineticheskogo uravnieniya slozhnoy kataliticheskoj reakcji* / Yablonskiy G., Yevstignieyev V., Noskov A., Bykov V. // *Kinetika i kataliz*, **1981**, 22 (3), p. 738.

[7] Snagovsiy Yu., Ostrovskiy G. *Modelirovanije kinetiki heterogennykh kataliticheskikh processov*, Moskva: Khimija, **1976**, 248 p.

[8] Yablonskiy G., Bykov V., and Gorban' A. *Kineticheskije modeli kataliticheskikh reakcij*, Novosibirsk: Nauka, **1983**, 253 p.

[9] *Chemical Applications of Topology and Graph Theory.* Edited by R. B. King, University of Georgia, Athens, and USA, New York, **1983**.

[10] Cepalov V. *Kinetika cepnogo prevrashcheniya mnogokomponientnykh system* // *J. Phys. Chem.*, **1961**, 35 (5), p. 1086.

[11] Aris R. *Prolegomena to the rational analysis of systems of chemical reactions* // *Arch. Rational Mech. Anal.*, **1965**, 19 (1), p. 81.

[12] Aris R. *Prolegomena to the rational analysis of systems of chemical reactions* II. *Some addenda* // *Arch. Rational Mech. Anal.*, **1968**, 27 (3), p. 356.

[13] Aris R. *Mathematical aspects of chemical reactions* // *Ing. Eng. Chem.*, **1969**, 61 (6), p. 17.

[14] Horiuti J. Stoichiometrische Zahlen und die Kinetik der Chemischen Reactionen // *J. Res. Inst. Catal.*, Hokkaido University, **1957**, 5 (1), p.1.

[15] Horiuti J., Nakamura T. *Stoichiometric number and the theory of steady reaction // Zhurnal Phys. Chem.*, Neue Folge, **1957**, II, p. 358.

[16] Horiuti J., Nakamura T. *On the theory of heterogeneous catalysis* // In: *Advances in Catalysis*, 17, New York, Acad. Press, **1967**, p. 1.

[17] Medvedevskikh Yu., Kucher R. *Marshruty i kineticheskaya model kataliticheskikh reakcij v slozhnykh sistiemakh // Dokl. AS USSR*, **1984**, (1), p. 42.

[18] Berlin A., Medvedevskikh Yu., Kucher R. *Kineticheskaya model marshrutov nierazvietvlionnykh radikalno-cepnykh processov // Dokl. AS USSR*, **1986**, (6), p. 32.

[19] Walling C., Mc Elhill E. *The reactivities of benzaldehydes with perbenzoat radicals // J. Chem. Soc.*, **1951**, 73 (6), p. 2927.

[20] Mayo F., Walling C. Copolymerization // *Chem. Rev.*, **1950**, 46 (2), p. 191.

Chapter 3. Problems of stationary kinetics of polymerization up to the high conversion state

3.1. General and different characteristics of 3-D and linear polymerization up to the high conversion state

Three-dimensional (3-D) polymerization of polyfunctional monomers, in contrast to linear polymerization of monofunctional monomers, is accompanied by the cross-linking of polymeric chains at the expense of double bonds, which do not react in the linear chain propagation. That is why in the case of 3-D polymerization the final product is a polymer, the structure of which forms a united network [1, 2].

From the point of view of elementary reaction steps, free-radical cross-linking polymerization does not differ from linear polymerization of monovinyl compounds and involves initiation, propagation and termination. However, the investigation of mono- and multifunctional monomer polymerization kinetics up to late in the conversion process shows both common and special features of these processes that do not follow the classical kinetic scheme of polymerization which is satisfactory for the initial state of polymerization.

The first common feature is the S-like shape of kinetic curves (see Figure 3.1), which reflects the dependence of polymerization state on time. Such a dependency is characteristic for autocatalytic reactions. The maximum rate of polymerization corresponds to the transition point on the curve, indicating the presence of an autoacceleration state in the process. The maximum also occurs in the case of a relative polymerization rate, here indicating the prescence of an autodeceleration state at which the process rate is decreasing faster than the decrease in monomer concentration [3–8].

The second feature is based on the pronounced "post-effect", i.e., on the "dark" (after turning off the UV illumination) post-polymerization process taking place from the beginning of the autoacceleration state and continuing until full conversion in some cases. The kinetic curve of the post-polymerization consists of two sections (see Figure 3.2): the first is fast and short, and the second is slow and prolonged [9–16]. This suggests that the postpolymerization process is led by two types of radicals, sharply differing in life time. The molecular mass of the polymer increases during the postpolymerization process.

Sometimes it is reflecting the existence of a second maximum on the molecular mass distribution curve. Incorporation of a chain0transformation agent reduces the polymer molecular mass both in the light and in the dark polymerization regimes.

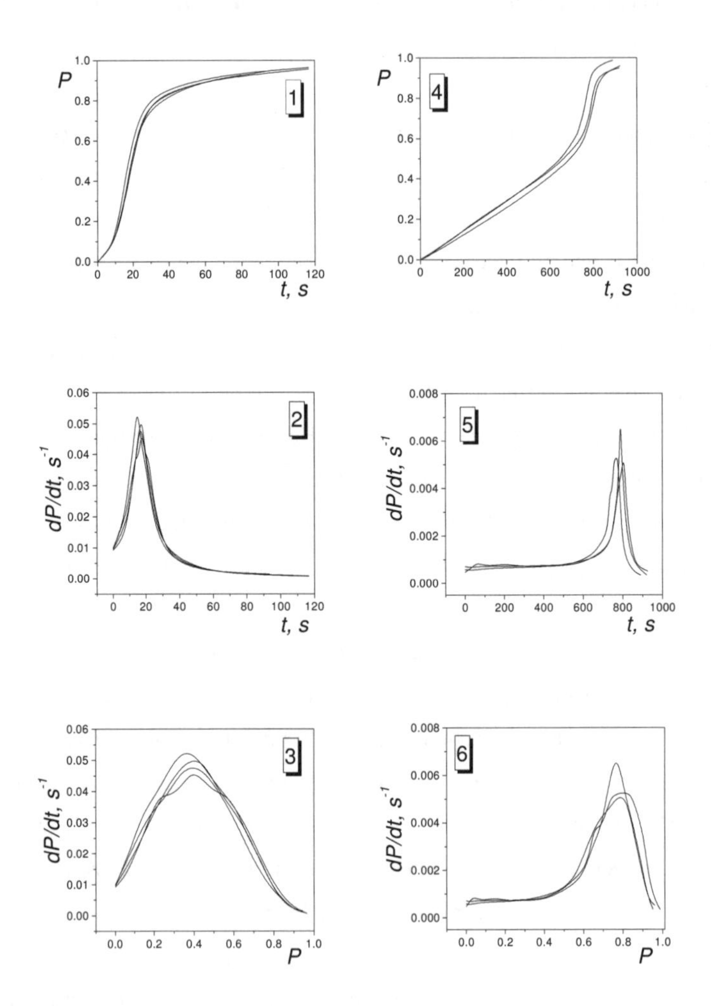

Figure 3.1. Typical kinetic curves of 3-D (1–3) and linear (4–6) polymerizations in different coordinates.

According to observations of the authors [9] the polymerization rate depends on the state of conversion corresponding to the beginning of the dark period of the process. The same dependence is observed also in another study [10]: after the UV-illumination is stopped in the initial states of triethyleneglicoltrimethylacrylate (TGM-3) polymerization the post-effect lasts for 10 min, but in the late states decreases to 4 min.

From the analysis of methylmethacrylate postpolymerization under different depths of the conversion it was concluded [11, 12] that from the moment the UV-illumination was stopped until the moment of the autoacceleration state, chain termination in the postpolymerization state is described by the bimolecular chemical mechanism.

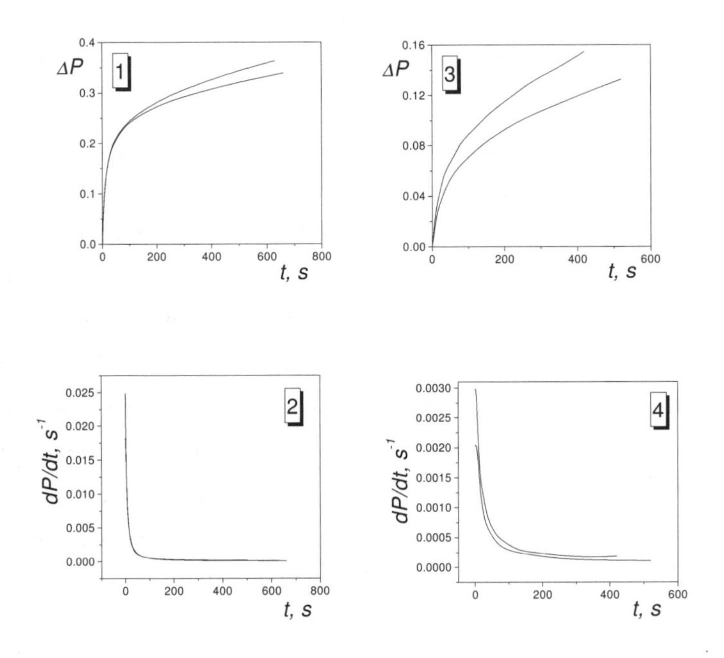

Figure 3.2. Typical kinetic curves of 3-D (1, 2) and linear (3, 4) post-polymerizations in different coordinates.

At the same time, after the UV-illumination was stopped in the field of process autoacceleration, simultaneously with the bimolecular chain termination chemical mechanism, the monomolecular mechanism is also observed, the presence of which is explained by the authors [11, 12] as a "stacking" (or trapping) of the radicals in the polymeric matrix, resulting in the loss of their diffusion mobility; in such a case these radicals can not continue the kinetic chain. At conversions higher than the autoacceleration state (20–80%) chain termination is realized only in accordance with the monomolecular chemical mechanism, which from the authors' point of view [11, 12] leads to the accumulation of radicals in the polymerizate. These radicals do not participate in the propagation reactions.

Results of several investigations [9–13] led to the conclusion that there are at least two types of radicals in the postpolymerization state. These radicals differ in characteristic life times and in the kinetic role in the process, respectively.

In other studies [14, 15] the conclusions of in the reasearch [11, 12] of the monomolecular chain termination on the autoacceleration state is simplified, and the presence of trapped radicals is assumed only in the glass-like polymerizate in the limited conversion states, where diffusion processes of macroradicals and monomers are completely stopped.

The third feature of the linear and 3-D polymerizations is a high (up to 10^5 mol/m^3) radical concentration accumulation, as determined by EPR spectroscopy *in situ* at the polymerization end [8, 9, 12, 16–22]. Such an accumulation of radicals is observed in the autodeceleration state. In the stationary light process regime, detectable accumulation of radicals starts from the autoacceleration state. In the postpolymerization the radical conception decreases insignificantly or remains constant. Radical decay is detectable only at relatively high temperatures, namely >60°C, and the kinetics of the process in some cases are described as first order. Thus, the radical characteristic periods of life determined by EPR-spectroscopy *in situ* at the end of the polymerization process are significantly higher than the periods of time of the stationary light and the non-stationary dark processes. Hence, they do not lead the polymerization and are radicals of the third type [9].

Note that in the autodeceleration state of a polymerization process, the concentration of radicals practically does not change *via* a conversion and depends on the conditions under which the experiment is carried out, namely temperature and initiation rate [9] and also on the nature of a starting monomer [16, 17]. Stationary concentration of radicals discovered by EPR-spectroscopy in samples of cured oligoesteracrylates at a conversion degree Γ=80% achieved a high value: 10^{15} spin/g [8]. After the introduction of a plastifying agent in the polymerization system, the concentration of radicals does not reach the stationary value and is not stabilized after reaching a certain value, but begins to decrease [9].

Radical trapping into polymeric matrix becomes visible only at relatively high temperatures. For example, at 60°C the radical-trapping process in the polymeric matrix of

TGM-3 proceeds for 8 h; when the temperature is increased to 90°C the radical trapping reaction is practically finished in 1 h [16].

Oliva *et al.* [23] describe the EPR characterization of radicals trapped during the photoinitiated polymerization of trimethacrylate monomer (TMA), which leads to a highly cross-linked polymer structure; also, vinylmethacrylate monomer (VMA), a difunctional monomer carrying different reactive groups, has been studied. The EPR spectra obtained at room temperature from photoinitiated samples of both trimethacrylate monomer TMA and vinylmethacrylate monomer VMA consist of a nine-line EPR pattern, typical of methacrylate propagating radicals (Structure **I**).

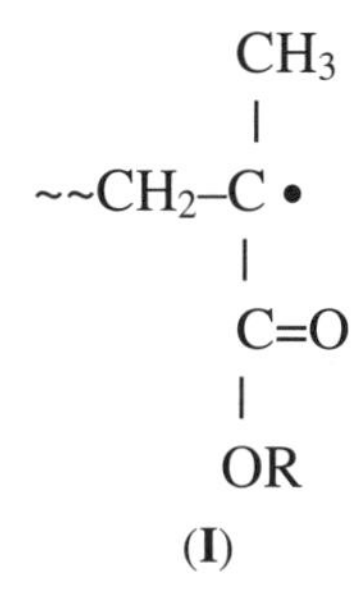

(**I**)

Very good fittings of the experimental spectra (see Figure 3.3) were obtained by attributing the nine lines to three equivalent hydrogen atoms with hyperfine coupling constants (hfcc) a_{3H}=2.22±0.02 mT for pre-irradiated TMA, a_{3H}=2.25±0.01 mT for pre-irradiated VMA, and to two equivalent hydrogen atoms with hfcc a_{2H}=1.17 ± 0.02 mT and 1.22 ± 0.01 mT, respectively, for the two pre-irradiated systems. The intrinsic line width of the EPR patterns is ΔW_0=0.77±0.01 mT and 0.70±0.01 mT in the two cases. An additional broadening contribution ΔW_{exch}=0.57±0.01 mT and 0.51±0.02 mT, respectively, is measured for the central line of the hydrogen multiplet characterized by a_{2H} of pre-irradiated TMA and VMA, respectively.

At such high temperatures much better fittings of the experimental spectra can be obtained by assuming that a single-line pattern, with the same *g*-value, overlaps the nine-line signal. The ratio of the concentration of radicals giving the single-line pattern over that of the species giving a nine-line pattern increasing with temperature, being 0.5/100 for TMA at 383 K and nearly 10 times higher for VMA at the same temperature, while it attains the same value (50/100) in both systems at 443 K.

The increase in concentration of the single-line carrier is accompanied by an increase in both line intensity and line width of this EPR pattern. At 383 K, the last parameter is 0.28±0.03 mT (TMA) and 0.48±0.02 mT (VMA), while at 443 K it is 1.95±0.06 mT (TMA) and 1.54±0.04 mT (VMA). Also, a variation of ΔW_{exch} is seen to occur with temperature. In fact ΔW_{exch}=0.54±0.01 mT (TMA) and 0.511±0.009 mT (VMA) at 383 K, while ΔW_{exch}=0.37.±0.02 mT (TMA) and 0.41±0.01 mT (VMA) at 443 K.

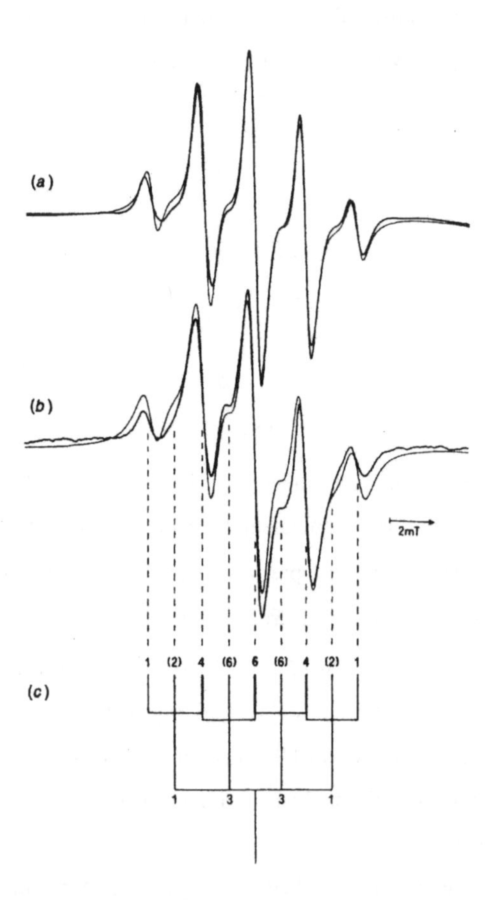

Figure 3.3. Experimental (noisy trace) and computer-simulated spectrum of pre-irradiated (a) TMA, (b) VMA at room temperature and (c) interpretation scheme. Thick lines are due to the overlapping of lines unaffected by the motional broadening; in parentheses the degeneration of the motional broadened lines is given [23].

The spectral shape does not change with temperature in the range 153–373 K, while major changes occur at temperatures above 373 K for both pre-irradiated systems (see, for example, Figure 3.4 for TMA).

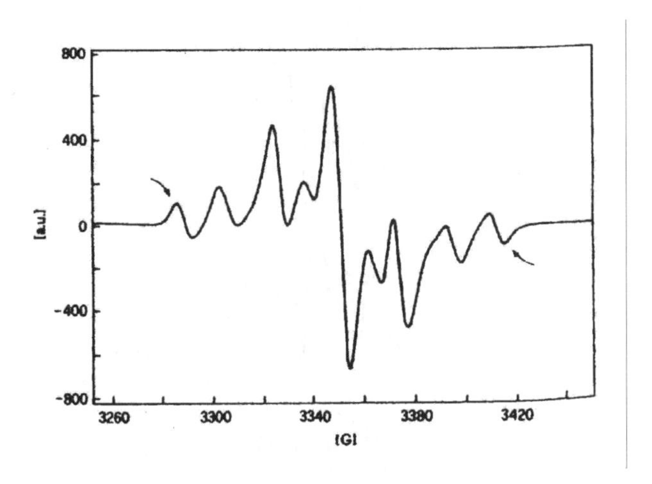

Figure 3.4. EPR spectrum of pre-irradiated TMA at 443 K. The new EPR lines are indicated by arrows [23].

Moreover, two new lines appear on both sides of the EPR spectrum of pre-irradiated TMA at temperatures above 383 K (Figure 3.4), which do not disappear after cooling the sample back to room temperature. These two features are not observed with photopolymerized VMA.

The EPR spectral shape undergoes marked modifications during radical decay kinetic runs in the range 363–423 K, showing the same trend for both photopolymerized systems. An example of this behaviour is given in Figure 3.5.

According to the results of the fitting procedure, the integrated area of the single-line spectrum continuously increases during radical decay, while the integrated area of the whole EPR pattern decreases at different rates, depending on temperature.

Many studies have been devoted to the determination of the structures of radicals and to the changes in their concentrations during cross-linking photopolymerizations. Most of these studies focused on the structures of the trapped radicals. In the case of multiacrylates, generally a 3-line spectrum was observed, which was consistent both with the chain-end structure (**II**) as well as with a mid-chain structure (**III**) obtained after tertiary hydrogen abstraction from the polymer chain:

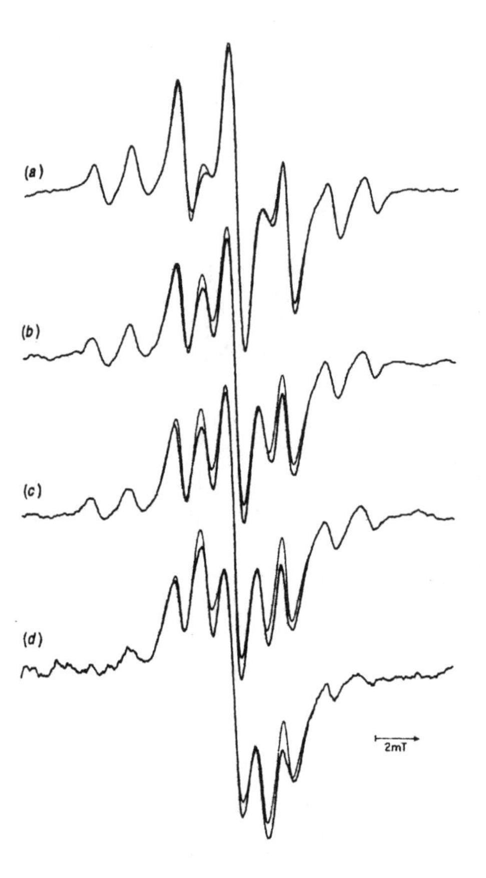

Figure 3.5. EPR spectra of pre-irradiated TMA at various times during the radical decay at 423 K (noisy trace) and least-square fittings of their central parts. Maximum spectral intensity was normalized. (a) 0, (b) 30, (c) 60, (d) 310 min [23].

$$\begin{array}{ccc}
\begin{array}{c} CH_3 \\ | \\ \sim\sim CH_2\text{–}C\bullet \\ | \\ C{=}O \\ | \\ OR \\ \textbf{(I)} \end{array} &
\begin{array}{c} H \\ | \\ \sim\sim CH_2\text{–}C\bullet \\ | \\ C{=}O \\ | \\ OR \\ \textbf{(II)} \end{array} &
\begin{array}{c} \bullet \\ \sim\sim CH_2\text{–}C\text{–}CH_2\sim\sim \\ | \\ C{=}O \\ | \\ OR \\ \textbf{(III)} \end{array}
\end{array}$$

By selective deuteriation at both α positions, the 3-line spectrum has been definitively attributed to structure **II**, by assuming that only one of the β-protons located on the CH_2 group contributes to the hyperfine splitting. Propagating chain-end radicals (structure **II**) could not be observed because they terminated by combination or hydrogen abstraction [24]. Further studies [25–28] showed that the spectrum observed was the superimposition of one 3-line and one single-line pattern, both attributed to the same radical in different environments. The 3-line spectrum corresponds to radicals present in a more fluid phase (so-called $R^{\bullet}$ radicals), while the single line corresponds to radicals in denser, cross-linked regions (so-called $S^{\bullet}$ radicals). Based on the various rates of decay for the EPR-spectra of the two types, it was the 3-line pattern that decayed faster. Such a 3-line spectrum was observed for hexane–1,6-diol diacrylate (HDDA) with one hyperfine coupling constant $\alpha_{2H} \cong 24$ G [24, 29], tetraethyleneglycol diacrylate [29] and butane-diol diacrylate [26–28, 30].

On the other hand, di- and trimethacrylates give a 9-line EPR-spectrum for the propagating radical **I**. This results from the interaction of the unpaired electron of the propagating radical with the three completely equivalent hydrogen atoms of the methyl group and with the two β-methylene hydrogens, which undergo a fast chemical exchange between two different conformations. In the case of EGDM, the hyperfine structure resulted from three hydrogen nuclei of the CH_3 group with α_{3H}=22.7 G and the two β-methylene hydrogen atoms characterized by a mean value of α_{2H}=10.6 G [23, 28, 31]. For methacrylates, no superimposed single EPR line was observed at room temperature during sample irradiation or immediately after and the spectrum was totally determined by radicals of the type $R^{\bullet}$. At higher temperatures and further cross-linking in the dark, the single broad band appeared in the EPR spectrum, which overlapped the 9-line signal. The single-line spectrum also originated from the same propagating radicals. The lack of hyperfine splitting is a consequence of spin–spin interaction between radicals trapped within the rigid matrix [23, 27, 30, 31].

So, the characteristic life times of a radical, which are registered by EPR-spectroscopy *in situ* at the end of a polymerization process, greatly exceed the characteristic times of the stationary light and non-stationary dark processes; thus, they do not cause polymerization and represent radicals of the third type. The difference between spectra of radicals trapped into the polymeric matrix (9 lines of the spectrum), which were observed under the EPR-spectroscopic investigations of the methylmethacrylate samples at the limited conversion degree *via* polymerization, and the spectra of radicals of the methylmethacrylate liquid phase (13-line spectrum) was also found [18, 19].

Two types of conditions for the process proceeding are distinguished by the authors of Refs [16, 17, 20–22]; the first type, with a rapid and short routine, and a second type, which is slow and long (up to 1500 h for some samples) [18]. By evidence of two routines of radical trapping in Ref. [19] the integral dependency $1/[R]–t$, which is approximated by two straight lines with different slopes, is found. This means that radical trapping in the polymeric matrix is insufficiently described by the kinetic equation of the bimolecular chemical mechanism.

It was also noted in Ref. [19] that the rate of radical trapping is a function of the structure of a 3-D network. For example, in the homological range of the oligoesteracrylates differing one from other only by a number of oligomeric links, at the longer transverse cross-linking the rate of a radicals trapping is higher.

Results of the investigations by Berlin [8] showed that autoacceleration *via* the polymerization processes of polyester(meth)acrylates is observed on the conversion depth $\Gamma \leq 1\%$ and at $\Gamma \approx 20–50\%$, a system is near to solid state. In some cases it is observed that the polymerization process finishes, even though half of the double bonds did not react [5, 8]. Berlin considers these characteristics as characteristic only for 3-D polymerization and gives reasons for the autoacceleration of the linear polymerization starting at conversion $\Gamma \approx 20–50\%$.

Experimental data and conclusions of Berlin are confirmed by the results obtained in Refs. [6–8]. For example, it was shown in Ref. [6] that with increasing glycoldimethacrylate content in the methacrylate/glycoldimethacrylate system the autoacceleration process starts earlier. At the polymerization of the low-molecular monomers the autodeceleration begins far earlier than in the case of the high-molecular oligomers [8].

In Ref. [22] data are presented on the limited conversion $\Gamma_{lim.}$, at which the photopolymerization process of the diethylenglycoldimethacrylate (DEGDA) is finished. At the illumination light equal to 0.02 MW/m^2 and a photoinitiator concentration of 1%, the limited conversion reaches $\Gamma_{lim.}$=57.2%. Increasing the illumination light by 10 times leads to the conversion increasing to $\Gamma_{lim.}$=97.0%. According to Ref. [22] even the highest illumination intensity does not lead to a limited conversion value $\Gamma_{lim.}$=100.0%. It was determined by Notley [7] that in the photoinitiated polymerization of triethylenglycolmethacrylate (TGM-3) in the solution of an acethylcellulose $\Gamma_{\text{lim.}}$ is sharply decreased at the increasing solution degree and is proportional to the concentration of the photoinitiator and the illumination intensity to the degree equal to 0.24. In Ref. [8] the dependence $\Gamma_{\text{lim.}}$ on the molecular weight of the monomer was also investigated.

Investigations of a range of dimethacrylic esters, glycols and other polyfunctional oligomers containing (meth)acrylic end-groups with the use of the thermometric express method [23] allowed to determine that the polymerization kinetics of all investigated oligomers on the autoacceleration state can be described by the empirical dependence:

$$W/[M]=W_0 + \alpha\Gamma \qquad (3.1)$$

Here, W and W_0 are current and starting rates of the polymerization, respectively, [M] is current concentration of a monomer, Γ is the depth of polymerization and α is an empirical constant.

It was determined that the empirical value of α depends on the nature of a monomer.

An essential characteristic of the oligomeric polyfunctional (meth)acrylates is the efficient decreasing of the strong inhibitors in the polymerizing medium of these oligomeric (meth)acrylates. It was shown in Ref. [8] that the polymerization of polyfunctional oligo(meth)acrylates in the presence of even an effective inhibitor as benzohinone is not fully inhibited, but proceeds with a visible rate until the late states of the conversion.

A correlation of a reactive ability has been determined for the (meth)acrylates in the late states of a conversion, as observed for a range of polymerhomologous di(meth)acrylates with the same chemical structure but different flexibility of oligomeric molecules. In all cases the polymerization rate increased with increasing flexibility of the starting oligomeric molecule [32]. Interesting data on the analysis of polymeric products at different conversion degrees in the polymerization of some poly(meth)acrylates and diallyl esters were also presented in Ref. [32]. It was noted that during (meth)acrylates polymerization in the bulk, even at a low conversions degrees, all of the product consists of the non-soluble branched or 3-D polymer. In the case of diallyl esters non-cross-linked polymers, which dissolve well in organic solvents, are formed up to the high conversion states. The maximal rate of the poly(meth)acrylates polymerization is higher than the rates for the diallyl esters [32]. In accordance with Ref. [33] it was found that acrylates polymerized 3–7-times faster than the respective methacrylates.

3.2. Conceptions of the diffusion-controlled reactions

Two conceptions are used for the quantitative explanation and description of the regularities of a polymerization and the postpolymerization of mono- and polyfunctional vinyl monomers (mainly (meth)acrylates).

The first conception (or microheterogeneous model) is based on the assumption that the main contribution to the kinetics of the process in the initial state does not lead to homophaseous polymerization in the liquid monomer/polymeric solution, but the heterophaseous one, proceeding on the solid polymer–liquid monomer boundary under the gel-effect conditions.

The second conception follows from the assumption that the polymerization process proceeds mainly in the monomer/polymeric solution, that is in the homogeneous medium and the character of the elementary reactions is determined by diffusion control; that is why their kinetic constants depend upon the composition of the monomer/polymeric solution and upon the depth of polymerization, respectively. In this conception of the diffusion-controlled reactions (DCR) a classical kinetic scheme of the polymerization process and, following from this, its kinetic equation are mainly used. However, the parameters of the above-mentioned kinetic equation are not constant and represent a function of the monomer/polymeric solution current state.

There are different variants of the DCR conception, differing one from another by the set of physical parameters, which are taken into account, and by the details of the analytical description of diffusion control on the rate of an elementary reaction. As a rule, the main attention focuses on bimolecular chain termination, the constant rate of which is considered as a function of macroradicals mobility, depending on their length [34–46], free volume [47–53] or characteristic viscosity [54–57] of the monomer/polymeric solution. In a range of studies [58–61] the initiation efficiency and the constant rate of the chain propagation are also used as a function of macroradicals mobility.

The processes of the macroradicals diffusion in the reactive medium according to models of DCR are described mainly with the use of three mechanisms, namely:

1. The first chemical mechanism considers the diffusion of two macroradicals from one into another. The kinetics of such a process are described mainly in the framework of the modified Smolukhovskiy's theory and are characterized by the coefficient of the translational diffusion D_n.
2. The second chemical mechanism of the diffusion follows from the model of a ball-shaped combination of two macroradicals with the characteristic sphere radius σ. The rate of such a process is determined by the constant of the diffusion segments of polymeric chains D_c [62, 63].
3. The third chemical mechanism is observed in the late states of the polymerization, when the polymerizate is mainly in the glass-like state and the relative mobility of the active links of macroradicals is determined only by the propagation reaction and can be described with the use of the coefficient of reactive diffusion D_p [64–66].

There are various approaches to the description of a polymerization process in the above-mentioned conception of diffusion-controlled elementary reactions. The first approach to the description of diffusion-controlled reactions of radical polymerization follows from the assumption that the main parameter which is changed with the polymerization state is the rate constant of the bimolecular chain termination k_t, depending on translational and segmental mobility of the macroradicals, but only in the autoacceleration state. It is assumed that k_t is constant in the initial state ($\Gamma < \Gamma^*$, where Γ^* is a conversion corresponding to the autoacceleration start) and in the final state ($\Gamma > \Gamma_{max}$, where Γ_{max} is the state of the polymerization corresponding to the beginning of the autodeceleration). Initiation efficiency and the constant rate of chain propagation in these states are also constant during the process.

The basis of such an approach to the description of the diffusion-controlled reactions of radical polymerization has been laid in the works of Nourish and Smith [67]. In 1942, they explained the phenomenon of autoacceleration by a decrease of the constant rate of chain termination with increasing system viscosity. After that a series of investigations has been published in which the influence of the medium viscosity on the constant rate of chain termination has been studied. A whole range of the empirical models [54–56, 62, 68] was considered, in most of which constant chain termination is a reverse function of the medium viscosity η: k_0=const/η. However, it has been noted [54] that such a dependence can describe the polymerization process only in the case when it is limited by translational diffusion. The direct connection between k_t and the viscosity of the medium is not observed in the case when the viscosity of a system is increased until the moment that the step-by-step diffusion of tangled macromolecules is sharply slowed but the mobility of the chains segments is yet sufficiently high.

A range of models in the diffusion-controlled reactions conception takes into account the dependence of a constant chain termination rate k_t on the current concentration of monomer (that is, on the conversion) and on the quantity of polymer formed *via* the polymerization process [54, 56].

A model for methylmethacrylate polymerization has been proposed by Benson and North [62] in 1959. According to this model the constant rate of chain termination is limited by the segmental diffusion and depends upon the length l of an interactive radical. A characteristic value L is introduced in the model and it is assumed that at the small lengths l of the chain the segmental diffusion depends on l and at the achievement of this characteristic value it becomes independent from l.

According to Pravednikov's conception [69] in a case, when the average degree of a macromolecule's chain polymerization is small in comparison with some characteristic value P, the process can be described by the usual equation of classic kinetics. Radicals, the length l of which is more than this value of the polymerization degree P, sharply lose their mobility in comparison with the radicals of shorter chain length and in this case the kinetics are described by a dependence of the constant chain termination rate on the chain length.

Pravednikov's ideas were developed in Ref. [34]. In this work a number of empiric functions taking into account the dependence of the constant rate of bimolecular chain termination on the length l of interacting macroradicals and a characteristic value P are introduced.

Gardenas and O'Driscoll [35] considered three chemical mechanisms of macroradical termination depending on their lengths n and a value n_c, close to the characteristic value L from Ref. [34]:

$$R_r^{\bullet} + R_s^{\bullet} \xrightarrow{k_t}, \qquad \text{where } r, s < n_c$$

$$R_r^{\bullet} + R_s^{\bullet} \xrightarrow{k_{tc}}, \qquad \text{where } r < n_c, s > n_c$$

$$R_r^{\bullet} + R_s^{\bullet} \xrightarrow{k_{tl}}, \qquad \text{where } r > n_c, s > n_c \qquad (3.2)$$

In accordance with Ref. [**35**], in the gel-effect state, chain termination mainly proceeds during the interaction between relatively short ($n < n_c$) and long radicals ($n > n_c$). A constant rate of this termination k_{tc} is proposed to be the geometric average of the constant rate of chain termination up to the moment of the beginning of the gel-effect and the constant rate of chain termination in the late states of the autodeceleration.

Ito's model [68] bears resemblances to the model of Ref. [35], but is different by two aspects. Firstly, it assumes that the constant rate of the chain termination k_{mn} depends on the number of monomeric units (so-called polymerization degree) of m and n radical chains taking part in the termination reaction and represents the sum of the independent contributions of m and n. Secondly, the dependence of the chain termination constant on the length of chains under two types of conditions is described: the first condition is $n < n_c$, controlled by segmental diffusion, and the second one is $n > n_c$, controlled by the reptation diffusion. In the reptation chemical mechanism of diffusion in the deep states of conversion the macroradicals move snake-like between the network joints. De Gennes connected a reptative moving of macroradicals with the dynamic properties of the medium with the use of scaling ratios [37–40] as applied in Refs. [41–46] for the description of constant chain termination in the late conversion state.

In Ref. [44] a model considering the reactive medium as a physical network with the average number of monomeric links between side couplings n was proposed. After the macroradicals reach length n_c (so-called coupling scale), the character of the diffusion movement of the macroradicals changes from the segmental to reptation one. At this time, it is considered [44] that the constant chain termination rate k_{tij} depends on the concentration of polymer and the lengths i and j of the reacting macroradicals. A diffusion of radicals characterized by a relatively short length ($n < n_c$) in comparison with the couplings scale is described with the use of Rauz's m odel for concentrated solutions of polymers in which a scaling ratio for constants of the macroradicals diffusion is true: $D_n \sim n^{-1}c^{-1/2}$, where n is a length of the radical and c is a concentration of the polymer. A diffusion of the macroradicals with length $n > n_c$ is described by the reptative chemical mechanism, according to which $D_n \sim n^{-2}c^{-7/4}$.

In Refs. [43–46] the autoacceleration model according to which a visible gel-effect is started when the majority of the macroradicals diffuse in accordance with the reptation chemical mechanism is postulated, and such a model is accompanied by the following characteristics:

1. a greater influence of side reactions of chain transfer and inhibition;
2. widening of the curve of molecular-weight distribution;
3. change of character of the dependence of the process rate on the rate of initiation.

In Ref. [36] the characteristics of a polymerization up to the high conversion state are explained by the change of chemical mechanism of chain termination, that is, by the transition from the bimolecular chemical mechanism to the monomolecular one. The last is considered as a physical excluding of active radicals resulting from their trapping by solid polymeric matrix.

Two types of functions in the models for the description of the polymerization processes at moderate and high conversions have been described in the literature [47–53, 57–61, 63, 70–75]. For example, in Ref. [63] until the depth of conversion Γ=0,6, the kinetics of the polymerization process are described by a change of the bimolecular chain termination rate constant; parameters of the propagation and initiation efficiency are assumed to be constant values. Decreasing the chain termination constant rate *via* especial empirical dependence has been introduced for values of $\Gamma > 0,6$. Decreasing the chain termination constant rate for $\Gamma < 0,6$ is determined *via* the coefficients of the translational and segmental diffusion of the macroradicals:

$$1/k_t = 1/k_R + 1/k_C \tag{3.3}$$

here:

k_R is a chain termination constant rate limited by the translational diffusion and is calculated *via* the coefficients of the self-diffusion of the macroradicals D_n and the size R of their balls;

k_C is a chain termination constant rate limited by the segmental diffusion. Besides the segmental diffusion coefficient D_c and a segment size, the empirical sterical factor is introduced, that takes in to account the side effects in the cell [47–51, 70]. The segmental diffusion coefficient is assumed as a function of segment number of the macroradicals and the characteristic viscosity of the solution.

A change of the diffusion-controlled constant k_t and k_p *via* conversion [47–53, 58, 70] is expressed as a function of free volume.

Specific free volume, in accordance with the model of free volume [47, 49–51], is a function of conversion and, at the glass transition temperature, equals 0.025; this volume changes linearly with increasing temperature of the polymerization medium. According to the above, the following expressions are used for the calculation of specific volumes of the reactive medium:

$$V_{fp}=0.025 + \alpha(T-T_{gp})$$
$$V_{fm}=0.025 + \alpha(T-T_{gm})$$
$$V_f=V_m\Phi_m + V_{fp}(1-\Phi_m) \quad (3.4)$$

Here

m and p are subscripts indicating monomer and polymer, respectively;

T is the temperature of the polymerizing medium;

T_g is the glass transition temperature;

Φ_m is a volumetric concentration of the monomer;

α is a contraction coefficient;

V_f is a specific free volume of the reactive system.

On the basis of the free volume model in Ref. [**70**] the diffusion-controlled chain termination constant is described *via* all processes as follows:

$$k_t = \left(\frac{M_W}{M_{WC}}\right)^{-\alpha} k_{t0} \exp\left\{-B\left(\frac{1}{V_f}-\frac{1}{V_{f_{0t}}}\right)\right\} \quad (3.5)$$

Here

M_w is the molecular weight of the polymer;

M_{wc} is the molecular weight of the polymer in the starting point of the diffusion control on the chain termination reaction rate;

k_{t0} is the rate constant of the bimolecular chain termination at the starting point of the diffusion control on the chain termination reaction rate;

B is an empiric constant;

V_f is the specific free volume of the reactive system;

V_{f0t} is the specific free volume of the reactive system at the starting point of the diffusion control on the chain termination reaction rate;

α is an empiric constant.

In accordance with the model [70] the chain termination rate constant depends on the molecular weight only in the early conversion states, when the translational diffusion is the limiting factor.

The constant of the chain propagation rate is described in Ref. [70] by the following dependence:

$$k_p = k_{p0} \exp\left\{ C\left(\frac{1}{V_f} - \frac{1}{V_{f_{01}}} \right) \right\} \qquad (3.6)$$

Here

k_{p0} is the chain propagation rate constant at the starting point of the diffusion control on the chain propagation reaction;

V_f is the specific free volume of the reactive mixture;

V_{f01} is the specific volume of the reactive system at the starting point of the diffusion control on the chain propagation reaction;

C is an empiric constant.

The proposed model [70] is in good agreement with the experimental results of the polymerization of methylmethacrylate in all conversion states. However, the conditions of the diffusion control starting with the elementary reactions of the chain propagation and its termination are not discussed and that is why the change of chemical mechanism of diffusion control with increasing conversion state cannot be understood.

In Ref. [49] a thesis of Ref. [70] for a description of the polyfunctional 3-D polymerization processes of (meth)acrylates is used.

The main diffusion-controlled parameter in Refs. [52, 53, 58, 76] is the constant rate of chain propagation depending on the translational mobility of the monomer, determined by the function of the free volume of monomer/polymeric solution. It is considered that the chain termination is limited by the chain propagation, even in the middle states of the conversion as a result of the formation of a viscous network system. In this case only one type of diffusion exists, namely a diffusion of the polymer chain propagation, and the coefficient of this diffusion D_p [52] is determined as a function on the bond length l taking into account that all lengths of bonds in the macromolecule are the same:

$$D_p = \frac{2}{3} \ell^2 k_p [M] \tag{3.7}$$

From this a constant of the chain termination limited by the diffusion of a propagation is determined as follows:

$$k_t = \frac{32}{3} \pi \, r \ell^2 k_p [M] \tag{3.8}$$

where r is the reactive radius of the radical.

Assuming a direct dependence of the macroradicals termination constant rate on their propagation constant rate, the presence of 'frozen" radicals in the matrix is proven using the following arguments: if the diffusion is decreased until the value at which the propagation acts cannot proceed, then the chain termination connected with the propagation act also cannot be realized and the radicals are 'frozen" by the matrix of a po lymer.

Typical of the model proposed in Ref. [58] is the introduction of the initiation efficiency f_i as a function of conversion depth into the equation of the general process rate. The initiation efficiency is expressed in the model *via* empiric dependence of the diffusion parameter on the polymerization state and is estimated taking into account the model of the cell effect and the following approximations:

- diffusion of all radicals is the same;
- radii of the cell for all recombined radicals are also approximately the same.

In Ref. [77] the dependence of the chain propagation rate constant and the initiation efficiency upon the volumetric concentration of polymer *via* the processes of radical polymerization in the late conversion states was considered. With the aim of describing such functions, the models of the cell effect and a side cell effect have been used.

The following kinetic equation for the initial state of the process is used most often for the quantitative description of the photoinitiated polymerization [78–84]:

$$-\frac{\mathrm{d[M]}}{\mathrm{d}t}=\frac{k_{\mathrm{p}}}{k_{\mathrm{t}}^{1/2}}[M]\left\{\gamma_{\mathrm{i}}I_0(1-e^{-\varepsilon[\mathrm{In}]\ell})\right\}^{1/2} \tag{3.9}$$

Here

γ_i is the quantum yield of the photoinitiation;

I_0 is the light intensity of the UV illumination on the surface of the photocomposition;

ε is the molar extinction coefficient;

[In] is the photoinitiator concentration;

ℓ is the thickness of the photocomposition layer;

k_p and k_t are the propagation and termination rate constants, respectively, which change in the process and have been described earlier.

At low photoinitiator concentration and assuming a quasi-steady state for the radicals' concentration, the evolution of $k_p/k_t^{1/2}$ can be expressed by:

$$\frac{\gamma_{\mathrm{i}}^{1/2}k_{\mathrm{p}}}{k_{\mathrm{t}}^{1/2}}=-\frac{\mathrm{d[M]}}{\mathrm{[M]d}t}\left\{I_0(1-e^{-\varepsilon[\mathrm{In}]\ell})\right\}^{1/2} \tag{3.10}$$

The influence of temperature and conversion degree on the values $\frac{\gamma_{\mathrm{i}}^{1/2}k_{\mathrm{p}}}{k_{\mathrm{t}}^{1/2}}$ for the photoinitiated polymerization of the dimethacrylate by structure (**IV**) was investigated in Ref. [78]. 2,2-Dimethyl-2-hydroxyacetophenone (Darocur 1173: 0.15% (w/w), i.e., 10^{-2} mol· l^{-1}) was used as photoinitiator, which was dissolved in the oligomer under stirring at room temperature for 3 h.

$$CH_2{=}C(CH_3){-}C({=}O){-}O{-}({-}CH_2{-}CH_2{-}O{-})_x{-}C_6H_4{-}C(CH_3)_2{-}C_6H_4{-}({-}O{-}CH_2{-}CH_2{-})_y{-}O{-}C({=}O){-}C(CH_3){=}CH_2$$

(**IV**)

Experimental results, shown in Figure 3.6, are explained as follows by the above-mentioned authors. At the beginning of the reaction, the high increase in viscosity reduces the segmental or translational diffusion of the reactive species and the collision between two radicals is more and more difficult. As a result k_t decreases faster than k_p and $k_p/k_t^{1/2}$ increases. Then, the termination mechanism becomes a diffusion-controlled reaction and can be assimilated to a propagation mechanism. From the top of the curve and because of the great decrease in the mobility of species, first propagation and termination constants and then $k_p/k_t^{1/2}$ decrease quickly.

Equation (3.9) has been derived on the assumption of bimolecular chain termination. Furthermore, this equation does not take into account the presence of gradients of photoinitiator and monomer concentration, and conversion depth on the layer of a photocomposition. These gradients are caused by the illumination gradient. That is why the presented equation will be true for the description of the process under ideal conditions.

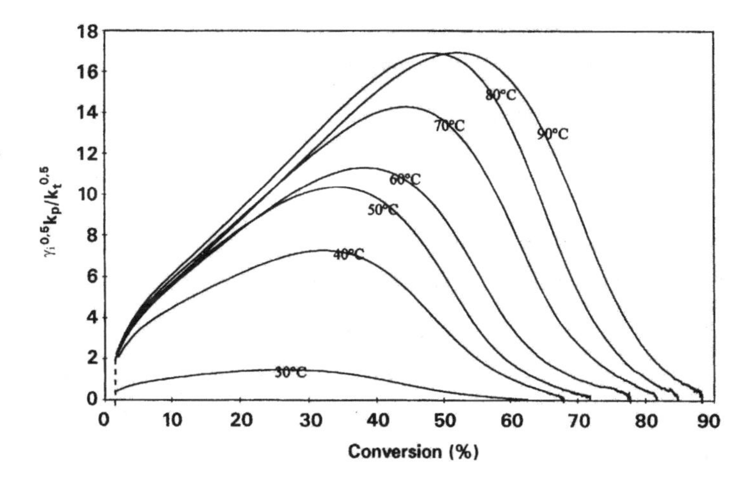

Figure 3.6. Variation of $\dfrac{\gamma_i^{1/2}k_p}{k_t^{1/2}}$ *versus* conversion for different reaction temperatures [78].

Thus, although all variants of the diffusion-controlled reactions conception can quantitatively describe the stationary kinetics of the polymerization up to the high conversion state, they forced us to divide the whole process into unfounded intervals and to choose different chemical mechanisms of the diffusion control and analytical dependences, including the empiric ones, for each interval.

Among the numerous models in the framework of the diffusion-controlled reactions also the conception of a semi-empirical model (the so-called autocatalytic one) is used for the description of the 3-D polymerization kinetics.

In general an autocatalytic effect is due to the formation of intermediate species that markedly accelerate the reaction. For instance, the curing process of epoxy-amine thermoset resins is autocatalysed by hydroxyl groups formed in the reaction. Processes of this type are often described using the phenomenological model developed by Kamal [85, 86]. Generally, phenomenological models are based on two kinetic relations [87–90]. The general phenomenological model proposes that, when the initial rate of reaction is not negligible, the expression becomes:

$$\frac{\mathrm{d}p}{\mathrm{d}t} = (k_1 + k_2 p^m)(1-p)^n \tag{3.11}$$

Here

k_1 and k_2 are the Arrhenius-type rate constants;

n is the reaction order exponent;

m is the autocatalytic exponent;

p represents the relative conversion;

k_1 is the rate constant of the uncatalyzed process;

k_2p^m represents the autocatalytic nature of the process.

Several simplifications have been suggested in the literature, among which a widely employed one makes use of the observation that, in general, the curing reaction of epoxy resins is a second-order process, that is, $m+n=2$ [91]. The sum of m and n represents the overall reaction order.

With this assumption, equation (3.11) reduces to

$$\frac{\mathrm{d}p}{\mathrm{d}t} = (k_1 + k_2 p^{2-n})(1-p)^n \tag{3.12}$$

In this form, the Kamal equation does not take into consideration the fact that, as the conversion increases and the T_g of the material approaches the reaction temperature, the process becomes diffusion-controlled. Thus, the reaction rate decreases considerably and reaches zero before reaching full conversion. In general, the final conversion is strongly dependent on the reaction temperature. This effect could be accounted for by substituting the term ($1-p$) on the right side of equations (3.11) and (3.12) with the term ($p_{max}-p$), where p_{max} represents the final conversion reached at the investigated temperature. Thus,

$$\frac{\mathrm{d}p}{\mathrm{d}t} = (k_1 + k_2 p^{2-n})(p_{max} - p)^n \tag{3.13}$$

The above-discussed kinetic models were applied to the TGDDM/DDS system cured at 140°C and to the TGDDM/MNA system cured at 120°C to evaluate their ability to simulate the real process.

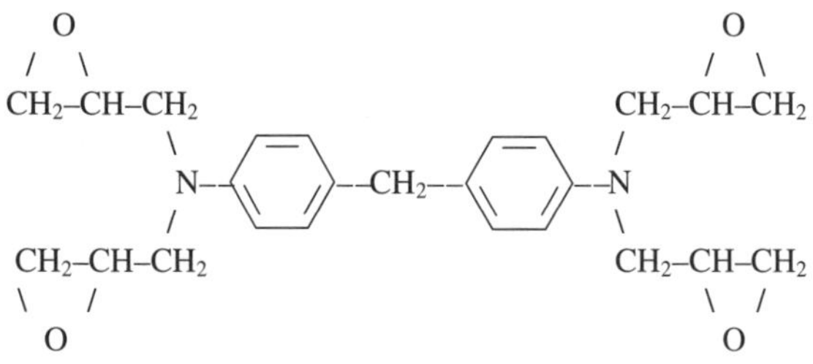

```
      O                                              O
     / \                                            / \
CH2–CH–CH2                                    CH2–CH–CH2
         \                                    /
          N––(benzene ring)––CH2––(benzene ring)––N
         /                                    \
CH2–CH–CH2                                    CH2–CH–CH2
  \    /                                          \   /
    O                                               O
```

Tetraglycidyl 4,4′-diamino diphenyl methane (TGDDM)

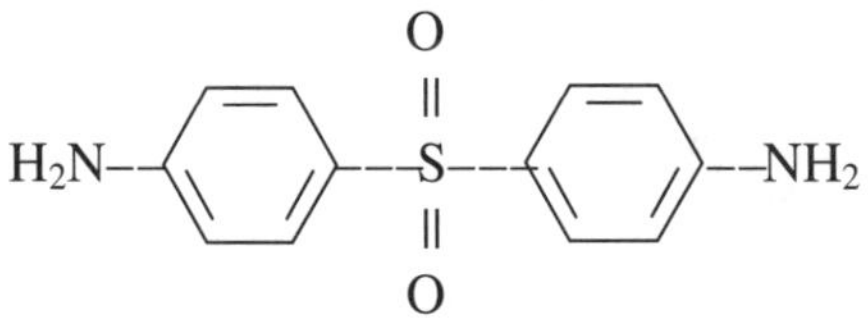

```
                          O
                          ||
H2N––(benzene ring)––S––(benzene ring)––NH2
                          ||
                          O
```

4,4′-diamino diphenyl sulfone (DDS)

Methyl nadic anhydride (MNA) Benzyl dimethyl amine (BDMA)

The results of the simulations are presented in Figure 3.7 and in Table 3.1.

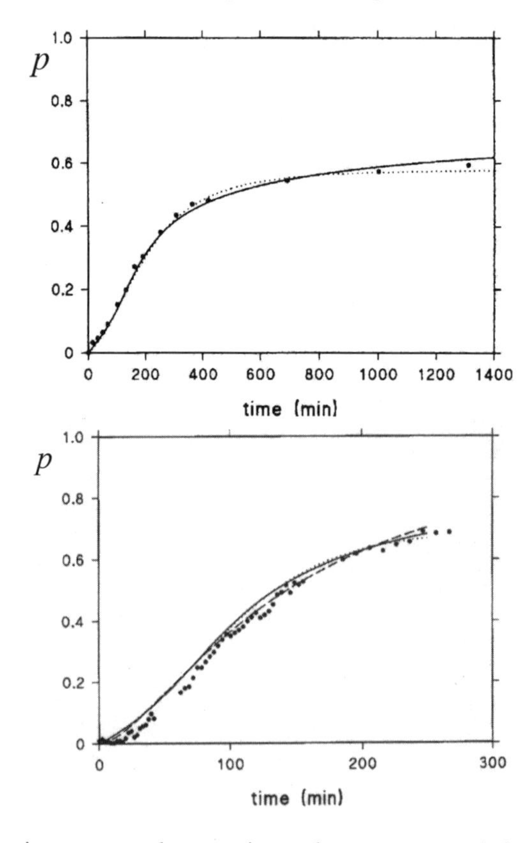

Figure 3.7. Comparison among the experimental p–t curves and the predictions of the various kinetic models, relative to the TGDDM/DDS system cured at 140°C: (•) experimental data points; (continuous line) model (3.11); (dashed line) model (3.12); (dotted line) model (3.13) [91].

If the initial rate of the reaction is negligible, equation (3.11) can be reduced to:

$$\frac{\mathrm{d}p}{\mathrm{d}t} = kp^m(1-p)^n \qquad (3.14)$$

The temperature dependence was assumed to reside in the rate constant k, and was expressed as an Arrhenius relation [92]; thus, equation (3.14) takes the form [93]:

$$\frac{\mathrm{d}p}{\mathrm{d}t} = A\exp\left\{-\frac{E}{RT}\right\}p^m(1-p)^n \qquad (3.15)$$

Here:

A is the pre-exponential factor;

E is the apparent overall activation energy;

R is the universal gas constant.

Table 3.1. Kinetic parameters evaluated according to the different model equations [91]

Parameter	System	
	TGDDM/DDS 100:30	TGDDM/MNA/BDMA 100:80:0.56
T_{cure}	140°C	120°C
p_{max}	0.581	0.600
$k_1{}^a$	0.001282	0.063340
$k_2{}^a$	0.220259	2.0565
m^a	2.00	1.86
n^a	7.50	6.38
$k_1{}^b$	–	0.04928
$k_2{}^b$	–	0.002595
n^b	–	2
$k_1{}^c$	0.001417	0.105027
$k_2{}^c$	0.014274	0.035403
n^c	1.30	1.18

[a] Equation (3.11); [b]equation (3.12); [c]equation (3.13).

The above-mentioned phenomenological model is usually used to describe the kinetics of autocatalytic reactions that are characterized by a maximum reaction rate between 20 and 40% conversion. The shape of the photoinitiated polymerization rate curves is the same as that of the autocatalytic reaction (see Figure 3.8). Although photoinitiated polymerization is autoaccelerated and not autocatalyzed, this model can be applied to describe these reactions in a purely mathematical way.

Autocatalyzed reactions often appear to show an "induction" period during which no apparent reaction occurs. This can be interpreted as a manifestation of the parallel or seeding reaction. The "duration" of the induction period, thus, depends critically on the concentration of the seed or on the rate constant of the parallel reaction. A consequence is that the zero origin of the reaction time-scale can be chosen arbitrarily, and it may be convenient to choose this origin to be different from the actual start of the experiment [87].

The autocatalytic model has been shown to predict the cure of thermoset resins accurately [92]; some of the more recent studies in this area are cited in a number of references [90, 91, 93–96].

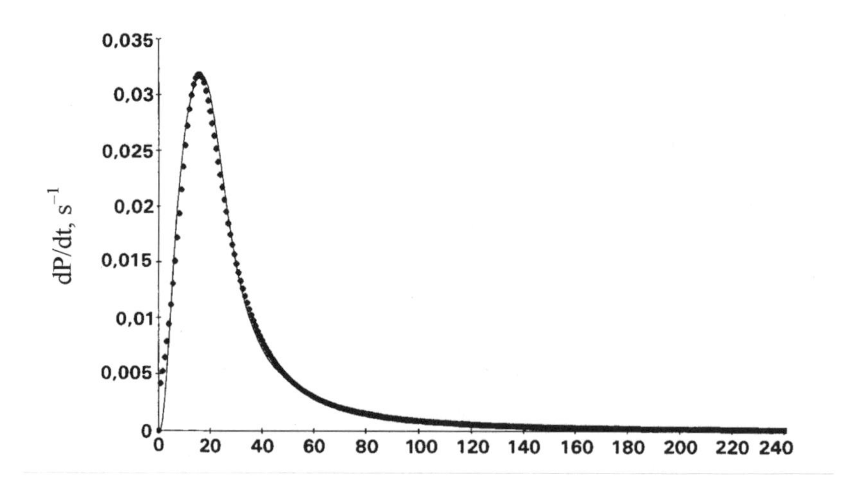

Figure 3.8. Typical photoinitiated polymerization rate *versus* time curve at 50°C. (---) Experimental curve and (♦) modelization according to equation (3.14) [78].

The autoaccelerated process kinetically resembles an autocatalytic reaction. In the case of cross-linking polymerization, the increase in the polymerization rate is caused by restricted diffusion of the reactive chain ends connected to the network being formed [97]. As a result, the shape of the polymerization rate *vs.* conversion curves for autoaccelerated processes is the same as that for autocatalytic reactions; thus, equation (3.14) can also be applied to describe the cross-linking polymerization. However, only a limited number of studies have been devoted to the description of photoinitiated cross-linking polymerizations by the autocatalytic model [78, 88, 89, 97–102] and all of these were based on DSC measurements.

Waters and Paddy [87] analyzed DSC curves registered for autocatalysed reactions and found that such curves exhibited a maximum at the value of p_{Rm} given by:

$$\frac{1-p_{\mathrm{Rm}}}{p_{\mathrm{Rm}}}=\frac{n}{m} \tag{3.16}$$

Thus, the determination of exponents n and m was based on the linearization of equation (3.14) to the form:

$$\log\frac{\mathrm{d}p}{\mathrm{d}t}=\log k+n\ \log[p^{m/n}(1-p)] \tag{3.17}$$

and adjusting the ratio m/n estimated by $p_{Rm}/(1-p_{Rm})$ to get a linear relation with the slope n [88, 89, 100].

Chandra and Soni analyzed the photo(co)polymerization of di(meth)acrylate oligomers. The values of exponents m and n were 0.58 and 0.72, respectively, at 50°C (for a dimethacrylate oligomer) [88]. No clear tendency with temperature rise was observed, although the temperature affected the exponents, more so for n than for m [88, 100]. The results obtained showed a slight decrease of the rate constant k with increasing temperature [100]. The authors found that the autocatalytic model fitted well the data obtained from the start of the reaction at various concentration degrees, which depended on the system polymerized. In these conversions, the data fitted an nth order model rate equation, which described the diffusion limited reactions better [88, 100]:

$$\frac{\mathrm{d}p}{\mathrm{d}t}=k(1-p)^{n} \tag{3.18}$$

An analogous analysis of the kinetics of the photopolymerization of (ethoxylated Bisphenol A)dimethacrylate [78, 98] gave constant (independent on temperature) values for the m and n exponents equal to 0.8 and 2, respectively. The rate constant k increased with the reaction temperature from 30 to 80°C, with a low activation energy of 2.2 kcal*mol^{-1}.

An increase of k with temperature for a series of analogous di(meth)acrylates was also observed in another work [97] and an activation energy (in the range 30–90°C) of about 7 kJ· mol^{-1} was found.

Preliminary results of the analysis of the photopolymerization kinetics according to the autocatalytic model of a series of alkylenediol dimethacrylates (structure **V**) with an increasing number of CH_2 groups in the spacer have been presented recently [101]. The parameters k, m and n were calculated from a best fit between model prediction and the data points of the experimental polymerization rate *vs.* conversion degree curves (see Figure 3.9).

$$CH_2{=}C(CH_3){-}C({=}O){-}O{-}(CH_2)_n{-}O{-}C({=}O){-}C(CH_3){=}CH_2$$

V

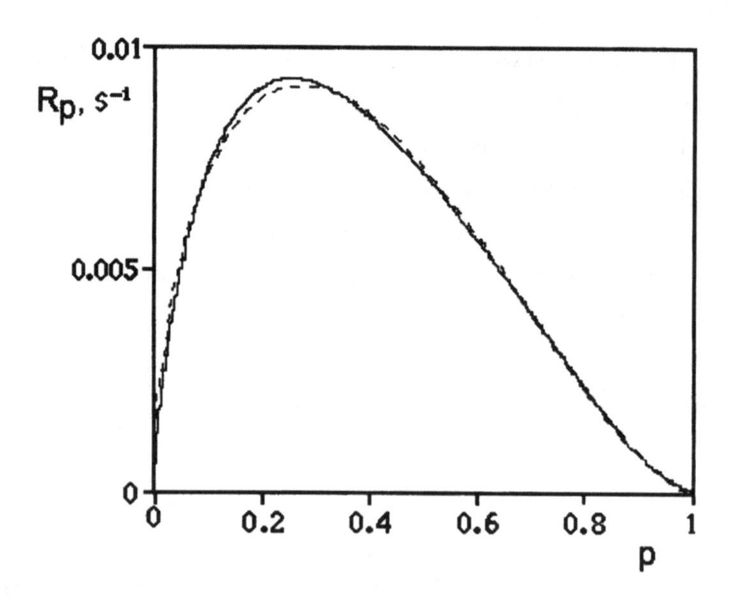

Figure 3.9. Polymerization rate *vs.* relative conversion for ethylene glycol dimethacrylate polymerization in Ar at 50°C. Solid line, experimental curve; dashed line, prediction of autocatalytic model [101].

The k values increased with the number of CH_2 groups in the monomer molecule and with the polymerization temperature (between 40 and 100°C). This reflected the resulting increase in the polymerization rate. The values of exponents m and n showed a tendency to decrease with increasing temperature. The values for the parameters for all of the monomers studied were within the range of about 0.025–0.06 s^{-1} (k), 0.45–0.65 (m) and 1.4–1.8 (n).

In one study [102], the expression for the autocatalytic reaction was modified to model diffusion effects (autoacceleration and vitrification) as well as the light intensity effect on the photopolymerization kinetics of a commercial acrylic resin. The parameters of the model were calculated using non-linear regression analysis. Expression of the rate constant k as $k_0(T)I_0^b$, where b is a fitting parameter and $k_0(T)$ is temperature-dependent kinetic constant, enabled the authors to determine to the light intensity exponent. A value of about 0.7 was found, showing that the termination occurred following both monomolecular and bimolecular pathways.

The autocatalytic model was also applied to compare the effects of various photoinitiators on the polymerization kinetics [99].

On the other hand, despite the fact that equations (3.11)–(3.15) are widely used for the description of the polymerization of polyfunctional monomers up to the high conversion state and satisfactory describe the experimental data, their physical sense has not been discovered yet. That is why results obtained *via* the autocatalytic model do not offer the possibility to explain the chemical mechanism of a process and theoretical grounds of the constants k_1 and k_2.

Thus, kinetic modeling of the curing process can be approached in two ways: mechanistic (in accordance with the classical equations) or phenomenological (autocatalytical model). The mechanistic approach consists of considering the complete set of reaction steps that constitute the overall mechanism and in determining the single-rate equations for each step. A complex system of simultaneous differential equations is obtained, which, whenever possible, is solved numerically. The solutions yield the absolute values of the kinetic constants of each step, which allow the simulation of the concentration profiles of all the reactive species involved in the process. If the system of differential equations is not amenable to numerical integration, a number of simplifying assumptions are adopted. This approach is complicated from a computational point of view, especially when side reactions play an important role in the process.

3.3. Conception of the microheterogeneity of the polymerization system

Studying 3-D polymerization foresees an investigation of the total of all chemical and physical processes, useful properties of the polymeric material and phenomena responsible for the formation of a polymer and, as a result, also for the formation of a complex. Therefore, knowledge of the structural and physical aspects of the microheterogeneous chemical mechanism of the 3-D radical polymerization is very important.

In 1962, Gallacher [103] observed, for the first time, the origin and propagation of solid particles in the liquid phase of the polymerizate on the basis of acrylic and (meth)acrylic monomers using the light-scattering method (in the variant of spectra turbidity). From this time, at least for polyfunctional monomers, the micro-unheterogeneity of a polymerization system is a fact proven by experimental methods [2, 33, 104–119]. In this case micro-unheterogeneity represents the presence in the polymerization system of micrograins of a solid polymer in the liquid monomeric phase, or after the phase inversion, the microdrops of the liquid monomer, distributed in the solid polymeric matrix.

It was determined by the spectrum turbodimetry method that at the polymerization of polyfunctional oligoesteracrylates a number of micrograins of the solid polymer formed on the initial state of a process does not change at the simultaneous propagation of their sizes [104]. A hypothesis of the propagating grains has been confirmed by experimental data of TGM-3 polymerization described *via* ranges of models of micro-particles propagating *via* the process [105].

Analysis of the reactive medium *via* the process of TGM-3 polymerization has been done with the use of the light-scattering method [106]. According to data presented in Ref. [106], a system is stable up to 6–8% conversion. At 20% conversion, a sharp increase both of scattering centers concentration and the light-scattering intensity is observed. At the same time, it was determined with the use of ultra-microscopy [107] that at 20% TGM-3 conversion the concentration of scattering centers is increased by one order of magnitude.

Very important are results of the investigation of the phase-inversion state, when in a certain step of a polymerization process discrete micrograins of the solid polymer in the homogeneous liquid phase of monomer are propagated into a united polymeric matrix, in which microdrops of the liquid monomer are distributed [2, 33, 104–112]. According to data represented in Ref. [104], a phase inversion is observed in the state of a process when the volume of polymeric grains is approximately half of the volume of the reactive system.

Phase inversion, or a monolytization state, is accompanied by a microsyneresis [107–109] that occurs as result of the thermal incompability of the densely cross-linked polymeric formations with the liquid phase of a monomer on this state of the polymerization process and leads to an elimination of the liquid-phase excess from the cross-linked polymer.

In Ref. [108] the phase inversion phenomenon has been investigated by the light-scattering method. In Ref. [108] a negative temperature coefficient of the light-scattering at the initial conversion states of the TGM-3 polymerization is observed. At such a negative temperature coefficient of the light-scattering, the temperature decrease of a process causes the increase of light-scattering. This phenomenon of a negative temperature coefficient of the light-scattering is characteristic for solutions of solid polymer in liquid monomer. In a certain determined transformation state, a change of the negative temperature coefficient to the positive one is observed. At the same time, increasing the temperature of a process causes an increase of light-scattering and is characteristic of the polymeric matrix. The change of the negative temperature coefficient to a positive one is explained by phase inversion in the reactive medium and by microsyneresis of the system [108].

The phase inversion state on the kinetic curves of polymerization corresponds to the section of transition from the autoacceleration state to the autodeceleration state and is characterized by a sharp sudden leap-like change of properties of the cross-linked polymeric material [33, 111]. Thus, for TGM-3 the module elasticity is increased 3.5-fold and the value of the diffusion coefficient is decreased 1000-fold [111]. The structures of grains and an inter-grains space were investigated with the use of the paramagnetic probe sorption method. 2,2′,6,6′-Tetramethylpiperedyne-1-oxyl was used as a probe [112, 113]. The above-mentioned method is based on the difference between the diffusion coefficients of the 2,2′,6,6′-tetramethylpiperedyne-1-oxyl at sorption of the probe into polymeric grains and into the interface-grains space [114]. The concentration of 2,2',6,6'-tetramethylpiperedyne-1-oxyl in the polymeric matrix has been determined by ESR spectroscopy. It was shown in Ref. [113] that in the cross-linked polyacrylates of different structure sorption of 2,2′,6,6′-tetramethylpiperedyne-1-oxyl takes place only in the interface-grains space, a part of which is decreased *via* process depth. Investigations of the sorption into the inter-grain space in the 2,2′,6,6′-tetramethylpiperedyne-1-oxyl spectrum at different temperatures discovered a glass-transition-like phenomenon, and this allowed to draw conclusions on structural un-homogeneity of the interface-grains space.

Investigation results of the acrylic oligomers TGM-3 and MGPh-9 by sol–gel analysis and IR-spectroscopy [107, 116, 120] show that in all conversions states the sol phase is a starting oligomer with the conversion degree Γ=0 and the gel phase the non-soluble densely-cross-linked polymer with limited conversion degree. On the basis of these studies [107, 116, 120] it was assumed that the process was localized on the surface of the polymeric grains.

The structure of the polymeric grains was investigated in Ref. [115] by the scanning calorimetric method and by warm measurement. Samples for the investigations have been prepared by polymerization of TGM-3. A glass transition temperature was determined by sharply increasing the heat flow $d\Delta H/d\tau$ or a specific heat C_p of the polymer by heating the sample related to the polymer transition from the glass-like state to the high-elastic state. A great jump of the $d\Delta H/d\tau$ was fixed for a sample at the converstion degree Γ=10%. At the same time, it was shown in Ref. [115] that grains, similar to interface-grains layers, are structurally un-homogeneous, that is, the core of the grain is in a glass-like state up to 190° C, while the structure of the outer shells of the polymeric grains can transit to the high-elastic state, even at low temperature increases.

An attempt to give a complete interpretation of the chemical mechanism of the oligomeric acrylates transformation process into cross-linked polymers has been done *via* the framework of the microheterogeneous model of polyunsaturated compound polymerization. In this relation, works corresponding to the main states of the microheterogeneous structure of polymer formation are considered, namely the origin of polymeric grains, their propagation and growth into the monolith in the late states of the polymerization.

A microheterogeneous model of the radical polymerization process of polyunsaturated compounds at the present time is based on the investigations in which the kinetics and chemical mechanism of acrylic oligomer polymerization, structure and properties of the forming polymers are complexly studied. This model draws a general conclusion about the most essential regularities of inter-connecting chemical and structural-physical phenomena proceeding during the formation of polymeric solids. The microheterogeneous conception consisting of the model base allows the best interpretation of all experimental data presented thus far. At the same time this model is the most promising for the search of new ways of improvement of the properties of polymeric acrylates. This model helps researchers in investigating of the nature of processes and phenomena accompanying the polymerization with the aim of elucidating the scientific principles of rational technologies for obtaining densely-cross-linked polymeric materials.

The main principles of the microheterogeneous model can be formulated as follows. The densely-cross-linked polymer formation process is characterized by a parallel proceeding of inter-connected chemical and physical phenomena. Chemical transformations (or interaction of the reactive-able groups) are accompanied by physical phenomena, namely chain aggregation, phase division (or syneresis) with the elimination of the starting monomer from the sufficiently dense polymeric networks, local glass transition of the densely-cross-linked microvolumes, re-distribution of the inter-molecular interactions in connection with phase surface division, etc.

The so-called microheterogeneous model of 3-D radical polymerization, based on an image of the microheterogeneous character of 3-D polymerization, has been proposed for the first time by Korolev in the 1970s.

The main principles of the above-mentioned microheterogeneous model were formulated in Refs. [2, 71]. According to these works the process of 3-D polymerization is considered to be a total combination of chemical processes and structural-chemical transformations in the reactive medium. From Korolev's point of view, this reactive medium represents the chain aggregation, phase division, at which a starting monomer is separated from the sufficiently dense polymers (the so-called syneresis phenomenon), and also local glass transition of the densely-cross-linked microvolumes. In accordance with this model one of the main kinetic reasons causing the proceeding of structural-physical transformations is the dependence of the chain termination rate on the viscosity of the reactive medium. This dependence is explained by the fact that the local viscosity propagation in the macromolecular balls formed as a result of polymeric chain aggregation leads to a decreasing chain termination rate and, in turn, to increasing viscosity, i.e., the process has the autocatalytic character of the gel-effect type. The total of chemical and structural transformations firstly leads to appearance of the kinetic un-homogeneity of a system and after that to the formation of micro-particles with the conversion states higher than the average-volumetric one.

A thermodynamical reason for the chemical-structural transformations proceeding is thermodynamical incompability of the densely-cross-linked polymers with a liquid phase of a reactive system. In accordance with the Flory–Renner ratio, syneresis is started simultaneously with the elimination of the liquid phase from the cross-linked polymer at a time when the density of polymeric network in a micro-volume of the reactive system will be higher than the value corresponding to equilibrium swelling conditions [113]. The process of syneresis in 3-D systems proceeds as microsyneresis and is the result of grave diffusion complications, and distribution processes of the other components of the reactive system between the micro-phases increasingly intensify the micro-un-heterogeneity of a system. At some state of the transformation micro-gel particle glass transition is started with the formation of glass-like polymeric grains. The final polymeric product consists of densely-cross-linked polymeric grains and liquid layers between them.

Via the framework of the microheterogeneity model of the 3-D radical polymerization, topological reasons for the microheterogeneity appearing also have been discussed. Among them is increasing the probability of macromolecules cyclization processes that promote the formation of the micro-gel particles. According to this model a totality of kinetic, thermodynamical and topological reasons causes the structural-physical transformations, which together with the polymerization processes lead to the above-mentioned chemical mechanism *via* the 3-D polymerization process.

Kinetic reasons cause the formation of a micro-gel particle formation in which a local gel-effect proceeds. Further polymerization is carried out at the expense of growing new outlying layers on the surface of polymeric grains that are micro-reactors in which the polymerization process is localized. At the state of the transformation when the volumetric part of the propagating discrete particles is sufficient for their interaction, these particles are joined. The growth into a uniform structure in the monolytization state proceeds at the expense of polymerization transformations in the joint zones.

The microheterogeneous model foresees a sudden leap-like change of the properties of the cross-linked polymeric material in the monolytization state in sufficiently narrow depths of the transformation. This model also assumes the existence of two topologically different types of polymerizing process on the monolytization state, which are essentially different in their contribution to the monolith formation.

The first process is localized indirectly in the narrow zones of the grains joint, and the second process in the wide cracks between them. The contribution of the polymerization in the narrow zones to the formation of a monolith from joint grains is maximal, and polymerization in a wide range leads only to decreasing cracks between them.

Development of the theoretical basis for the dense-cross-linked systems formation stimulated a new wave of (meth)acrylates polymerization investigations, which were carried out in the 1980s. The results of these studies showed that in all cases experimentally determined regularities are in good agreement with the model prognosis and allow essentially to detail and support this model.

At the qualitative level the microheterogeneous model does not take the place of the diffusion-controlled reactions conception; however, its quantitative interpretation causes complications in the description of complicated processes proceeding in the microheterogeneous reactive medium [2]. The kinetic equation of the polyfunctional oligoesteracrylates 3-D polymerization process is presented in Refs. [2, 113]. The quantitative model [2, 113] starts from the assumption of the microheterogeneous model of the 3-D radical polymerization that the reactive mediums consists of solid polymeric grains with limited conversion degree and inter-grain space, in which the primary monomer is located.

Thus, the polymerization process can be simply represented by the following transformations, namely: origination of polymeric grains from the fluctuations of polymer concentration → microdistribution → polymeric grains propagation → monolytization (see Scheme 3.10). Furthermore, as it was mentioned above, even at the initial states of the process a system consists of a polymer with practically limited conversion degree and the starting monomer with the conversion degree equal to zero. Thus, two reactive zones appear: first is the inter-grains layer or liquid homogeneous medium and the second is the surface layer on the solid polymer–liquid monomer boundary division, making a maximum contribution to the polymerizing process [107, 116].

The next subsections give short characteristics of the above-mentioned states of the polymerization.

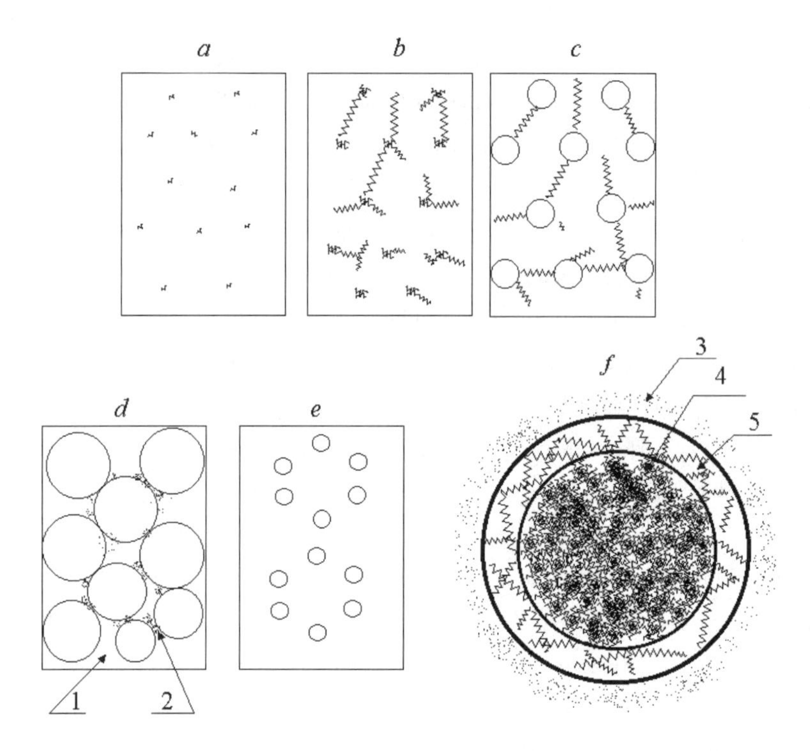

Scheme 3.10. States of microheterogeneous structure polymer formation. (a) Origination of polymeric grains from the fluctuations of polymer concentration; (b and c) micrograins propagation; (d) polymer at the state of monolytization (polymeric grains co-growing); (e) formed microheterogeneous polymer; (f) polymeric grain; (1) contact and co-growing zone; (2) inter-grain space; (3) monomer; (4) core of a grain; (5) surface layer (shell) of a grain.

3.3.1. Origination of polymeric grains in the initial states of a polymerization

At the radical chain polymerization of oligomer containing n vinyl groups per molecule, the length of a primary polymeric chain will be equal to [121]:

$$\nu = \frac{K_{\mathrm{p}}[M]_0}{(\omega_{\mathrm{i}} K_{\mathrm{t}})^{0.5}}, \qquad (3.19)$$

where $[M]_0$ is the general concentration of vinyl groups, K_p and K_t are rate constants of chain propagation and chain termination, respectively, and ω_i is the initiation rate.

Such a primary chain contains (*n*–1) un-reacted vinyl groups, since the molecules of the monomer joined in the primary chains react only with one from the *n* available vinyl groups. At the start of the process the quantity of primary chains in the system is increased at the rate ω_i. At time $\tau_{cr.}$ such a quantity of the un-reacted vinyl groups $[M']_{cr}$ is stored, that each propagating polymeric chain reacts at least with the one of vinyl groups.

If we assume, that the reactive ability of 'hanging" vinyl groups M' and vinyl groups in the monomer M is the same, then the value $[M']_{cr}$ can be determined from the expression

$$[M]_{cr}/[M']_{cr} = 1/\nu \quad (3.20)$$

taking into account that $[M]_{cr} \approx [M]_0$ is (*n*–1) times higher than the concentration of reacting vinyl groups $[M]_0 \times \Gamma$ we will obtain

$$[M]_0 \Gamma_{cr}(n-1)/[M]_0 = 1/\nu\ , \quad (3.21)$$

from which

$$\Gamma_{cr} = 1/(n-1)\nu \quad (3.22)$$

Taking into account equation (3.19) we obtain

$$\Gamma_{cr} = \frac{(\omega_i K_t)^{0.5}}{(n-1)K_p[M]_0} \quad (3.23)$$

Starting from the polymerization depth Γ_{cr} the formation of new polymeric chains will stop, since each new chain in its propagation will be attached compulsory to at least one link of the former formed macromolecule with the formation of the joint with its particle. It can be assumed that each primary chain is a nucleus of the isolated particle in the solution of an oligomer. The maximum quantity of the isolated grains N_{max} that can be accumulated in the system until the moment Γ_{cr} is reached is equal to

$$N_{\max} = \frac{\Gamma_{\text{cr}}[\text{M}]_0}{\nu} = \frac{[\text{M}]_0}{(n-1)^2} \tag{3.24}$$

Taking into account equation (3.22) we obtain

$$N_{\max} = \frac{\omega_{\text{i}} K_{\text{t}}}{(n-1)K_{\text{p}}^2[\text{M}]_0} \tag{3.25}$$

At the constant initiation rate ω_i and at rather little polymerization degrees Γ:

$$\Gamma = \omega_{\text{i}} \nu \tau / [\text{M}]_0 \tag{3.26}$$

Achievement time Γ_{cr} and N_{max} are determined from equation (3.23):

$$\tau_{\text{cr}} = \frac{[\text{M}]_0 \Gamma_{\text{cr}}}{\omega_{\text{i}} \nu} = \frac{K_{\text{t}}}{(n-1)K_{\text{p}}^2[\text{M}]_0} \tag{3.27}$$

3.3.2. Polymeric grain propagation

The identification of polymeric grains as microreactors, in which the polymerization process is localized from the moment of grain formation and until the moment of their contact and propagation into a microheterogeneous polymeric body, has been carried out by sol–gel analysis [122] and thermometrical methods [123] in studies of oligoesteracrylates polymerization kinetics.

As result of investigation the composition of reactive system as a function of the polymerization depth Γ_t for oligoesteracrylates with 2–6 functional groups the following dependence has been determined:

$$\Gamma_{\text{t}} = \Gamma_{\text{d}} (1 - V_{\text{s}}) \tag{3.28}$$

where Γ_d is the polymerization depth in the gel-phase and V_s is the volumetric part of the sol-fraction in the polymer.

At this time, Γ_d is not changed by the polymerization and is equal to the maximum value Γ_t at the respective temperature:

$$\Gamma_{\text{d}} = (\Gamma_{\text{t}})_{\max} = \text{constant} \tag{3.29}$$

It has been noted [124–126] that in all cases the dependence on Γ_t of V_s has a linear character (see Figure 3.11). This fact agrees with equation (3.28) at Γ_d=constant. From this, it follows that the polymerization degree Γ_d in the non-soluble polymer (gel) does not change with the propagation of the total polymerization degree Γ_t.

Generally, the dependence $\Gamma=f(S)$ is described by the expression

$$\Gamma_t=\Gamma_s V_s + \Gamma_d(1-V_s) \qquad (3.30)$$

where Γ_s and Γ_d are transformation depths in the sol- and the gel-phase, respectively.

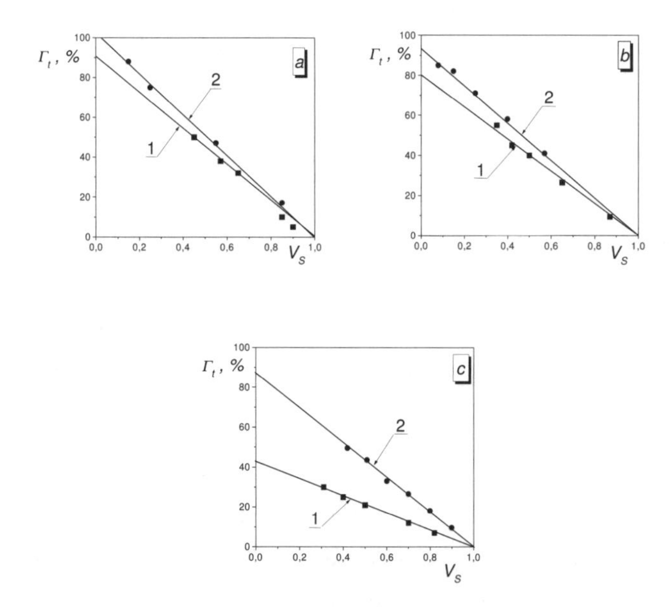

Figure 3.11. Dependence of the sol yield on a depth of the polymerization for oligoesteracrylates with different functionality. (a) TGM-3 (n=2); (b) tetramethacrylate (bis-ethylene glycol)adipinate (n=4); (c) hexamethacrylate (bis-pentaeritrite)adipinate (n=6). Temperature: (1) 60° C; (2) 90° C.

We can conclude from equations (3.28)–(3.30) that Γ_s=0 and the sol-fraction consists of the starting monomer [124–126]. Correspondly, the reactive system of the polymerizing process proceeding consists of only two components, namely the non-soluble densely-cross-linked polymer (so-called gel) with maximum conversion depth and the starting monomer.

A correlation between the sol-fraction yield and 3-D polymerization depth obtained using equation (3.28) allows us to propose the following chemical mechanism of grain propagation [127]. During the polymerization process grain formation takes place due to the propagation of new polymeric layers on the outer area of particles. At this time, in the outer area there is a radial gradient of conversion depths, from $(\Gamma_d)_c$ on the inside of the outer area of layer to $(\Gamma_d)_e=0$ on the outside. The average polymerization depth in the propagating outlying area layer (shell) is $\overline{\Gamma}_{sh}$ in the interval $[\Gamma_{max}-0]$ and in the case of a linear gradient $-\overline{\Gamma}_{sh}=0.5(\Gamma_t)_{max}$. Then the average polymerization depth of the total volume of a grain $\overline{\Gamma}_d$, taking into account its differentiation on the core and surface layer, can be expressed by the following ratio:

$$\overline{\Gamma}_d = \Gamma_{core}(1-V_{sh}) + \overline{\Gamma}_{sh}V_{sh} \tag{3.31}$$

where V_{sh} is the volumetric part of the outer area layer shell and Γ_{core} is the conversion depth in the core of the grain ($\Gamma_{core}=(\Gamma_t)_{max}$).

It is evident that at the end of a polymerizing process, the end of the polymerization means that the outer area layers have disappeared and $V_{sh}=0$. Thus, at the end of the polymerizing process equation (3.31) transforms into equation (3.29). In the earlier states of the polymerization, at $V_{sh} > 0$, equation (3.31) will be satisfied, from which follows

$$\overline{\Gamma}_d < \overline{\Gamma}_{core} \tag{3.32}$$

Since a constant value of Γ_d is observed in all states of the polymerization, we can conclude that the conditions of equations (3.26) and (3.32) agree with one another only at $V_{sh} \approx 0$ at the conversion during the polymerization. The physical meaning of this condition is that the propagation of polymeric grains proceeds in the thin outer layer practically on the surface of grains. A possible model of such a process is a horizontal organization on the surface of the outer area layer of polymeric chains with the unfinished structure propagating from the volume of the liquid layers. An analogy of such a model can be “spooling” of propagating polymeric chains on the densely-packed ball (polymeric grain). These opinions have been postulated elsewhere [57, 72, 121].

A kinetic model of the process has been elaborated based on the following assumptions. According to Refs. [2, 57, 113], a polymerization process proceeds in two zones, namely in the inter-grain space and at the surface layer of grains, in the range of which the kinetic parameters are stable and constant at all conversion degrees.

The contribution of the inter-grain space (so-called liquid oligomer) to the polymerization process is essentially rapidly decreased and at the initial states of a process is insufficient in comparison with the contribution of the other zone, namely the surface layer of grains. The autoacceleration state of a polymerizing process in the surface layer of grains is explained [2] by a sharp increase of the volume of effective reactors as a result of increasing total surface area of the grains in accordance with the law:

$$S=(4\pi n)^{1/3}\,(3\Gamma/\Gamma_{\mathrm{lim}})^{2/3} \qquad (3.33)$$

where n is the concentration of grains;

Γ is current polymerization depth;

Γ_{lim} is limited polymerization depth.

Total polymerization rate in this model [2] change sin accordance with the law:

$$W=3m\,(\Gamma/\Gamma_{\mathrm{lim}})W_r \qquad (3.34)$$

where Wr is a polymerization rate in the surfaced layer, which is described by the standard kinetic equation:

$$W=k_{\mathrm{p}}[\mathrm{M}](W_i/k_{\mathrm{t}})^{1/2} \qquad (3.35)$$

where k_{p} is the chain propagation constant;

k_{t} is the chain termination constant;

[M] is monomer concentration;

W_{i} is the initiation rate.

3.3.3. Propagation of the polymeric grains into the monolith in the late states of polymerization

The formation of densely-cross-linked polymeric particles with their following propagation at the expense of polymerization transformation of the starting monomer leads to a range of consequences. Of these consequences the most important is the existence of a state of polymerization propagation of discrete grains into monolith.

A monolytization state should be characterized by a sharp change of the properties of formed polymer. Grain propagation in the polymerization system proceeds *via* at least two states. The first state is reached when the size of grains is sufficiently small. In this case the system is dispersed, in which the disperse medium is formed by grains and the liquid starting monomer is the interruptive phase. Properties of dispersions are determined by the properties of the interruptive phase, which is why the strength or elasticity modulus should be rather small and penetration should be higher.

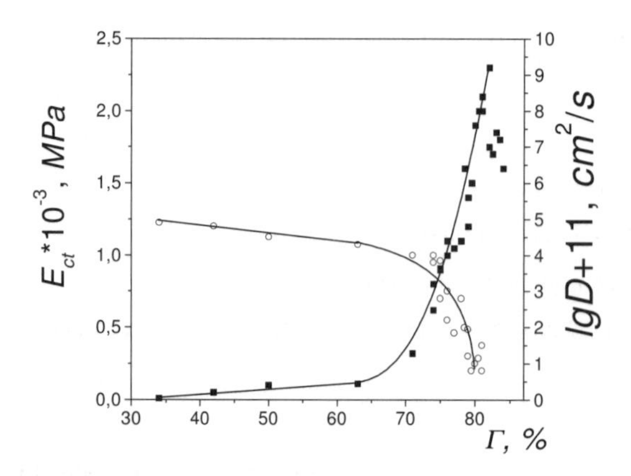

Figure 3.12. The dependence of the elasticity modulus with compressing $E_{compr.}$ (1) and diffusion coefficient D of acetone (2) on conversion degree in the polymerization of TGM-3. Conditions of polymerization are presented in Tables 3.2 and 3.3.

The second state of a system will be reached in the case when the size of polymeric grains after propagation will be such that the grains will be in contact with each other. The oligomer, which was earlier by the interruptive phase divided in zones of grain propagation on discrete microvolumes, becomes a disperse phase. In contrast, polymeric grains, *via* propagation from a disperse phase, are transformed into the interruptive phase. *Via* the polymerization process the critical conversion depth Γ_{cr} is achieved, which can be considered as the point of phase inversion. After the Γ_{cr} a structured framework, that consists of the solid densely-cross-linked grains, should obligatory lead to a sudden leap-like change of the properties of the polymeric material in the field of the polymerization depth Γ_{cr}. At the same time, the strength and the elasticity modulus are sharply increased and the diffusion penetration is sharply decreased (see Figure 3.12) [4, 128]. Diffusion ways appearing at $\Gamma < \Gamma_{cr}$ on interruptive, easy penetrated, phase (oligomer) at $\Gamma > \Gamma_{cr}$ are discovered partied of barriers.

Table 3.2. The elasticity module under polymers of TGM-3 compressing

Conditions of the polymerization	Γ (%)	E_{compr} (MPa)
50°C, 240 min	35.8	5.4
20°C, 150 days	38.8	10.0
60°C, 180 min	42.0	13.3
60°C, 240 min	53.3	50.3
100°C, 25 h	68.0	297.8
60°C, 61 h	75.0	564.0
60°C, 180 min+ 100°C, 24 h	77.8	928.0
60°C, 365 min+ 100°C, 24 h	77.0	1047.6
60°C, 180 min+ ^{60}Co, 1 Mrad. 5 h	80.2	1736.0
60°C, 180 min+ 100°C, 24 hours + ^{60}Co, 1 Mrad. 5 h	79.3	1933.7
60°C, 61 h+100 °C, 24 hours + ^{60}Co, 1 Mrad. 5 h	80.4	2079.0
20°C, 150 days + ^{60}Co, 11.5 Mrad. 10 h	79.0	1679.2
20°C, 150 days + ^{60}Co, 20 Mrad. 100 h	81.4	2230.6

Polymerization under the conditions indicated in this table was carried out using the initiator DAC ([DAC]=2×10^{-2} mol/l) and inhibitor trinitrotoluene ([TNT]=5×10^{-2}mol/l).

The monolytization state has been experimentally determined for the first time in a wide range of varied conditions in the investigation of the polymerization of TGM-3 [111, 129]. The monolytization degree has been estimated from the elasticity modulus of a polymer under compressing (E_{compr}) and the coefficient of diffusion (D) into the polymer of a range of the solvents. We can clearly see from Figure 3.12 and from the data in Tables 3.2 and 3.3 the presence of a sufficiently narrow field of conversion depths near $\Gamma_{cr} \approx 75\%$, in which a sudden leap-like change of properties of the cross-linked polymeric material takes place: the elasticity module is increased almost 3.5 times, from 700 to 2300 MPa, and the value of diffusion D is decreased approximately 1000 times under the same conditions. The variation in polymerization conditions (namely, the method of an initiation (substantial and radiational), the initiation rate, temperature and introducing the inhibitors) does not change the indicated character of the dependence of properties on the state of polymerization. We can assume that the described regularity has an universal character and should be observed also in radical polymerization of polyunsaturated compounds of other types.

It is necessary to note that the observed sudden leap-like change of polymer properties in the narrow range of the conversion depths cannot be explained by gel formation as usual, since the gel-point under the chosen conditions for the TGM-3 polymerization is located in the field of low conversion depths (at $\Gamma < 1\%$). The sudden leap-like change of polymer properties is a result of monolytization, growth of supra-molecular formations (polymeric grains), being the result of the second 'gel-point" at the supra-molecular level.

Table 3.3. Diffusion coefficients for acetone in the samples of TGM-3 polymers (at T=20° C) under different polymerization conditions

Polymerization conditions	[DAC]×10^2 (mol/l)	[TNT]×10^2 (mol/l)	Γ (%)	D×10^8 (cm^2/s)
20° C, 730 days	–	–	78.4	0.982
	0.5	–	80.7	0.052
	1.31	–	77.7	0.047
	3.27	–	79.8	0.019
	3.27	1.00^a	75.7	3.736
	3.27	0.94	71.3	6.140
	3.27	2.36	71.3	9.502
20°C, 400 days+100° C, 5 h	–	–	82.4	0.474
	0.65	–	85.2	0.014
	1.31	–	83.9	0.019
	3.27	–	86.5	0.022
	3.27	1.00*	80.5	0.050
	3.27	0.94	80.0	0.422
20°C, 400 days+^{60}Co, 1 Mrad, 5 h	–	–	85.1	0.007
	0.65	–	80.1	0.004
	1.31	–	82.5	0.009
	3.27	–	83.9	0.011
	3.27	1.00^a	83.9	0.028
	3.27	0.94	8.9	0.065

^{a} Benzohinone was used as inhibitor instead of TNT..

The monolytization state, which is a direct result of the microheterogeneous mechanism of the 3-D radical polymerization, has a decisive influence on the formation of physical-mechanical and other properties of densely-cross-linked polymeric materials. Sometimes in practice we can observe a sharp deterioration of the properties of polymeric materials on the basis of oligoesteracrylates at a visible constant conversion depth. Evidently, this fact can be explained by the sudden leap-like change of polymer properties at the monolytization state.

According to the kinetic interpretation of the microheterogeneous model, the autodeceleration is caused by reduction of the total surface area of the grains in the monolytization state *via* joining of grains. In this case, the polymerization rate is not described by equation (3.34) due to the loss of a reactive area part at the grains' joint.

This is why the autodeceleration state was considered in the model described in Ref. [2], taking into account the assumption, that at a sufficiently late polymerization state, the form of the inter-grain space with the starting oligomer will be near to spherical; at this time, phase inversion will be observed in the system, that is, the disperse phase (a polymer grain) becomes as interruptive one, and an interrupted phase (the so-called starting oligomer) is transformed into the dropped micro-dispersion. In connection with these assumptions the total polymerization rate is described in the autodeceleration state as

$$W = W_I (4\pi n)^{1/3} h(1 - \Gamma/\Gamma_{\lim})^{2/3} \qquad (3.36)$$

However, a comparison of the kinetic equations (3.34) and (3.36) with the experimental data shows, that they cannot quantitatively describe the process both in the autoacceleration state and in the autodeceleration states and do not agree in the section of maximal polymerization rate, since in the variant of the microheterogeneous model [2] the regularities of the polymerization at the interface layer are considered to be the same as the regularities in the volume of liquid monomer. Analysis of the experimental data shows that this assumption is not true.

REFERENCES

[1] Irzhak V., Rosenberg B., Enikolopian N. *Sietchastyje polimery* (*syntez, struktura, svojstva*); Moscow: Khimija, **1983**, 232 p.

[2] Berlin A., Korolev G., Kepheli T. *Akrilovye oligomery i materialy na ikh osnovie;* Moscow: Khimija, **1983**, 232 p.

[3] Bemford K., Barb U., Jenkins A., Onion P. *Kinetika radical'noj polimierizacji vinilovyh soedinienij*; Moscow: Izd-vo. Inostr. Lit., **1961**, 348 p.

[4] Bagdasarjan X. *Teoriya radical'noj polimierizacji*; Moscow: Nauka, **1966**, 300 p.

[5] Berlin A., Kepheli T., Korolev G. *Khimicheskaya promyshliennost'*, **1962**, 12, p. 870.

[6] Hayden P., Melville H. *J. Polym. Sci.*, **1960**, XLIII (141), p. 215.

[7] Notley N. *J. Phys. Chem.*, **1962**, 66 (9), p. 1577.

[8] Berlin A., Kepheli T., Korolev G. *Poliefiracrylaty*; Moscow: Khimija, **1967**, 72 p.
[9] Efimov A., Bugrova T., Dyachkov A. *Vysokomol. soed. (A)*, **1990**, 32 (11), p. 2296.
[10] Babich A., Lazarenko E., Tokaryk Z. *Zhurnal nauchnoj i prikladnoj photografii i kiniematografii*, **1979**, 24 (3), p. 204.
[11] Dyachkov A., Efimov A., Efimov L. *Vysokomol. soed. (A)*, **1983**, 25 (10), p. 2176.
[12] Guzieeva E., Efimov A., Dyachkov A. *Vysokomol. soed. (A)*, **1986**, 28 (8), p. 587.
[13] Efimov A., Dyachkov A., Kuchanov S. *Vysokomol. soed. (B)*, **1982**, 24 (2), p. 83.
[14] Zhu S., Tian Y., Hamieles A. *Macromolecules*, **1990**, 23, p. 1144.
[15] Doetschman D., Mehlenbacher R. *Macromolecules*, **1996**, 29, p. 1807.
[16] Broon E., Ivanov V., Kaminskiy V. *Doklady AN USSR*, **1986**, 291 (3), p. 618.
[17] Broon E., Ivanov V., Kaminskiy V. *Vysokomol. soed. (A)*, **1992**, 34 (4), p. 40.
[18] Lazarenko E., Tokaryk Z., Kenig V. *Zhurnal nauchnoj i prikladnoj photografii i kiniematografii*, **1982**, 27 (2), p. 122.
[19] Smirnov B., Korolev G., Berlin A. *Kinetika i kataliz*, **1966**, 7 (6) p. 990.
[20] Garrett R., Hill D., O'Donnell I. *Polym. Bull.*, **1989**, 22, p. 611.
[21] Shen I., Tian Y., Wang G. *Macromol. Chem.*, **1991**, 192, p. 2669.
[22] Kurdikar D., Peppas N. *Polymer*, **1994**, 35 (5), p. 1004.
[23] Oliva C., Selli E., Di Blas S., Termignone G. *J. Chem. Soc. Perkin Trans*, **1995**, 2, p. 2133.
[24] Kloosterboer J., Lijten G., Greidanus F. *Polym. Commun.*, **1986**, 27, 268.
[25] Selli E., Bellobono I. *Photopolymerization of multifunctional monomers: kinetic aspects. In: Radiation curing in polymer science and technology*, **1993**, London: Elsevier Applied Science, 1, p. 1-32.
[26] Selli E., Oliva **C.,** Galbiati M**.,** Bellobono I. *J. Chem. Soc., Perkin Trans*, **1992**, (2), p. 1391.
[27] Selli E., Oliva C. *Macromol. Chem. Phys.*, **1995**; 196, p. 4129.
[28] Selli E., Oliva C. *Macromol. Chem. Phys.*, **1996**; 197, p. 497.
[29] Bellobono I., Oliva C. Morelli R., Selli E., Ponti A. *J. Chem. Soc., Faraday Trans.*, **1990**, 86, p.3273.
[30] Oliva C**.,** Selli E., Bellobono I., Morelli R. *Phys. Chem., Chem. Phys.*, **1999**, 1, p. 215.
[31] Selli E., Oliva C**.,** Termignone G. *J. Chem. Soc., Faraday Trans.*, 1994, 90, p. 1967.
[32] Korolev G., Berlin A. *Vysokomol. soed.*, **1962**, IV (11), p. 1654.
[33] Golikov I., Berezin M., Mogylevich M. *Vysokomol. soed. (A)*, **1979**, 21 (8), p. 1824.
[34] Kuchanov S., Povolotskaya E. *Doklady AN USSR*, **1976**, 227 (5), p. 1147.
[35] Gardenas I., O'Driscoll K. *J. Polym. Sci.*, **1976**, 14, p. 883.

[36] Boots H. *J. Polym. Sci.*, **1982**, 20, p. 1695.

[37] De Gennes P. *J. Chem. Phys.*, **1982**, 76 (6), p. 3322.

[38] De Gennes P. *J. Chem. Phys.*, **1971**, 55 (2), p. 372.

[39] De Gennes P. *Ideji skejlinga v phisikie polimerov*; Moscow: Myr, **1982**, p.252.

[40] Tulig T., Tirrell M. *Macromolecules*, **1981**, 14, p. 1501.

[41] Tulig T., Tirrell M. *Macromolecules*, **1982**, 15, p. 459.

[42] Broon E., Kaminskiy V. *Book of Abstr. of the III^td^ All-Union Conf. "Mathematical methods for investigation of polymers"*; Pushchino, **1988**, p. 85.

[43] Ivanov V., Kaminskiy V., Broon E. *Vysokomol. Soed. (A)*, **1991**, 33 (7), p. 1442.

[44] Broon E., Kaminskiy V., Gladyshev G. *Doklady AN USSR*, **1984**, 278 (1), p. 134.

[45] Kaminskiy V., Ivanov V., Broon E. *Vysokomol. Soed. (A)*, **1992**, 34 (9), p. 14.

[46] Budtov V., Revnov B. *Vysokomol. Soed. (A)*, **1994**, 36 (7), p. 1061.

[47] Bowman C., Peppas N. *Macromolecules*, **1991**, 24, p. 1914.

[48] Anseth K., Bowman C., Peppas N. *J. Polym. Sci.(A)*, **1994**, 32 (1), p. 139.

[49] Anseth K., Wang C., Bowman C. *Macromolecules*, **1994**, 27, p. 650.

[50] Anseth K., Kline L., Walker T. *Macromolecules*, **1995**, 28, p. 2491.

[51] Kurdikar D., Peppas N. *Macromolecules*, **1994**, 27, p. 4084.

[52] Kurdikar D., Somvarsky J., Dushek K., Peppas N. *Macromolecules*, **1995**, 28, p. 5910.

[53] Kurdikar D., Peppas N. *Macromolecules*, **1994**, 27, p. 733.

[54] Gladyshev G., Gibov K. *Polimeryzaciya pri glubokikh stepenyakh prevrashcheniya i metody jejo issliedovaniya*; Alma-Ata.: Nauka, **1968**, 144 p.

[55] Gladyshev G. *Radikalnaya polimeryzaciya pri glubokikh stadiyakh prevrashcheniya*; Moscow: Nauka, **1974**, 243 p.

[56] Fujita H., Kishimoto A. *J. Chem. Phys.*, **1961**, 34 (2), p. 393.

[57] Budtov V., Podosenova N., Zotikov E. *Vysokomol. Soed. (A)*, **1983**, 25 (4), p. 237.

[58] Litvinenko G., Lachinov M., Sarkisova E. *Vysokomol. Soed. (A)*, **1994**, 36 (2), p. 327.

[59] Revnov B., Podosenova N., Ivanchov S. *Vysokomol. Soed. (A)*, **1988**, 30 (3), p. 184.

[60] Shen J., Tian Y., Wang G. *Science in China (B)*, **1990**, 33 (9), p. 1046.

[61] Kaminskiy V., Broon E., Ivanov V. *Doklady AN USSR*, **1985**, 282 (4), p. 923.

[62] Benson S., North J., Simple A. *J. Amer. Chem. Soc.*, **1959**, 81 (6), p. 1139.

[63] Marten F., Hamielec A. *High Conversion Diffusion-Controlled Polymerization*; Washington, DC, **1979**, 104, p. 43.

[64] Chern C., Poehlein G. *Polymer plastic technology*, **1990**, 23 (5/6), p.577.

[65] Russell G., Napper P., Gilbert R. *Macromolecules*, **1988**, 21, p. 2133.

[66] Buback M., Huckerstein B., Russell G. *Macromol. Chem. Phys.*, **1994**, 195, p. 539.

[67] Bagdasaryan H. *Teoriya radikal'noj polimeryzacji*; Moscow: Nauka, **1966**, 300 p.
[68] Ito K. *J. Polym. Sci.*, **1970**, A1(8), p. 1313.
[69] Pravednikov A. *Doklady AN USSR*, **1956**, 108 (4), p. 923.
[70] Anseth K., Decker C., Bowman C. *Macromolecules*, **1995**, 28, p. 4040.
[71] Korolev G. *Book of Abstr. of the first All-Union conf. on chem. and physico-chem. of polymerizing oligomers*, Chernogorlovka, **1977**, p. 144.
[72] Lazarenko E. *Photokhimicheskoje formirovanije pechatnykh form*; L.: Vyshcha shkola, **1984**, 151 p.
[73] Vatsulyk P. *Khimija monomerov*; Moscow: Inostr. lit., **1960**, 1738 p.
[74] Shybanov V., Bazyluk K., Marshalok I. *Vysokomol. Soed. (A)*, **1993**, 35 (2), p. 115.
[75] Karnaukh A., Gudzera S., Grishchenko V. *Vysokomol. Soed. (A)*, **1982**, 14 (10), p. 2227.
[76] Kurdikar D., Peppas N. *Vysokomol. Soed. (A)*, **1994**, 36 (11), p. 1852.
[77] Korolev B., Lachinov M., Dreval'V. *Vysokomol. Soed. (A)*, **1983**, 25 (11), p. 2430.
[78] Lecamp L., Youssef B., Bunel C., Lebaudy P. *Polymer*, **1999**, 40 (6), p. 1403.
[79] Wrzyszczynski A., Filipiak P., Hug G., Marciniak B., Paczkowski J. *Macromolecules*, **2000**, 33, p. 1577.
[80] Lecamp L., Youssef B., Bunel C. *Polymer*, **1997**, 38 (25), p. 6089.
[81] Lecamp L., Lebaudy P., Youssef B., Bunel C. *Polymer*, **2001**, 42, p. 8541.
[82] Diertz E., Peppas N. *Polymer*, **1997**, 38 (15), p. 3767.
[83] Decker C., Decker D. *Polymer*, **1997**, 38 (9), p. 2229.
[84] Takacs E., Emmi S., Wojnarovits L. *Radiat. Phys. Chem.*, **1999**, 55, p. 621.
[85] Kamal M., Sourour S. *Polym. Eng. Sci.*, **1973**, 13, p. 59.
[86] Kamal M. *Polym. Eng. Sci.*, **1974**, 14, p. 230.
[87] Waters D., Paddy J. *Anal. Chem.*, **1988**, 60, p. 53.
[88] Chandra R., Soni K. *Polym. Int.*, **1993**, 31, p. 239.
[89] Chandra R., Soni K. *Proc. Polym. Sci.*, **1994**, 19, p. 137.
[90] Stutz H., Mertes J., Neubecker K. *J. Polym. Sci. Polym. Chem.*, **1993**, 31, p. 1879.
[91] Musto P., Martuscelli E., Rragosta G., Russo P., Villano P. *J. Appl. Polym. Sci.*, **1999**, 74, p. 532.
[92] Strombeck L., Gebart R. *Thermochim. Acta*, **1993**, 214, p. 145.
[93] Auad M., Arangueren M., Elicabe G., Borrajo J. *J. Appl. Polym. Sci.*, **1999**, 74, p. 1044.
[94] Vyazovkin S., Sbirrazzuoli N. *Macromolecules*, **1996**, 29, p. 1867.
[95] Flammersheim H. *Thermochim. Acta*, **1997**, 296, p. 155.

[96] Lee J., Shim M., Kim S. *Kor. J. Mater. Res.*, **1998**, 8, p. 797.

[97] Andrzejewska E., Linden L., Rabek J. *Polym. Int.*, **1997**, 42, p. 179.

[98] Lecamp L., Youssef B., Bunel C., Lebaudy P. *Nucl. Instr. Methods Phys. Res.*, **1999**, B(151), p. 285.

[99] Alien NS, Hardy S, Jacobine AF, Glaser DM, Yang B, Wolf D. *Eur. Polym. J.*, **1990**, 26, p. 1041.

[100] Chandra R, Soni RK. *Polym. Int.*, **1993**, 31, p. 305.

[101] Andrzejewska E, Andrzejewski M, Bogacki M. *World Polymer Congress, IUPAC Macro* 2000, Warsaw, Poland, 9–14 July **2000**, Book of Abstracts, p. 237.

[102] Maffezzoli A, Terzi R. *Thermochim. Acta.*, **1998**, 321, p. 111.

[103] Gallacher L., Bettelheim F. *J. Polym. Sci.*, **1962**, 58 (166 (F2)), p. 697.

[104] Roshchupkin V., Ozerkovskij B., Kalmykov Yu., Korolev G. *Vysokomol. Soed.* (*A*), **1977**, 19 (4), p. 699.

[105] Roshchupkin V., Ozerkovskij B., Karapetian Z. *Vysokomol. Soed.* (*A*), **1977**, 19 (10), p. 2239.

[106] Volkova M., Belgovsky I., Golikov I. *Vysokomol. Soed.* (*A*), **1987**, 28 (3), p. 435.

[107] Korolev G., Mogylevich M., Golikov I. *Sietchatyje poliakrylaty. Mikroheterogennyje struktury, phizicheskije sietki, deformacionno-prochnostnyje svojstva*; Moscow: Khimija, **1995**; 276 p.

[108] Vasylev D., Belgovsky I., Golikov I. *Vysokomol. Soed.* (*B*), **1990**, 32 (9), p. 678.

[109] Dushek K. *Kompozicionnyje polimernyje matierialy*; K.: Naukova dumka, **1975**, p. 26.

[110] Berezin M., Korolev G. *Vysokomol. Soed.* (*A*), **1980**, 22 (8), p. 58.

[111] Berezin M. *Formation of the microheterogeneous structure via* 3*D radical initiated polymerization of oligoesteracrylates*; Ph. D Thesis, **1985**, 130 p.

[112] Lagunov V., Berezin M., Golikov I. *Vysokomol. Soed.* (*A*), **1981**, 23 (12), p. 2747.

[113] Siemiannikov V., Prohorov A., Golikov I. *Vysokomol. Soed.* (*A*), **1989**, 31 (8), p. 1602.

[114] Wassermann A., Kovarsky A. *Spinovyje mietki i zondy v phiziko-chimiji polimerov*; Moscow: Nauka, **1986**, 246 p.

[115] Berezin M., Lagunov V., Bakova G. *Vysokomol. Soed.* (*A*), **1984**, 23 (2), p. 422.

[116] Kloosterboer I., Lijten G. *Cross-linked Polymers: Chemistry, Properties, Application*; Washington, DC, ACS Symp. Ser., **1988**, 367, p. 409.

[117] Allen P., Bennet D., Hagias A. *Eur. Polym. J.*, **1989**, 25 (7/8), p. 785.

[118] Scranton A., Bowman C., Klier J. *Polym. J.*, **1992**, 33 (8), p. 1683.

[119] Louie B., Carratt G., Soong P. *J. Appl. Polym. Sci.*, **1985**, 30, p. 3985.

[120] Pfejpher P. *Fractals in physics*; Moscow: Myr, **1988**, p. 72.

[121] Debsky V. *Polymethylmetacrylate*; Moscow: Khimija, **1972**, p. 69.

[122] Semenov N. *Tsepnyje reakcji*; Moscow: Nauka, **1986**, p. 533.

[123] Anishchuk T., Bernstein V., Gal'perin V. *Vysokomol. Soed.* (*A*), **1981**, 23 (5), p. 963.

[124] Bernstein V., Nikitin V. and Razgulajeva L. *Thermodinamika i structurnyje svojstva granichnykh slojov polimerov*; Kiev: Naukova dumka, **1976**, p. 66.

[125] Smolanskij A. *Mol. Spectrosc.*, **1963**, p. 254.

[126] Lagunov V., Polozenko V. *Zavodsk. Laboratory*, **1977**, 43 (8), p. 947.

[127] Chifferi A., Word I. *Svierkhvysokomodul'nyj e polimery*; Leningrad: Khimija, **1983**, 272 p.

[128] Macknight W., Mackenna L., Read R., Stein R. *J. Phys. Chem.*, **1968**, 72, p. 1122.

[129] Berezin M., Korolev G. *Vysokomol. Soed.* (*A*), **1980**, 22 (8), p. 1872.

Chapter 4. Non-stationary (postpolymerization) polymerization kinetics problems

Postpolymerization, or the non-stationary dark process of the photopolymerization, is carried out after the completion of ultraviolet (UV) irradiation and gives important information about the mechanism of the process, particularly the elementary break reactions. The kinetic curve of the post-polymerization process has two regions: one quick and short and the other slow and long. These two regions indicate that two active radicals with different life times are involved in the process of postpolymerization.

Two basic concepts are used for the interpretation of the postpolymerization process: (1) diffusion-controlled reactions (DCR) and (2) the microheterogeneous model. With regard to DCR, the basis for the analysis of experimental data is the famous kinetic equation of the initial state of the stationary process. The parameters of this equation are the function of the mobility of macroradicals.

The second concept, the microheterogeneous model, is based on the assumption about the microheterogeneity of the system.

Below we will be considering both of these approaches and also those in which postpolymerization kinetics are discussed from the formal kinetic positions of variation of mono- and bi-molecular chain termination.

4.1. Bimolecular chain termination in the DCR conception

In general, DCR is used for the description of the photoinitiated post-polymerization.

In accordance with this conception (i) the chain termination and the chain propagation rate contstants represent functions of the viscosity of the polymerization system; (ii) the microheterogeneity of the system is considered only as additional diffusive complications; (iii) the main focus is on bimolecular chain termination.

As it was mentioned earlier, in Chapter 3, the chain termination rate constant depends on the translation and segmental diffusion coefficients of a monomer and a macroradical, and can be rewritten by the equation:

$$1/k_{tD}=1/k_{SD} + 1/k_{TD} \qquad (4.1)$$

Reactive diffusion is an alternative to classical (translation and segmental) diffusion mechanisms.

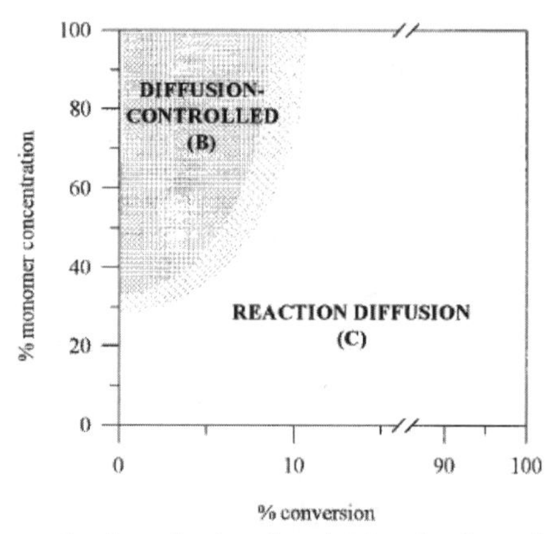

Figure 4.1. Termination mechanisms in the photoinitiated polymerization of tetrafunctional methacrylic monomers in an SBS matrix as a function of monomer concentration and double-bond conversion [2].

In bulk polymerizations of multifunctional monomers, reaction diffusion starts to control the termination reaction early, at approximately 10% double bond conversion [1–3] (see Figures 4.1 and 4.2).

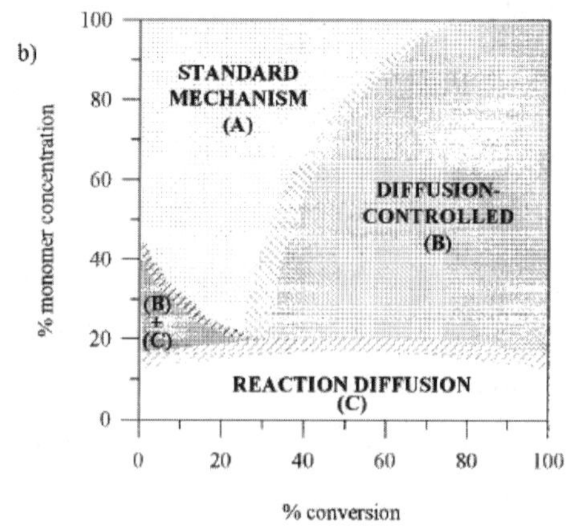

Figure 4.2. Termination mechanisms in the photoinitiated polymerization of difunctional methacrylic monomers in an SBS matrix as a function of monomer concentration and double-bond conversion [2].

Until that conversion is reached, autoacceleration takes place; the termination kinetic constant, k_t, decreases from the start to approximately 10% double-bond conversion. Above this conversion, k_t values maintain a plateau until high double-bond conversions are reached.

The notion of diffusion-controlled termination reaction was first proposed by Schulz in 1956 [4] (Figure 4.3). Several investigators have shown experimentally that when reaction diffusion is controlling the termination mechanism, the termination kinetic constant becomes proportional to the propagation frequency, as shown below [5–8]:

$$k_t = R\, k_p\, [\mathrm{M}] \tag{4.2}$$

where k_t is the termination kinetic constant, R is the reaction diffusion parameter, k_p is the propagation kinetic constant and [M] is the concentration of double bonds.

Another possible termination mechanism is unimolecular trapping of radicals which has been demonstrated [9, 10] using electron spin resonance spectroscopy (ESR). Anseth and Bowman [9] have shown that bimolecular termination through reaction diffusion is generally the dominant mechanism even in highly crosslinked methacrylates.

A model for linear polymerizations proposed by Russell *et al.* [11, 12] suggests the following equations for predicting the kinetic termination constant when reaction diffusion is the dominant termination mechanism:

$$k_t(\mathrm{res}) = \frac{4\pi}{3} p k_p [\mathrm{M}] a^2 r_a \tag{4.3}$$

where k_t(res) is the residual termination constant (diffusion-controlled termination reaction constant), p is the probability of reaction when two radicals come within capture distance of each other, a is the root-mean-square end-to-end distance per square root of the number of monomer units and r_a is the radius of interaction. The parameter p has values between 0 and 1 due to the effects of spin multiplicity and steric hindrance [12, 13]. It is thought that p is 1 at high conversions and 0.25 at low conversions, or even 0 in systems that are sterically hindered [12, 13]. The equation for k_t(res) was derived from the Smolukhovskij equation for the prediction of diffusion-controlled bimolecular rate coefficients [14] and Einstein's relation between the diffusion coeffici ent and the frequency of propagation [11, 15].

To complete the derivation, the radius of interaction must be specified. Russell *et al.* [11] defined two extremes, the rigid and flexible limits. At the rigid limit, the chain end cannot move on the time-scale of propagation and, thus, the radius of interaction is represented by half the Lennard–Jones diameter of the monomer, ($\sigma/2$). At the flexible limit, the chain can sample the entire volume around its last entanglement point. Thus, its radius of interaction is approximated as $a\sqrt{j_c}$ [16] where j_c is the distance between entanglements. Consequently, two limiting equations are derived for the diffusion-controlled termination reaction constant [11], which can be rearranged to predict minimum and maximum reaction diffusion parameters:

$$R_{\min} = \frac{k_t(\text{res}, \min)}{k_p[\text{M}]} = \frac{2\pi}{3} pa^2\sigma \tag{4.4}$$

$$R_{\min} = \frac{k_t(\text{res}, \max)}{k_p[\text{M}]} = \frac{4\pi}{3} pa^2 j_c^{1/2} \tag{4.5}$$

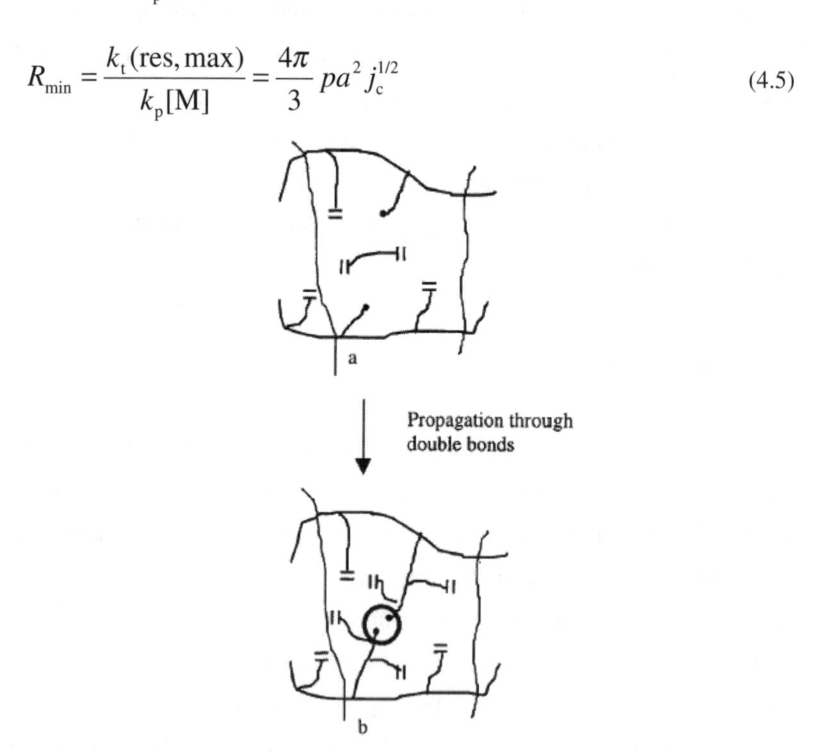

Figure 4.3. Illustration of the diffusion-controlled termination reaction. In (a), the radicals are physically separated in space and are unable to terminate. In (b), the radicals have moved closer together by propagating through monomeric or pendent double bonds. The propagation reaction is a means for physical movement of the radical.

Diffusion reaction parameters have been measured by others for highly cross-linked acrylate and methacrylate systems and were remarkably consistent among systems. For methacrylates R is in the order of 2 l/mol and for acrylates R is 3–5 l/mol [17]. The equation for R_{min} (equation (4.4)) can fit these results with reasonable values of the model parameters. The constant value of R for highly cross-linked systems is not too surprising if we examine the ratio of R_{max}/R_{min}. This ratio is $2a\sqrt{j_c}\sigma$. Since the ratio (a/σ) is about 1 and the parameter j_c approaches 1 as the cross-linking density increases, the ratio R_{max}/R_{min} approaches 2 for a highly cross-linked system.

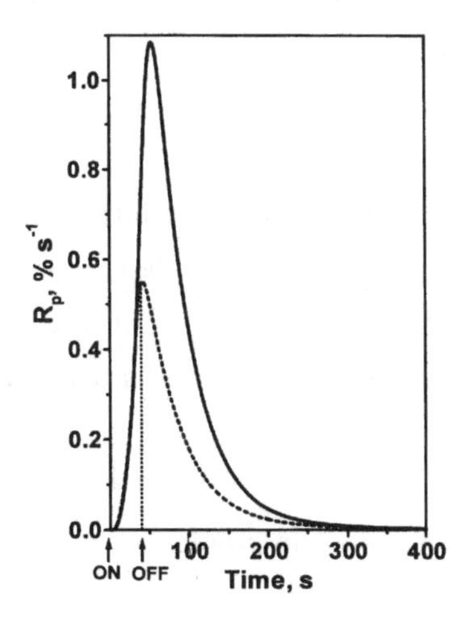

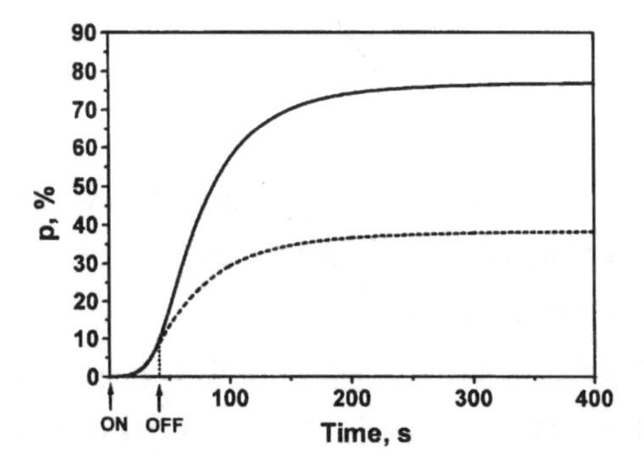

Figure 4.4. The kinetic curves of the polymerization under continuous illumination (solid lines) and after the light has been cut off (dashed lines). Diethylene glycol dimethacrylate, Ar, 40°C; initiator, 2,2-dimethoxy-2-phenyl acetophenone (Irgacure 651), 0.06 M [30].

For a loosely cross-linked network, polymerized above the glass transition temperature, R should approach the flexible limit, i.e., R_{max}, because the distance from the last attachment to the network is significantly larger and the overall mobility of the system is increased by polymerization above the T_g. As the polymerization temperature is lowered below T_g, the distance back to the last network attachment point (or entanglement) becomes less important, and the mobility of the radical chain end is reduced to the point where it is virtually immobile on the time scale of propagation. In this case, the rigid limit should be applicable, and R should approach R_{min}, just as it would for a highly cross-linked network.

It follows from this that the coefficient of a chain termination at free radical polymerization can be presented as a sum:

$$k_t = k_{t,D} + k_{t,RD} \tag{4.6}$$

where $k_{t,D}$ is the termination rate coefficient for diffusion controlled termination.

Postpolymerization process is usually used for the determination of numerical values of the rate constants of a chain termination (see typical kinetic curves in Figure 4.4).

The kinetic models given below are used for numerical descriptions of the postpolymerization process.

4.2. Formal models of the postpolymerization

4.2.1. Partly integrated models

The model that is most widely used was first revealed by Tryson and Schultz in 1979 [18]. This model analyses a non-steady-state kinetic experiment and allows one to calculate the propagation and termination rate coefficients. The model is based on classical equations and assumes that the only termination reaction is the reaction between two macroradicals, i.e., bimolecular termination.

The experimental procedure involves initiation of the polymerization by irradiation followed by cutting off the light after a certain time at a degree of conversion chosen, and monitoring the reaction in the dark. As experimental methods, both isothermal differential scanning calorimetry (photocalorimetry, photo-DSC) [2, 6, 7, 18–32] and real-time infrared spectroscopy (RTIR) [1, 33, 34] were used. Figure 4.4 presents the kinetic curves obtained for polymerization during continuous and interrupted irradiation.

The polymerization in the dark (post-effect) that occurs in the absence of initiation, may be described by following equations for reaction rates:

$$\text{propagation } R_\mathrm{P} = -\frac{\mathrm{d}[\mathrm{M}]}{\mathrm{d}t} = k_\mathrm{p}[\mathrm{M}]\cdot[\mathrm{P}^\bullet]$$

$$\text{termination } R_\mathrm{t} = -\frac{\mathrm{d}[\mathrm{P}^\bullet]}{\mathrm{d}t} = k_\mathrm{t}^\mathrm{b}[\mathrm{P}^\bullet]^2, \qquad (4.7)$$

where [M] is the concentration of double bonds; $[\mathrm{P}^\bullet]$ the radical concentration and k_t^b the overall termination rate coefficient calculated under the assumption of exclusively bimolecular termination.

Rearranging equation (4.7) to the form:

$$-\frac{\mathrm{d}[\mathrm{P}^\bullet]}{[\mathrm{P}^\bullet]^2} = k_\mathrm{t}^\mathrm{b}\mathrm{d}t\,. \qquad (4.8)$$

and integrating, we obtain:

$$\frac{1}{[\mathrm{P}^\bullet]} = 2k_\mathrm{t}^\mathrm{b}t + \frac{1}{[\mathrm{P}^\bullet]_0}, \qquad (4.9)$$

where $[\mathrm{P}^\bullet]_0$ is the concentration of radicals at the beginning of the dark period and t is time of reaction in the dark.

From equation (4.7) we have that

$$[\mathrm{P}^\bullet] = \frac{R_\mathrm{p}}{k_\mathrm{p}[\mathrm{M}]} \qquad (4.10)$$

and introducing this dependence into equation (4.9) we obtain a rate equation for the reaction in the dark:

$$\frac{[\mathrm{M}]_t}{[\mathrm{R_p}]_t} = \frac{2k_\mathrm{t}^\mathrm{b}}{k_\mathrm{p}} t + \frac{[\mathrm{M}]_0}{[\mathrm{R_p}]_0} \tag{4.11}$$

where the subscript 0 denotes the polymerization parameters at the beginning of the dark period, and subscript t denotes the parameters after a time t for the dark reaction (after the light has been cut off).

Expression (4.11) represents a partly integrated bimolecular termination model.

A plot of the ratio $[\mathrm{M}]_t/(\mathrm{R_p})_t$ *vs.* t should yield a straight line, which allows the calculation from its slope of the $2k_\mathrm{t}/k_\mathrm{p}$ ratio. However, just at the beginning of the dark period, often deviations from linearity are observed. These deviations are attributed to the response time of DSC and to sample conductivity [18]. Thus, this technique can be applied for the determination of the rate coefficients of those monomers which show an appreciable residual rate, well beyond the time of reaching linearity of $[\mathrm{M}]_t/(\mathrm{R_p})_t$ *vs.* t function. However, deviations may also result, in part, from the fact that the kinetic model used is inadequate, or that the rate coefficients change with the time of the dark reaction.

The analysis described above assumes that the rate coefficients are relatively constant over the chosen time increment t, where conversion changes are rather small. In practice, they are averaged over the conversion increment during the time t taken for calculations. Additionally, k_t values represent average propagation reactivities of both monomeric as well as pendant double bonds.

Because the quasi-steady-state approximation for radical concentration during autoacceleration is invalid [35], determination of individual rate coefficient becomes possible under the assumption of an instantaneous quasi-steady state at the point of the irradiation termination. It is assumed that the change in radical concentration during a single measurement at one reaction point is negligible, so the classical expression for the polymerization rate can be used:

$$R_\mathrm{p}^\mathrm{b} = \frac{k_\mathrm{p}}{(k_\mathrm{t}^\mathrm{b})^{0,5}} [\mathrm{M}] \phi^{0.5} \tag{4.12}$$

where $\phi = I_\mathrm{a}\Phi$, I_a is the absorbed light intensity in Einstein/l and Φ is quantum yield of initiation.

From equation (4.12), the $k_p/(k_t^b)^{0.5}$ ratio at the point of stopping the irradiation can be calculated. The k_p/k_t^b ratio (from equation (4.11)) together with the $k_p/(k_t^b)^{0.5}$ ratio enables the determination of individual values of k_p and k_t.

For the application of equation (4.12) Φ and I_a need to be known. The light intensity absorbed by the layer thickness d may be calculated from [36]:

$$I_a = I_0\,(1 - 10^{\varepsilon [A] d})/d \qquad (4.13)$$

where I_0 is the incident light intensity in mW cm^2, ε the extinction coefficient, d the layer thickness and [A] is the photoinitiator concentration.

To obtain I_a in Einstein/l one needs to divide I_a in mW cm^{-3} by the quantum energy of the monochromatic light used in kJ mol^{-1}. There are some problems with the Φ value. In the literature, the Φ value for 2,2-dimetoxy-2-phenylacetophenone (IRGACURE 651) has been estimated as 0.7 [7].

By terminating the initiation process at various states of the reaction (various conversions) and by monitoring the reaction rates in the dark, the conversion dependence of the rate coefficients can be determined.

It should be noted that, in the model considered, all processes that contribute to the termination process are represented by one termination rate coefficient, k_t^b.

There is another possible model for polymerization in the dark, namely a model in which the only means of termination is radical trapping. This type of termination was assumed by Batch and Macosko to be the only termination process during a vinyl ester resin polymerization starting from conversions near R_p^{max} [37]. At this reaction state bimolecular termination was assumed to be negligible.

The rate of termination when the radicals are eliminated in the monomolecular reaction is given by the following expression:

$$R_t^m = -\frac{d[P^\bullet]}{dt} = k_t^m [P^\bullet], \qquad (4.14)$$

where k_t^m denotes the overall termination rate coefficient calculated under the assumption of exclusive monomolecular termination.

Rearrangement and subsequent integration leads to the following relation:

$$[\mathrm{P}^\bullet] = \exp(-k_\mathrm{t}^\mathrm{m} t) \cdot [\mathrm{P}^\bullet]_0 \quad (4.15)$$

Substituting $[\mathrm{P}^\bullet]=(R_\mathrm{p})/(k_\mathrm{p}[\mathrm{M}])$ and rearranging we obtain a partly integrated monomolecular termination model:

$$\ln \frac{[\mathrm{M}]_t}{[\mathrm{R_p}]_t} = k_\mathrm{t}^\mathrm{m} t + \ln \frac{[\mathrm{M}]_0}{[\mathrm{R_p}]_0} \quad (4.16)$$

From the slope of the logarithm of the $[\mathrm{M}]_t/(R_\mathrm{p})_t$, ratio *vs*. reaction time in the dark k_t^m can be calculated.

The linearity of this function is obtained for postpolymerization processes when initiation has been discontinued at higher conversion degrees (near $R_\mathrm{p}^{\mathrm{max}}$) [32, 33]. The propagation rate coefficient is then determined from the corresponding steady-state equation for the polymerization rate [29, 30. 32, 33]:

$$R_\mathrm{p}^\mathrm{m} = \frac{k_\mathrm{p}}{k_\mathrm{t}^\mathrm{m}} [\mathrm{M}]\phi \quad (4.17)$$

Again, in this model, all processes that contribute to the termination process are represented by only one termination rate coefficient, namely k_t^m.

4.2.2. Fully integrated models

The two models presented above may be considered to be two extreme cases. However, the most probable situation is that both types of termination reaction, the usual bimolecular interaction of polymer radicals (bimolecular termination) and a first-order process involving only one polymer radical (monomolecular termination), occur in parallel. Thus, we can have not two, but three possible termination mechanisms:

1. monomolecular (when all the radicals become trapped), described by the rate expression (4.14)

2. bimolecular (usual reaction between two macroradicals, equation (4.7)) and

3. mixed (monomolecular and bimolecular processes occurring parallel):

$$R_{\mathrm{t}}^{\mathrm{mix}} = -\frac{\mathrm{d}[\mathrm{P}^{\bullet}]}{\mathrm{d}t} = 2k_{\mathrm{t}}^{\mathrm{b}}[\mathrm{P}^{\bullet}]^{2} + k_{\mathrm{t}}^{\mathrm{m}}[\mathrm{P}^{\bullet}] \tag{4.18}$$

Using the same procedure as detailed previously and then combining these equations with the expression for the polymerization rate, after full integration of equations, one obtains three termination models describing polymerization in the dark:

monomolecular termination model (model I):

$$-\ln(1-P_{\mathrm{d}}) = \frac{k_{\mathrm{p}}}{k_{\mathrm{t}}^{\mathrm{m}}}[\mathrm{P}^{\bullet}]_{0} \cdot \{1-\exp(-k_{\mathrm{t}}^{\mathrm{m}} \cdot t)\} \tag{4.19}$$

bimolecular termination model (model II):

$$-\ln(1-P_{\mathrm{d}}) = \frac{k_{\mathrm{p}}}{2k_{\mathrm{t}}^{\mathrm{b}}} \ln(1+2[\mathrm{P}^{\bullet}]_{0} k_{\mathrm{t}}^{\mathrm{b}} \cdot t) \tag{4.20}$$

mixed termination model (model III):

$$-\ln(1-P_{\mathrm{d}}) = \frac{k_{\mathrm{p}}}{2k_{\mathrm{t}}^{\mathrm{b}}} \ln\left(1+2[\mathrm{P}^{\bullet}]_{0} \frac{k_{\mathrm{t}}^{\mathrm{b}}}{k_{\mathrm{t}}^{\mathrm{m}}}(1-\exp(-k_{\mathrm{t}}^{\mathrm{m}} \cdot t)\right) \tag{4.21}$$

where $[\mathrm{P}^{\bullet}]_0$ is the macroradical concentration at the beginning of the dark period, t the time from the start of the dark reaction and P_{d} is the degree of double bond conversion in the dark.

Model II corresponds to the model introduced by Tryson and Schuiz but is fully integrated. Model I corresponds to the partly integrated monomolecular termination model. Note that models (4.19)–(4.21) are not linear; thus, because they do not require assumption of linearity, they are more accurate. From these models the following parameters can be calculated: k_{t}^{m}, $k_{\mathrm{t}}^{\mathrm{b}}[\mathrm{P}^{\bullet}]_{0.}$ $k_{\mathrm{p}}[\mathrm{P}^{\bullet}]_0$ and $k_{\mathrm{t}}^{\mathrm{b}}/k_{\mathrm{p}}$. These models require only the knowledge of the degree of double-bond conversion in the dark as a function of time.

These fully integrated models were first introduced by Timpe and Strehmel in 1991 [19] and were subsequently used and developed further [26–30. 32].

The corresponding steady-state equation for the determination of the rate coefficients of model III takes the form [29, 32]:

$$R_{\mathrm{p}}^{\mathrm{mix}} = \frac{k_{\mathrm{p}}}{4k_{\mathrm{t}}^{\mathrm{m}}}[\mathrm{M}]\{[(k_{\mathrm{t}}^{\mathrm{m}})^2 + 16\phi\, k_{\mathrm{t}}^{\mathrm{b}}]^{0.5} - k_{\mathrm{t}}^{\mathrm{m}}\} \tag{4.22}$$

It should be emphasized that the k_t^m thus calculated, describing the fraction of radicals immobilized and deactivated in the network per unit of time, corresponds to radicals that are registered as inactive on the time-scale measured by the experimental method used, for instance, photocalorimetry. It is possible that for longer reaction times they will react further but with very slow, undetectable rates due to very limited monomer diffusion or due to various relaxation processes [38–40].

4.2.3. Comparison of the formal models with the experimental data

Bimolecular termination model

The bimolecular termination model in its simplified, partly integrated form (equation (4.11)) has been extensively used for the estimation of the polymerization rate coefficients in cross-linking systems [2, 6, 7, 18, 20–25].

In their fundamental work, Tryson and Schultz observed a decrease in the propagation and termination rate coefficients with conversion for two multifunctional acrylates, with k_t decreasing more rapidly than k_p [18].

In a number of studies, Anseth *et al.* continued these investigations and determined propagation and termination rate coefficients as a function of double-bond conversion for a variety of multifunctional acrylates and methacrylates [6, 17, 20. 21], taking into account the effects of monomer type, functionality, temperature and initiation rates. Because of their higher reactivities the acrylates exhibited termination and propagation rate coefficients 2–3 orders of magnitude (at the plateau region k_p and k_t were of the order of 10^4 and 10^5 M^{-1} s^{-1}, respectively) greater than methacrylates (at the plateau region 10^2 and 10^3 M^{-1} s^{-1}, respectively) [6, 17, 20. 21]. As the number of acrylate and methacrylate groups in the monomer was increased (from 2 to 5), the rate coefficients decreased correspondingly. This was attributed to the greater viscosity of the higher-functionality monomers due to their higher molecular weights. For the higher-functionality monomers, propagation becomes diffusion-limited at much lower conversions than in the case of di(meth)acrylates. A plateau on the $k_p = f(p)$ and $k_t = f(p)$ plots was clearly observed only for di(meth)acrylate monomers; for higher functionality monomers, it was much less pronounced, being shorter and shifted to lower conversions (methacrylates) or even a continuous decrease occurred (acrylates). However, the ratio of both rate coefficients reached a plateau at a conversion degree indicating the dominance of reaction diffusion in the termination process. Increasing the number of functional groups decreased their average reactivity and the final conversion decreased as the functionality increased [20].

The plateau region was the most sensitive to polymerization temperature [17]. At higher temperatures, the mobilities of the reacting species increase and the onset of the reaction diffusion dominance over the segmental diffusion termination is delayed. The enhanced mobility and increased propagation rate coefficient retard the initial drop of k_t which enables k_p to reach higher values at the plateau region.

The results of the investigation of the photopolymerization of poly(ethylene glycol) dimethacrylates led to conclusions that the increase of the number of ethylene glycol units in the monomer (from 2 to 14) causes a shift of diffusion-controlled propagation and diffusion-controlled reaction termination to higher conversions [6].

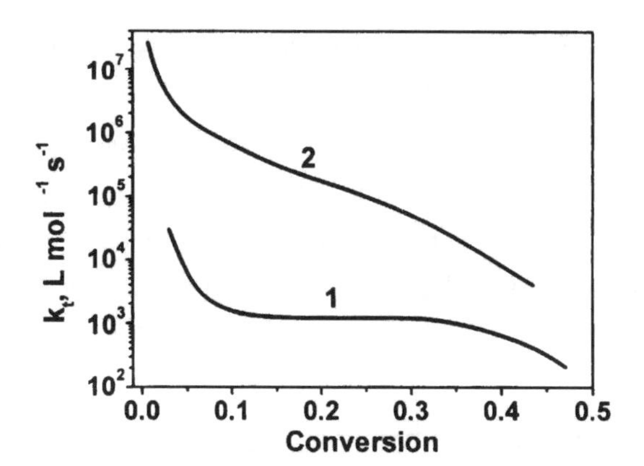

Figure 4.5. The comparison of the dependence of the termination rate coefficients of double-bond conversion for diethylene glycol dimethacrylate (1) and diethylene glycol dimethacrylate (2) at 30°C. Data taken from Ref. [17].

The polymerization of monoacrylates containing heterocyclic structures is very fast and leads to network formation due to enhanced chain-transfer reactions [41–45]. Determination of the rate coefficients for monomers of this type [33, 34] showed that the high reactivity of formulations based on these monomers appeared to be due to a poorly efficient termination process, with k_t values nearly 5-times lower than for diacrylate-based formulations.

The dependencies of rate coefficients on double-bond conversion for the polymerization of etoxylated bisphenol A dimethacrylate oligomer presented in Refs. [23, 24] exhibited similar shapes as in Figure 4.5 with a less pronounced plateau region for k_t. The k_p values reported in these studies were one order of magnitude higher than those for diethylene glycol dimethacrylate [17, 18].

Further insight into the effect of diffusion limitations on the polymerization rate coefficients was given by Mateo *et al.* [2, 7]. They studied the photopolymerization of di- (e.g. ethylhexyl methacrylate) and tetrafunctional (e.g. ethylene glycol dimethacrylate, EGDM) methacrylic monomers in polymeric media [2, 7]. The monomers, at concentrations of 12–13 wt%, were dissolved in polystyrene (PS), polybutadiene (PB) and styrene-butadiene-styrene (SBS) copolymer matrices. It was found that reaction diffusion was the only mechanism for termination from the onset of the polymerization until high double-bond conversions were reached. This is in contrast to the linear polymerization of difunctional monomers (diffusion control from *ca.* 40–50% double-bond conversion) and the bulk, cross-linking polymerization of tetrafunctional monomers (diffusion control from *ca.* 10% double-bond conversion). Reaction diffusion was the only termination mechanism for tetrafunctional monomers up to 30–40% of monomer content in the matrix. For higher monomer concentrations, reaction diffusion controlled the termination process starting from *ca.* 10% double-bond conversion. The values of the rate coefficients were of the order of 10^2 $M^{-1}s^{-1}$ (k_p) and 10^3–10^4 M^{-1} s^{-1} (k_t). The dominance of reaction diffusion in the polymerization of difunctional methacrylates right from the beginning of the reaction was observed only at low monomer concentrations in the matrix (10–15%). It is worth mentioning here that the SBS matrix participated appreciably in the polymerization process through the direct addition of the macroradicals or primary radicals to the double bond of the polybutadiene moiety and through hydrogen abstraction from the matrix with the formation of benzylic and allylic radicals [2, 46, 47].

In order to discern between the effects of monomer structure and viscosity on the polymerization kinetics, a two-component system, triethylene glycol dimethacrylate (TEGDM, **I**) and 2,2-bis[4-(2-hydroxy-3-methacryloyloxypropoxy)phenyl]-propane (*bis*-GMA, **II**) [25], was investigated and the rate coefficients were estimated.

$$CH_2=C(CH_3)-C(=O)-O-(CH_2-CH_2-O)_n-C(=O)-C(CH_3)=CH_2$$

I

$$CH_2=C(CH_3)-C(=O)-O-CH_2-CH(OH)-CH_2-O-C_6H_4-C(CH_3)_2-C_6H_4-O-CH_2-CH(OH)-CH_2-O-C(=O)-C(CH_3)=CH_2$$

II

The high viscosity of pure *bis*-GMA suppressed its polymerization rate coefficients by an order of magnitude when compared to TEGDM and its k_t values showed a plateau region from the very beginning of the reaction due to a diffusion-controlled termination reaction. The k_p values were relatively insensitive to compositions up to 50 wt% *bis*-GMA. Higher concentrations of *bis*-GMA shifted the diffusion control of propagation to lower conversions. The addition of a reactive diluent up to 50 wt% did not increase k_t values, not even at the beginning of the reaction, indicating that the high viscosity of the system forced the dominance of the diffusion-controlled termination reaction over the entire course of the polymerization. The initial termination during the polymerization of mixtures containing 0–25 wt% *bis*-GMA was not a diffusion-controlled reaction (k_t decreased with conversion increase), which was attributed to the relatively high mobility of polymerizing TEGDM compared to that of *bis*-GMA.

Under the assumption of bimolecular termination, the reaction diffusion parameter R (equation (4.2)) was determined for the bulk photopolymerization of multifunctional (meth)acrylates [1, 3, 6, 17, 48]. It was found that for methacrylates at the plateau level, R was in the order of 2 l mol^{-1} [6] and for acrylates R was 3-5 l mol^{-1} [17]. In general, reduction of network cross-linking density by co-polymerization with a difunctional reactive diluent (2-hydroxyethyl methacrylate (HEMA), octyl methacrylate) [3] increased the values of R due to increased mobility of the polymerization system until a plateau is reached. The addition of an inert diluent (polyethylene glycol 400 (PEG 400) [3, 6] or hydrogenated monomers [25, 53]) to a flexible monomer of low viscosity did not change R, not even when the solution contained 50–75% solvent. For *bis*-GMA, containing a stiff central core and hydroxyl groups that inhibit cyclization and cause diffusion-controlled termination to occur immediately upon exposure to light, dilution with hydrogenated TEGDM causes an increase in the reaction diffusion parameter from 3 to 9 as the amount of the diluent was increased [25, 48]. Increased system mobility also causes the reaction diffusion parameter to increase with a rise in polymerization temperature [3] above the glass transition point (T_g) for the methacrylate systems. However, R was also found to increase for loosely cross-linked systems when the reaction temperature dropped below T_g. It was proposed that chain transfer to monomer could cause this phenomenon.

In a recent study [49], Bowman' s group attempted to determine if the kinetic chain length of the macroradicals affects the termination process in cross-linking systems. They studied a chain-transfer reaction during the photopolymerization of multifunctional (meth)acrylates, assuming that chain transfer should influence the kinetic chain length ν, expressed as the ratio of the rate of polymerization, R_p, to the sum of the rate of termination, R_t, and the rate of chain transfer, R_{ct}, where R_t is equivalent to the rate of initiation, R_i, under the pseudo-steady-state assumption. It was found that chain transfer has a significant effect on the kinetics, supporting the hypothesis of chain length dependent effects in these systems. Additional support was obtained from non steady state *EPR*-experiments, which showed that there existed a fraction of the entire population of radicals terminating very rapidly upon extinction of the irradiation source: This fraction decreased when the polymerization shifted to higher conversions, as would be expected for a chain length dependent phenomenon.

Bimolecular and monomolecular termination models

In Ref. [33] Decker *et al.* analyzed the photopolymerization kinetics of polyurethane-acrylate resins using the bimolecular and monomolecular termination models given by equations (4.11) and (4.16). They considered ranges of conversion where the models show linearity and they differentiated three polymerization states with different termination types:

State I: dominance of bimolecular termination;

State II: mono- and bimolecular termination;

State III: dominance of monomolecular termination.

The conversion ranges of state I and state II depended on the structures and properties of the polymer network formed. At the beginning, the polymer radicals disappeared *via* bimolecular reactions according to the hyperbolic law (equation (4.9)) and then by radical trapping according to an exponential law (equation (4.15)). The relative importance of these two processes depended mainly on radical mobilities, monomer functionality and chain transfer reaction.

Monomolecular, bimolecular and mixed termination models

The three fully integrated termination models I–III ((4.19)–(4.21)) were used to find the termination types during the polymerization of acrylates in polymeric binders [19] and photocuring of acrylated silicones [26, 27]. In these studies discussion of the termination mechanism was based on uncoupled values of the k_t^b/k_p ratio and k_t^m values obtained for each model.

For acrylated silicones three cases were considered: photopolymerization under vacuum, in a laminate system and in the open air [26]. The results were interpreted in terms of the influence of oxygen, assuming that the reaction of radicals with oxygen leads to a monomolecular type of termination. For polymerizations under vacuum, the k_t^b/k_p ratios calculated from the bimolecular termination model II and the mixed termination model III were nearly the same and it was concluded that monomolecular termination was not dominant. In the laminate system, after a short exposure time (16 s, t_{Rm}=16 s), it was almost possible to describe the postpolymerization curve with the monomolecular termination model I; after a long exposure (more than 23 s), model II gave a better description of the results. This was explained by changes in the termination conditions during the polymerization. Residual oxygen, not consumed during the inhibition period, reacts with polymer radicals, leading to first-order termination at the beginning of the reaction. After their complete consumption, the bimolecular process begins to prevail. In the open-air system, both model I and model III can be used to describe the postpolymerization process, because the monomolecular termination rate coefficients for both models were nearly the same. It was concluded that bimolecular termination is not dominant in such systems due to the replenishment of oxygen by diffusion.

The termination mechanism during the photopolymerization of hexane-1,6-diol dimethacrylate in various polymeric binders depended on the type of matrix used [19]. In matrices with T_g values above room temperature, the first order mechanism dominated (the reaction was carried out at 25°C in air). For instance, the k_t^m values in poly(vinyl pyrrolidone) remained practically constant in the conversion range 11–17%. It was indicated that in this case the k_t^m rate coefficient corresponded more to radical trapping, because pseudo fist-order processes ($k_t^m = k_t'^{(m)}$ [radical scavenger]) are relevant only when diffusion of the component present in excess ($[P^{\bullet}]_0$ << [radical scavenger]) is not highly affected. In matrices with lower T_g values, mixed or bimolecular termination (at 10% double bond conversion) was observed.

The determination of termination mechanisms based on the termination models I–III was further developed in a numbers of studies [28–30. 32] using statistical techniques. Two-state statistical analysis (based on the F-Snedecor test) was applied to find the model that reproduced the experimental data best.

In one paper [28] the termination processes for a series of analogous heteroatom-containing monomers were considered. The investigated monomers included 2,2'-thiobisethanol dimethacrylate (TEDM) and diacrylate (TEDA) and 2,2′-oxybisethanol (diethylene glycol) dimethacrylate (OEDM) and diacrylate (OEDA, **III**).

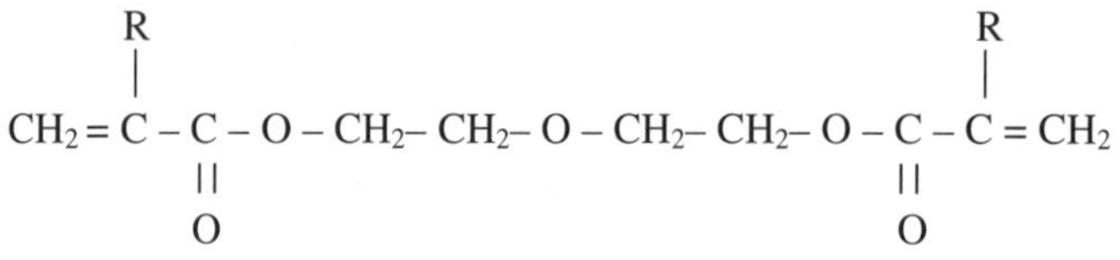

$$CH_2=C(R)-C(=O)-O-CH_2-CH_2-O-CH_2-CH_2-O-C(=O)-C(R)=CH_2$$

R = H; CH_3

III

The kinetic parameters were determined under various reaction conditions (light intensity, atmosphere). The statistical analyses showed that the experimental data in all cases were best described by the mixed termination model. The results obtained from this model indicated also that during the polymerization of sulfur-containing dimethacrylates, radical trapping occurred in a much lesser degree in comparison to the oxygen-containing counterpart and that the contribution of bimolecular reaction to the overall termination process was greater. This is clearly seen from k_t^b/k_p and k_t^m values obtained at R_p^{max} for TEDM and OEDM polymerization in argon (Table 4.1).

Table 4.1. Kinetic parameters at R_p^{max} calculated from models I–III and parameters of TEDM and OEDM polymerization in Ar and air

Model	Parameter	Argon		Air	
		TEDM	OEDM	TEDM	OEDM
I (equation (4.19))	$k_t^m \times 10^2$ (s^{-1})	0.696	1.794	1.829	2.148
II (equation (4.20))	k_t^b/k_p	3.2	8.0	22.8	72.8
III (equation (4.21))	$k_t^m \times 10^2$ (s^{-1})	0.231	1.518	1.052	1.536
	k_t^b/k_p	2.1	1.1	7.5	16.4
Additional parameters	R_p^{max} (% s^{-1})	0.50	0.47	0.28	0.09
	p_{Rm} (%)	19.3	23.5	7.6	6.1
	P_d (%)	37	32	10.0	5.0
	t_{Rm} (s)	82.6	114.7	91.0	187.0

I_0=1.4 mW/cm^2; initiator: IRGACURE 651, 0.02 M [30].

This result was explained by a chain-transfer reaction, which occurs during the polymerization of heteroatom-containing monomers (this subject will be discussed further). The chain-transfer constants for hydrogens in a CH_2 group attached to the sulfur atom are considerably higher than those for hydrogens in a CH_2 group attached to the oxygen atom [50]. Therefore, chain transfer is faster in the polymerization of sulfur-containing monomers and its influence on the polymerization course is stronger. Because chain transfer increases the mobility of radical sites attached to the network, it reduces the rate of monomolecular termination. The importance of chain transfer to monomer in increasing the rate of bimolecular termination was also indicated by Russel [51]. Similar calculations performed for TEDA and OEDA showed, however, that the difference in kinetic parameters between these two monomers is much smaller (k_t^m=0.924 and 1.113, respectively; k_t^b/k_p = 3.2 and 3.7, respectively; initiator concentration 0.008 M). This means that, although the contribution of radical trapping is still smaller the polymerization of sulfur-containing monomers, the effects of chain transfer on the CH_2–S group are less important due to the competition of chain transfer for tertiary hydrogens of the polyacrylate main-chain and its better flexibility.

A comparison of the results obtained using model II and model III showed that a discussion of the termination mechanism based on rate coefficients calculated under assumption of only bimolecular termination might lead to some misleading conclusions. For instance, whereas model II suggests that in TEDM polymerization termination (bimolecular) is slower than in OEDM polymerization (lower k_t^b/k_p ratio), model III shows, in contrast, that the bimolecular termination in TEDM polymerization is somewhat faster in comparison to OEDM (higher k_t^b/k_p ratio). The difference in the polymerization of these two monomers is caused mainly by a much lower monomolecular termination rate coefficient for TEDM.

The rate coefficients estimated for the polymerization in air additionally reflect the inhibiting effect of oxygen. The higher values for the ratio k_t^b/k_p for the polymerization under air suggested that oxygen accelerated bimolecular termination. This may occur due to oxygen-induced hydrogen transfer, which enhances the mobility of radical sites in the network [52], as well as to the fact that peroxy radicals formed by the addition of oxygen to growing macroradicals (although relatively unreactive in the re-initiation process) combine rapidly with other propagation radicals or with themselves, competing effectively with the bimolecular reaction between carbon-centered radicals (the rate constants in the reaction of $RO_2^\bullet$ with $R^\bullet$ and $RO_2^\bullet$ with $RO_2^\bullet$ are in the order of 10^8 and 10^4–10^8 $M^{-1}s^{-1}$, respectively [28]). Moreover, the k_t^b/k_p values, lower for methacrylates and for sulfur-containing monomers, confirm that methacrylates are less susceptible to oxygen inhibition and indicate that the presence of the sulfide group in the monomer molecule reduces oxygen effects on the polymerization kinetics.

In another study [29], the rate coefficients for each termination model were determined in relation to the degree of conversion of double bonds for OEDM polymerization. The rate coefficients were expressed as k_t^m, $k_t^b\phi$ and $k_p\phi$ where $\phi = I_a\Phi$. Statistical analysis showed that, from *ca.* 10% double-bond conversion, mixed termination dominated. Below this conversion degree the results of the statistical analysis were ambiguous. Determination of rate coefficients from model III is possible only when this model is indicated by statistical analysis to give the best fit to the experimental data. Thus, the initial range of conversion where the kinetic parameters change very quickly had to be excluded. The results of k_t^m calculations from models I and III are shown in Figure 4.6 and for $k_p\phi$ calculations from models II and III are shown in Figure 4.7.

A similarity in k_t^m (from model I) and $k_t^b\phi$ (from model II) behavior indicates that under conditions of continuous initiation, all the termination processes slow down as the reaction proceeds. The best-fitted, mixed-type model (model III) yields results that tie up the relationship obtained with model I and model II. The low $k_t^b\phi$ values, lower than $k_p\phi$ (not shown in Figures 4.6 and 4.7), suggested that the monomolecular reaction was the essential mechanism for termination in the polymerization of the multifunctional monomer under the experimental conditions applied.

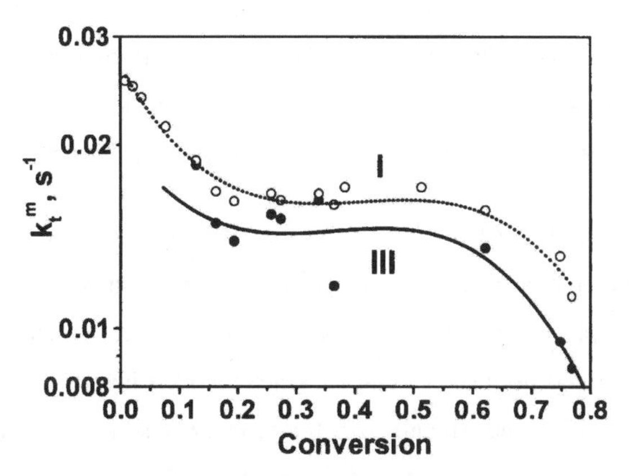

Figure 4.6. Effect of double-bond conversion on the monomolecular termination rate coefficient k_t^m calculated from the monomolecular termination model (I) and mixed termination model (III). Monomer, diethylene glycol dimethacrylate. Conditions: Ar, 40°C; initiator, Irgacure 651, 0.06 M. Data taken from Ref. [30].

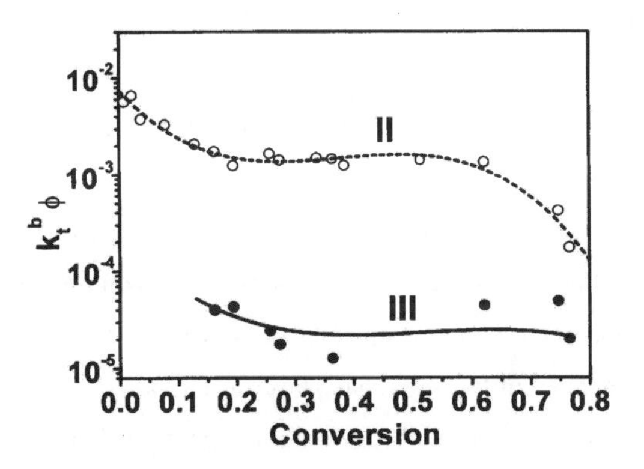

Figure 4.7. Effect of double bond conversion on the product of bimolecular termination rate coefficient k_t^b and absorbed light intensity I_a, multiplied by quantum yield of initiaton Φ, $\phi=I_a\Phi$, calculated from the bimolecular termination model (II) and the mixed termination model (III). Monomer and conditions as in Figure 4.6 [30].

It may be expected that during the post-effect, the rate coefficients as well as the termination mechanism will change due to drastic changes in the polymerization conditions. Within the post-effect, new short-chain and mobile radicals are no longer being generated and the entire living population is growing longer [53, 54]. In particular k_t^b should be strongly affected, due to its chain-length dependence. The changes of the kinetic parameters occurring with the time of the dark reaction (after switching off the irradiation) were modeled using the three termination models I–III [32]. Statistical analysis showed that the termination mechanism changed during the post-effect from a bimolecular mode to a mixed one when the light was cut off at low and medium double bond conversions. At higher starting conversions, the monomolecular termination mechanism dominated.

The approaches for the description of the postpolymerization kinetics considered above do not take into account the composition changes in a polymerization system composition and its phase state; that is why such approaches are formal. However, they are interesting because it has been shown that the kinetic models involving active radicals with one characteristic life time (mono- or bimolecular chain termination) correspond considerably worse to experimental data in comparison with the kinetic model characterized by two characteristic life times (mixed mono- and bimolecular chain termination). The same technique of widening the characteristic life times of active radicals from the position of a microheterogeneous model and monomolecular chain termination is proposed in Ref. [55].

4.3. Kinetic model of postpolymerization in the microheterogeneous conception

The microheterogeneous model is based on the following assumptions: (a) the rate of the 3-D polymerization is the sum of the rates of the homophase and heterophase processes, where the homophase process operates in the volume of liquid oligomer and thre heterophase process operates in the interphase layer at the boundary of the solid polymer and liquid oligomer in the regime of the gel effect. The cut-off rate is controlled by its growth rate; (b) the initiator is uniformly distributed between the phases and the interphase layer; and (c) clusters of the solid polymer in the liquid oligomer phase and clusters of the liquid oligomer in the solid polymeric matrix have the fractal structure.

4.3.1. Kinetic model of the chain propagation and monomolecular chain termination at an interface layer

For the explanation of all characteristics of a postpolymerization process proceeding, it was assumed that the 'frozen" radicals R_z are responsible for the long and slow postpolymerization. As a result, the following kinetic scheme of the process proceeding in the interface layer has been proposed [55]:

$$R_m + M_m \begin{cases} \xrightarrow{k_{p1}} R_m \\ \xrightarrow{k_{t1}} R_z \end{cases} \qquad (4.23)$$

$$R_z + M_m \xrightarrow{k_m} R_m' \qquad (4.24)$$

$$R_m' + M_m \begin{cases} \xrightarrow{k_{p2}} R_m'' \\ \xrightarrow{k_{t2}} R_z' \end{cases} \qquad (4.25)$$

In accordance with the above-presented scheme (4.23)–(4.25) a gel-effect in the interface layer is again considered as a control on the rate of chain termination. At this, acts of chain propagation (primary (4.23) and secondary chain (4.25)) and also termination represent two different interactions of active radicals R_m and R_m' with the functional group of a monomer, leading either again to the formation of active radicals (so-called chain propagation) or their 'freezing" (monomolecular chain termination), that is to the formation of non-active radicals R_z and R_z'.

Accordingly, reaction (4.24) represents the elementary reaction of the origination of secondary active radicals R_m' on the basis of which a secondary chain (4.25) with monomolecular termination is formed [56]. However, reaction (4.24) does not represent a chain-transfer reaction because this disagrees with the experimentally observed molecular weight of polymer propagation *via* the postpolymerization process. Thus, reaction (4.24) should be considered as an act of 'freezing" radical R_z "germination" [57] into an active zone of the interface layer. In this case, at primary chain lengths $v_1 = k_{p1}/k_{t1}$ and secondary chain lengths $v_2 = k_{p2}/k_{t2}$, the average length of the chain consisting of primary 'freezing" radicals will be equal to ν_1 and the length of the chain consisting of the secondary "freezing" radicals will be equal to $\nu_1 + \nu_2$. Exactly this fact explains the increase in average-number molecular weight of the polymer in the dark polymerization process at times longer than the life time of primary active radicals R_m: $\tau_1=(k_{t1}[M_m])^{-1}$.

4.3.2. Deduction of the kinetic model

A propagation act, for example, of the primary chain (4.23), relative to a radical with the average length ν_1, that is, R_z, represents the act of radical's death under observation of control of the monomolecular chain termination rate. It follows from this, that the ratio $k_{p1}[R_z]=k_{t1}[R_m]$ should be correct. Thus,

$$[R_z]=\frac{k_{t1}}{k_{p1}}[R_m]=\frac{[R_m]}{\nu_1} \quad (4.26)$$

On the basis of the above, we obtain the following expression (4.27) for the secondary chain (see equation (4.25)):

$$[R_z']=\frac{k_{t2}}{k_{p2}}[R_m']=\frac{[R_m']}{\nu_2} \quad (4.27)$$

According to equations (4.23)–(4.25) the kinetics of maintaining the active primary and secondary radicals in the interface layer is described by equations (4.28) and (4.29) [58]:

$$\frac{d[R_m]}{dt}=\nu_{im}-k_{t1}[M_m][R_m] \quad (4.28)$$

$$\frac{d[R_m']}{dt}=k_m[M_m][R_z]-k_{t2}[M_m][R_m'] \quad (4.29)$$

Since

$$\beta = k_{t1}[M_m],\ \gamma = k_{t2}[M_m] \quad (4.30)$$

let us re-write equation (4.28) and equation (4.29), taking into account equation (4.26), in the form of equations (4.31) and (4.32), respectively:

$$\frac{d[R_m]}{dt} = v_{im} - \beta[R_m] \tag{4.31}$$

$$\frac{d[R_m']}{dt} = \frac{k_m}{k_{p1}}\beta[R_m] - \gamma[R_m'] \tag{4.32}$$

For the first time, let us consider the stationary kinetics of a process in the interface layer, assuming that $d[R_m]/dt = d[R_m']/dt = 0$. Then, we obtain

$$[R_m] = \frac{v_{im}}{\beta}, \; [R_m'] = \frac{\left(k_m / k_{p1}\right) v_{im}}{\gamma} \tag{4.33}$$

The specific rate of polymerization in the interface layer at the expense of the primary and the secondary chains propagation will be equal to [59]:

$$w_m = k_{p1}\,[M_m][R_m] + k_{p2}\,[M_m][R_m'] \tag{4.34}$$

After the substitution of $[R_m]$ and $[R_m']$ from equation (4.33) into equation (4.34), and taking into account equation (4.30), we obtain:

$$w_m = \left(\frac{k_{p1}}{k_{t1}} + \frac{k_m k_{p2}}{k_{p1} k_{t2}}\right) v_{im} \tag{4.35}$$

It follows from this that the contribution of the process in the interface layer into total polymerization process, which is equal to $\varphi_m w_m$, will be again represented by the kinetic parameter k_2:

$$k_2 = \frac{h \dfrac{F_s}{F_v} \dfrac{f_m}{f_v} \left(\dfrac{k_{p1}}{k_{t1}} + \dfrac{k_m k_{p2}}{k_{p1} k_{t2}}\right)}{[M_0]\Gamma_0} \tag{4.36}$$

In accordance with equation (4.35), the effective length of the chain in the interface layer will be equal to $\nu_m = \nu_1 + (k_m/k_{p1})\nu_2$.

Next, let us consider the non-stationary kinetics of the postpolymerization in the dark. Since, as was noted before, the contribution of the homophaseous process into the total process under the postpolymerization process should be insufficient, we analyzed only the kinetics in the interface layer assuming that $w = \varphi_m w_m$. In the dark period $v_{im} = 0$, and equations (4.31) and (4.32) can be written as follows:

$$[R_m] = \frac{v_{im0}}{\beta} \exp\{-\beta\ t\} \qquad (4.37)$$

$$[R_m'] = \frac{\left(k_m / k_{p1}\right) v_{im0}}{(\beta - \gamma)} \left(\exp\{-\beta\ t\} + \beta / \gamma \exp\{-\gamma\ t\} \right) \qquad (4.38)$$

Here the time t is calculated from the start of the dark period ($t = 0$)and v_{im0} is the rate of initiation in the interface layer at the end of stationary light period, corresponding to the start of the dark period. At this, $[R_m]_0 = v_{im0}/\beta$ and $[R_m']_0 = (k_m/k_{p1})v_{im0}/\gamma$ are the concentrations of the radicals at the start of the post-polymerization period.

We obtainedhe following expression for the specific rate of the postpolymerization by substituting equation (4.37) and (4.38) into equation (4.34), taking into account equation (4.35):

$$w_m = \frac{w_{m0}}{\beta - \gamma} \left[\{(1-\alpha)(\beta-\gamma) - \alpha \cdot \gamma\} \exp\{-\beta\ t\} + \alpha\ \beta \exp\{-\gamma\ t\} \right] \qquad (4.35)$$

Here w_{m0} is the starting post-polymerization rate determined using equation (4.35) at the end of a stationary light period in which $v_{im} = v_{im0}$; α is a parameter characterizing the relative contribution of the secondary chain to the total kinetics in the interface layer and is determined by the expression

$$\alpha = \frac{k_m k_{p2} / k_{p1} k_{t2}}{k_{p1} / k_{t1} + k_m k_{p2} / k_{p1} k_{t2}} \qquad (4.40)$$

In accordance with the characteristic times of the rapid and short section and the slow and long part of the kinetic curve of the post-polymerization, which are determined by the life times of primary radicals R_m ($\tau_1=\beta^{-1}$) and the secondary radicals $R_{m'}$ ($\tau_2=\gamma^{-1}$), it will be correct to assume that $\beta >> \gamma$; It allows to write equation (4.39) in the simplified form:

$$w_{\mathrm{m}} = w_{\mathrm{m0}}\left[(1-\alpha)\exp\{-\beta\ t\}+\alpha\exp\{-\gamma\ t\}\right] \tag{4.41}$$

Neglecting the contribution of the homophaseous process in the post-polymerization, now we can write its rate expressed in units of the relative conversion P as follows:

$$\frac{\mathrm{d}p}{\mathrm{d}t} = k_2 p(1-p)v_{\mathrm{im0}}\left[(1-\alpha)\exp\{-\beta\ t\}+\alpha\exp\{-\gamma\ t\}\right] \tag{4.42}$$

Here k_2 is determined using expression (4.36).

In the case of photoinitiated post-polymerization v_{im0}, p and $\mathrm{d}p/\mathrm{d}t$ are functions not only of time, but also of layer coordinate x on the illuminated surface ($x = 0$); that is, there are differential characteristics of the process in the layer x, $x + \mathrm{d}x$.

Transition from the differential characteristics $p(x, t)$ and $\mathrm{d}p(x, t)/\mathrm{d}t$ to experimentally determined, that is, averaged on the photopolymerization composition layer, $P(t)$ and $\mathrm{d}P(t)/\mathrm{d}t$ are calculated *via* integral transformations of equation (4.42) according to equation (4.43).

$$W = \mathrm{d}P/\mathrm{d}t = \ell^{-1}\int_0^{\ell}(\partial P/\partial t)\mathrm{d}x \qquad\qquad P = \ell^{-1}\int_0^{\ell} p\mathrm{d}x \tag{4.43}$$

Let us again use the expression obtained in Refs. [60. 61] for the differential rate of photoinitiation as a function of layer coordinate x and time t:

$$v_{\mathrm{i}} = \gamma\varepsilon c_0 J_0 e^{\tau-y}[1+e^{-y}(e^{\tau}-1)]^{-2} \tag{4.44}$$

Here: $y = \varepsilon c_0 x$, $\tau_0=\gamma_v\varepsilon J_0 t'$;

c_0 is the start concentration of the photoinitiator for the light period ($t' = 0$);

J_0 is the intensity of the light falling on the surface of a composition;

ε is the molar extinction coefficient of the photoinitiator;

γ is the quantum yield of the photoinitiation;

t' is the duration of the light period.

At the end of a light period ($t' = t_0'$) the differential rate of photoinitiation becomes v_{i0}, determined by equation (4.44) at $\tau=\gamma_v \varepsilon J_0 t'$.

After integral transformation of equation (4.42), taking into account equation (4.44) at $\tau = \tau_0$, according to the main theorem we obtain:

$$\frac{dP}{dt} = k_2 \frac{P(1-P)\left[(1-\alpha)\exp\{-\beta\ t\}+\alpha\exp\{-\gamma\ t\}\right]\gamma_v J_0(1-\exp\{-y_0\})}{\ell(1+\exp\{-y_0\})(\exp\{\tau_0\}-1)} \quad (4.45)$$

where $y_0=\varepsilon c_0 \ell$ is a starting optical density of the layer.

At t = 0 (the start of a dark period) we will obtain that $P=P_0$ and

$$\frac{dP}{dt} \equiv W_{m0} = k_2 \frac{P_0(1-P_0)\gamma_v J_0(1-\exp\{-y_0\})}{\ell(1+\exp\{-y_0\})(\exp\{\tau_0\}-1)} \quad (4.46)$$

With the use of an equation (4.46) we simplify equation (4.45), namely:

$$\frac{dP}{dt} = W_{m0} \frac{P(1-P)}{P_0(1-P_0)}\left[(1-\alpha)\exp\{-\beta\ t\}+\alpha\exp\{-\gamma\ t\}\right] \quad (4.47)$$

4.3.3. Discussion of the experimental results and the estimation of the model parameters

To compare the results obtained using equation (4.47) with the experimental data (Figures 4.8–4.11), let us transform equation (4.47) into the integral form [59]:

$$\ln\frac{P(1-P_0)}{P_0(1-P)} = \frac{W_{m0}}{P_0(1-P_0)}\left[(1-\alpha)(1-e^{-\beta t})/\beta+\alpha(1-e^{-\gamma t})/\gamma\right] \quad (4.48)$$

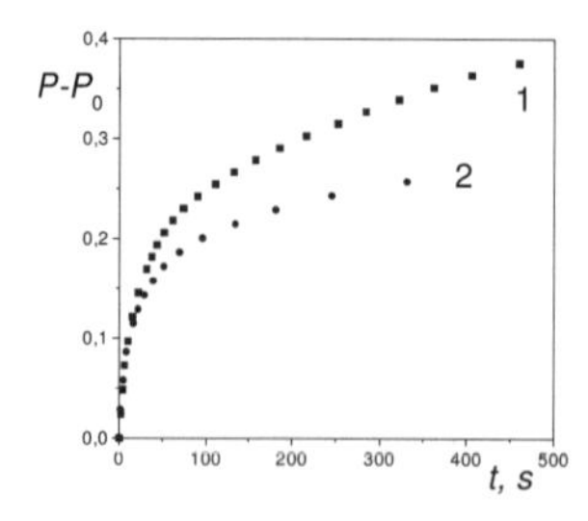

Figure 4.8. Experimental kinetic curves of dimethacrylate MDPh-2 post-polymerization at: (1) ℓ=2.5×10^{-4} m, P_0=0.42; (2) ℓ=2.7×10^{-4} m, P_0=0.46. E_0=40.6 W/m^2, c_0=0.06 mol/l.

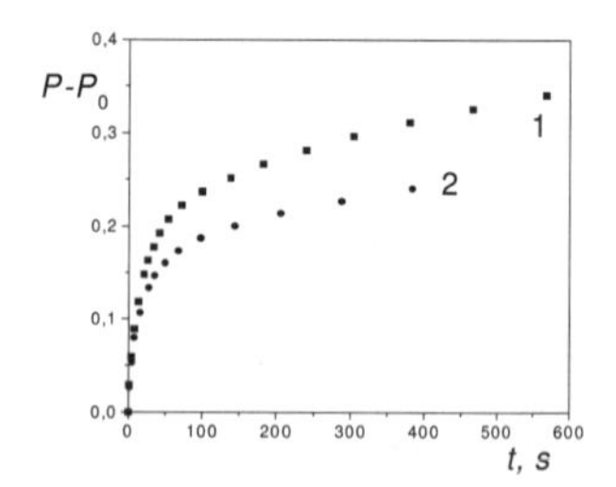

Figure 4.9. Experimental kinetic curves of dimethacrylate MGPh-9 post-polymerization at: (1) ℓ=1.0×10^{-4} m, P_0=0.52; (2) ℓ=1.4×10^{-4} m, P_0=0.20. E_0=40.6 W/m^2, c_0=0.10 mol/l.

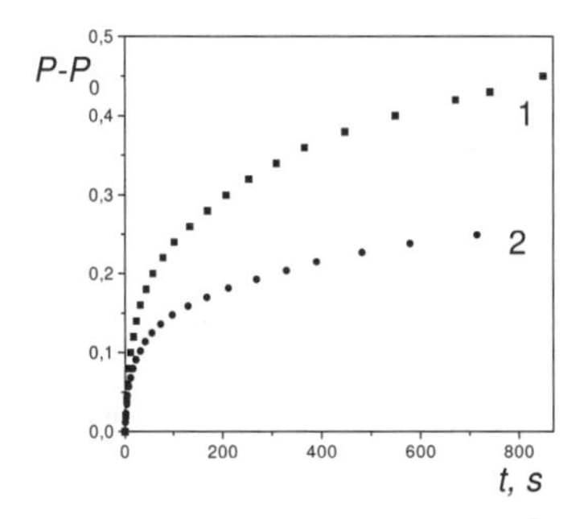

Figure 4.10. Experimental kinetic curves of dimethacrylate OCM-2 post-polymerization at: (1) ℓ=1.6×10^{-4} m, P_0=0.48; (2) ℓ=6.0×10^{-4} m, P_0=0.19. E_0=40.6 W/m^2, c_0=0.04 mol/l.

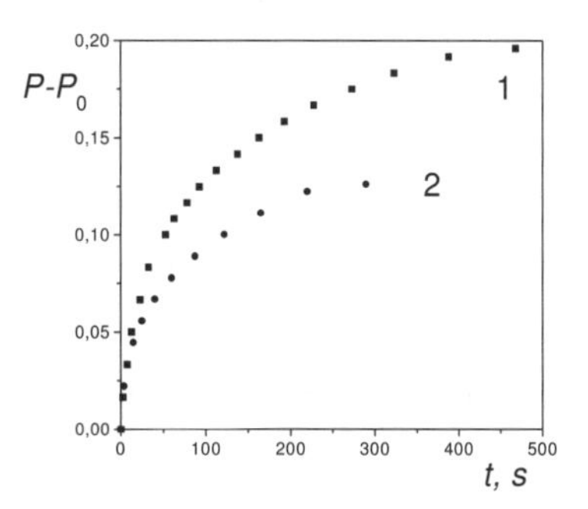

Figure 4.11. Experimental kinetic curves of dimethacrylate DEGDA postpolymerization at: (1) ℓ=1.0×10^{-4} m, P_0=0.47; (2) ℓ=1.4×10^{-4} m, P_0=0.14. E_0=40.6 W/m^2, c_0=0.06 mol/l.

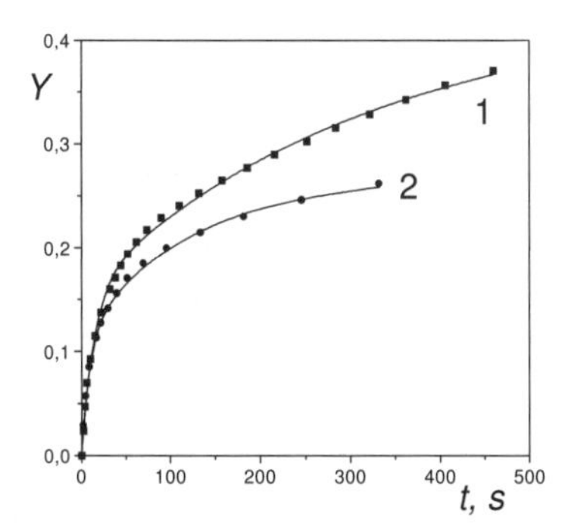

Figure 4.12. Experimental (see points) and calculated (see lines) accordingly to the equation (4.48) dependencies of the function $Y=P_0(1-P_0)\cdot \ln[P(1-P_0)/P_0(1-P)]$ on the post-polymerization time for MDPh-2: (1) ℓ=2.5×10^{-4}m, P_0=0.2; (2) ℓ=2.7×10^{-4} m, P_0=0.46. E_0=40.6 W/m^2, c_0 = 0.06 mol/l.

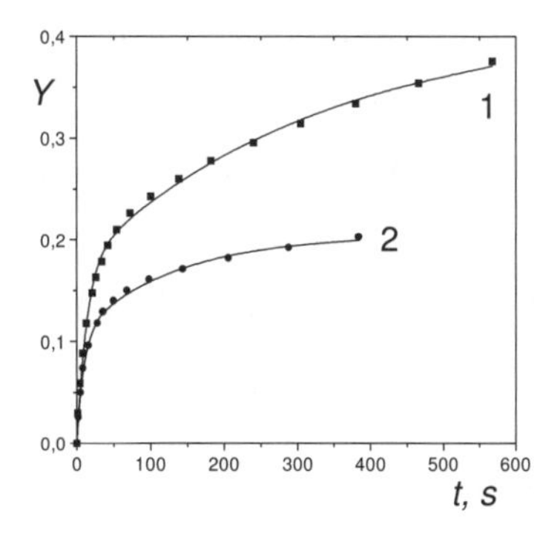

Figure 4.13. Experimental (see points) and calculated (see lines) accordingly to the equation (4.48) dependencies of the function $Y=P_0(1-P_0)\cdot \ln[P(1-P_0)/P_0(1-P)]$ on the post-polymerization time for MGPh-9: (1) ℓ=1.0×10^{-4} m, P_0=0.52; (2) ℓ=1.4×10^{-4} m, P_0=0.20. E_0=40.6 W/m^2, c_0=0.10 mol/l.

The value W_{m0} can be found both from the slope of the tangent angle of the post-polymerization kinetic curve at $t \to 0$ and from the calculation according to the stationary equation (4.46) for the end of a light period. However, both variants give a significant error at the estimation of the values W_{m0}. That is why equation (4.48) has been considered as four-parametric one and numerical values W_{m0}, α, β and γ were determined from the experimental curves of the post-polymerization by optimization in accordance with the program ORIGIN 5.0.

Calculations showed that the method used for optimization is true for a wide range of starting approximations of the parameters determined in a field $W_{m0.}$ β, $\gamma > 0$ and $0 \le \alpha \le 1$, which corresponds to their physical sense. A good agreement between the experimental dependence $Y=P_0(1-P_0)\cdot \ln[P(1-P_0)/P_0(1-P)]$ on time and the one calculated using equation (4.48) with a complete set of parameters, which are determined *via* the optimization method, is presented in Figures 4.12–4.15.

In all cases, that is, for each kinetic curve of the postpolymerization, the relation between the experimental and calculated dependencies is estimated by a correlation coefficient≥0.99. This means, that equation (4.48) correctly transforms the form of the post-polymerization kinetics curve and the experimental error in the construction of the individual (separate) kinetics curve is not significant. This can be also observed from the points location on the kinetic curves in Figures 4.8–4.15.

At the same time, a comparison of α, β and γ parameters calculated from the individual kinetics curves and presented in the Figures 4.16-4.18 as their dependencies on the starting polymerization depth in the dark period gives a great scattering of their numerical values, which greatly exceeds the error of an individual experiment.

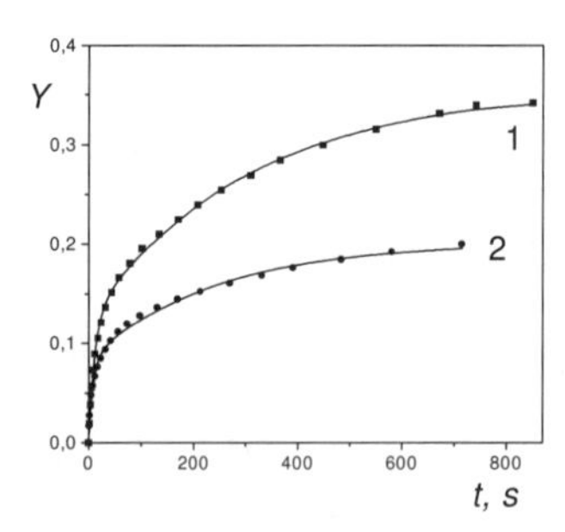

Figure 4.14. Experimental (see points) and calculated (see lines) accordingly to the equation (4.48) dependencies of the function $Y=P_0(1-P_0)\cdot \ln[P(1-P_0)/P_0(1-P)]$ on the post-polymerization time for OCM-2: (1) $\ell=1.6\times10^{-4}$ m, $P_0=0.48$; (2) $\ell=6.0\times10^{-4}$ m, $P_0=0.19$. $E_0=40.6$ W/m^2, $c_0=0.04$ mol/l.

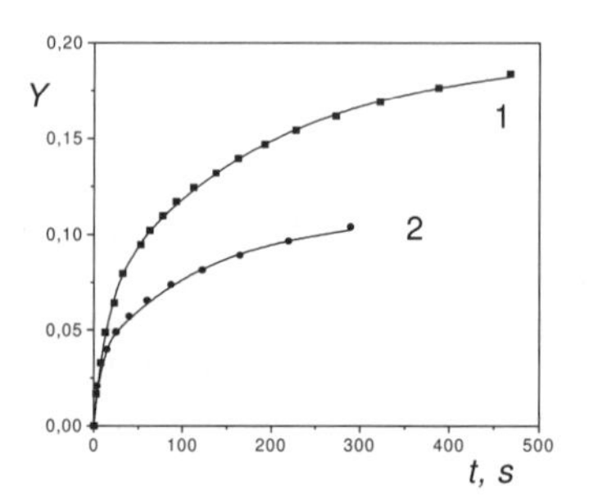

Figure 4.15. Experimental (see points) and calculated (see lines) accordingly to the equation (4.48) dependencies of the function $Y=P_0(1-P_0)\cdot \ln[P(1-P_0)/P_0(1-P)]$ on the post-polymerization time for DEGDA: (1) $\ell=1.0\times10^{-4}$ m, $P_0=0.47$; (2) $\ell=1.4\times10^{-4}$ m, $P_0=0.14$. $E_0=40.6$ W/m^2, $c_0=0.06$ mol/l.

The indicated characteristic of the 3-D polymerization is a direct proof of the microheterogeneity of a process, of an active role of the ‘liquid monomer–solid polymer” interface layer and also proof of the fluctuative mechanism of polymeric grain formation and propagation. This is reflected, first of all, in the kinetic constant k_2, the numerical value of which depends on the ratio of fractal characteristics of the surface and volume of the clusters of the solid polymeric phase into the liquid monomeric phase and liquid monomer into the solid polymeric matrix. Exactly that is why the calculations of W_0 according to stationary kinetic equation (4.46) cannot take into account the individual character of the postpolymerization curves.

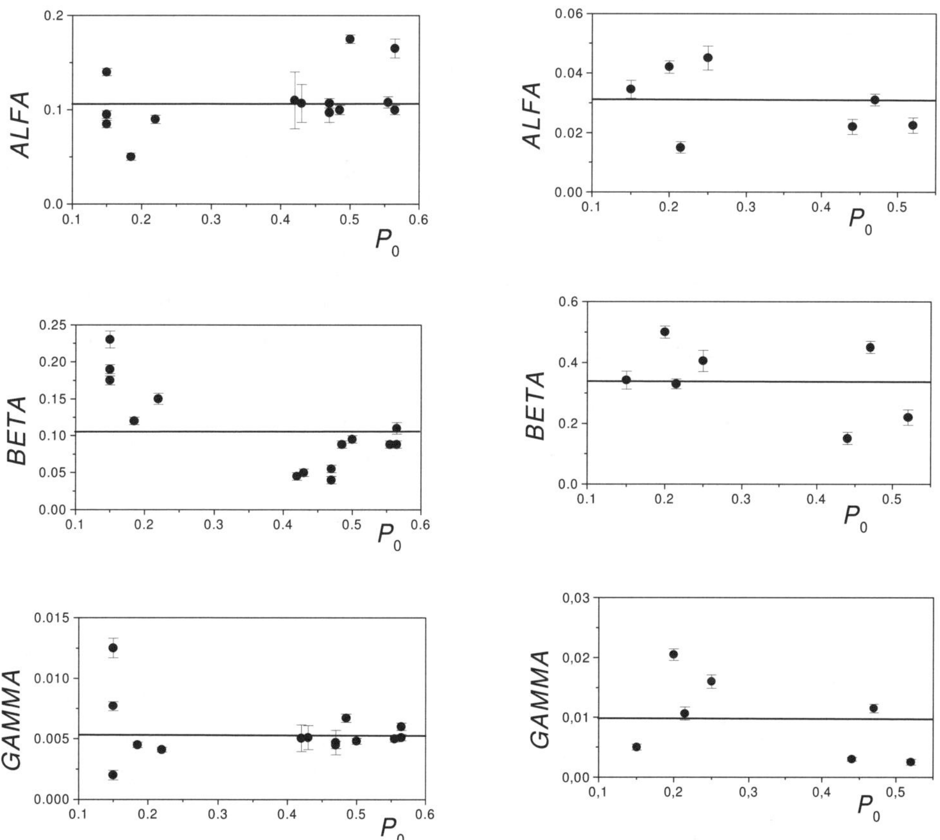

Figure 4.16. Values of parameters α, β and γ, calculated using equation (4.48) at different starting conversions of the OCM-2 post-polymerization.

Figure 4.17. Values of the parameters α, β and γ, calculated using equation (4.48) at different starting conversions of the DEGDA post-polymerization.

However, it should be noted that in the range of the same error determining the scattering of parameters α, β and γ in Figures 4.16–4.18, the W_{m0} values calculated in accordance with the postpolymerization kinetic curves and the stationary equation (4.48) are in good agreement between themselves. In spite of scattering of parameters α, β and γ, the statistical treatment allows to conclude that they do not depend on the starting conversion P_0 of the dark period and, thus, on the current value P *via* the postpolymerization process. This once more proves the agreement between the proposed kinetic model and the experimental data.

On the basis of all set of obtained experimental material standard values and standard deviations of parameters α, β and γ have been determined. They are presented in Table 4.2.

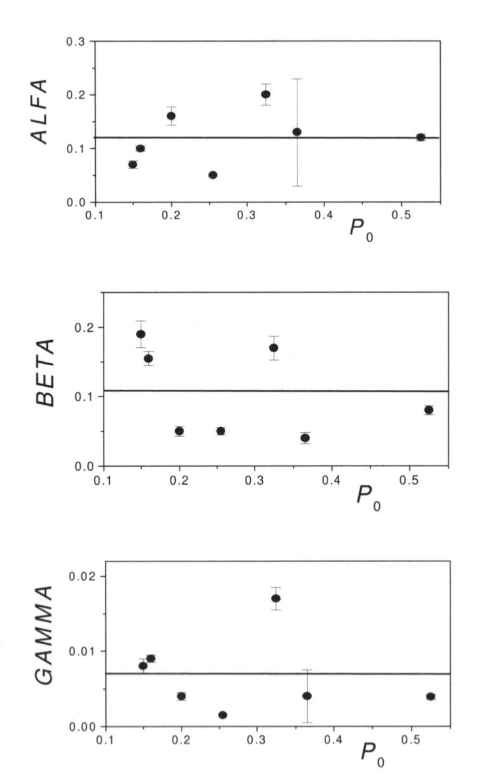

Figure 4.18. Values of parameters α, β and γ, calculated using equation (4.48) at different starting conversions of the MGPh-9 post-polymerization.

Determined values of the α parameter indicate that, for all dimethacrylates, the relative contribution of the secondary chain to total stationary kinetics in the interface layer is small and practically the same (about 10%) and for diacrylate this index is smaller (about 3%). Relaxation times of the primary radicals ($\tau_1=\beta^{-1}$) of dimethacrylates are practically the same and are in the order of 10 s; for diacrylates they are approximately 3-times less. Relaxation times of the secondary radicals ($\tau_2=\gamma^{-1}$) of dimethacrylates are in a range 100–200 s and are dependent on the molecular weight of a polymer.

However, if one extrapolates the values τ_2 of dimethacrylates on the molecular weight of the diacrylate, then τ_2 of the diacrylate will be essentially less than the τ_2 of dimethacrylate of the same molecular weight.

Table 4.2. Average values of parameters α, β and γ, calculated using equation (4.48), and their standard deviations

Monomer	$(\alpha\pm\delta\alpha)\times10^2$	$(\beta\pm\delta\beta)\times10^2$ (s^{-1})	$(\gamma\pm\delta\gamma)\times10^3$ (s^{-1})	No. of kinetic curves
MDF-2	11.0±1.0	11.5±0.8	8.6±0.5	2
MGPh-9	12.1±2.0	10.6±2.5	6.9±2.1	7
OCM-2	10.7±1.0	10.5±1.7	5.3±0.7	14
DEGDA	3.1±0.5	34.1±4.6	9.9±2.3	7

We can consider, that the values of parameters α, β and γ depend on the chemical nature of the functional group of a monomer. Since the concentration of a monomer $[M_0]$ in the interface layer is constant but unknown, then for the calculation of k_{t1} and k_{t2}, using β and γ parameters, the approximation $[M_m]=[M_0]$ has been used. The results of this calculation are presented in Table 4.3.

Table 4.3. Calculated values of rate constants for the primary and secondary chain termination in the interface layer

Monomer	$[M_0]$ (mol m^{-3})	$k_{t1}\times10^5$ (m^3 mol^{-1} s^{-1})	$k_{t2}\times10^6$ (m^3 mol^{-1} s^{-1})
MDF-2	1700	6.8	5.1
MGPh-9	2060	5.2	3.4
OCM-2	2890	3.6	1.8
DEGDA	8020	4.3	1.2

However, for the calculation of all parameters determining the total kinetic curves k_1 and k_2 data from the Tables 4.2 and 4.3 are insufficient. That is why for the first estimation of the unknown parameters in the k_1 and k_2 for dimethacrylates the polymerization degree has been used, which, in accordance with the numerical experimental data [62], is in the range 10^4–10^5.

Let us also take into account that the molecular weight of polymer corresponds to the second maximum on the molecular weight distribution curve in the final state of postpolymerization; this molecular weight is one order of magnitude higher than the molecular weight of the first maximum, corresponding to the initial state of the postpolymerization [63]. It allows to consider, that the sum v_1+v_2 is one order of magnitude than v_1 or $v_2=10v_1$. In accordance with this estimation let us assume, that $v_1=k_{p1}/k_{t1}=10^4$, $v_2=k_{p2}/k_{t2}=10^5$. Using the values k_{t1} and k_{t2} from Table 4.3, we found that $k_{p1}=k_{p2}=10^{-1}$ m^3 mol^{-1} s^{-1}. In accordance with Ref. [58], k_{pv} is characterized in the same order. Thus, we can conclude with the exactness of order that the chain propagation rate constants in the interface layer and in the volume of the liquid monomer are equal to $k_{p1}=k_{p2}=k_{pv}=10^{-1}$ m^3 mol^{-1} s^{-1}.

Taking into account that the α parameter is rather small, expression (4.40) can be approximately written as $\alpha=k_m v_2/k_{p1} v_1$. Now we can estimate: $k_m/k_{p1}=10^{-2}$ and $k_m=10^{-3}$ m^3 mol^{-1} s^{-1}. This means, that only 1% of the active radicals R_m formed after the initiator decomposition in the interface layer are transformed into the active R_m' radicals for the secondary chain.

Under the approximation that $\Gamma_0=1$ and taking into account a small value of the α parameter, let us simplify expression (4.38) for k_2: $k_2=h(F_s/F_v)\,(f_m/f_v)\,/\,v_1[M_0]$.

Setting $k_2=2\times10^{-1}$, $[M_0]=2.5\times10^3$, and assuming equality between the efficiencies of initiation of chains into interface layer and the liquid monomeric phase, that is, $f_m/f_v=1$, we obtain: $h(F_s/F_v)=5\times10^{-2}$. This parameter determines a free volumetric part of the interface layer φ_m *via* $\varphi_m=h(F_s/F_v)P(1-P)$. It follows from this that the maximal value φ_m, which is achieved at $P=0.5$, is characterized by the order 10^{-2} or 1%. This is physically a completely admissible value.

The estimations obtained for the kinetic constants allow us to discuss the question about diffusion control on the rates of the elementary reactions of chain propagation and chain termination. For this purpose let us use the Smolukhovkij diffusion equation:

$$k_{ij} = 4\pi(r_i + r_j)\,(D_i + D_j)\,N_A \tag{4.49}$$

Here: k_{ij} is the frequency of the interaction between particles having effective radii r and diffusion coefficients D; N_A is Avogadro's number which is introduced into equation (4.49) in order to obtain the general dimensionalities of the rate constants of bimolecular reactions.

Using the index "i" for indication of a macroradical and the index "j" for indication of a monomer let us set the condition: $r_i > r_j$, $D_i < D_j$. Then, we obtain for the chain propagation constant: $k_p = 4\pi r_i D_j N_A$. At rather low estimations of $r_i = 10^{-10}$ m and $D_j = 10^{-10}$ m^2 s^{-1}, we will obtain: $k_p = 10^5$ m^3 mol^{-1} s^{-1}. By six orders of magnitude this value exceeds the estimation obtained above on the basis of experimental data $k_p = 10^{-1}$ m^3 mol^{-1} s^{-1} and allows the conclusion that the chain propagation constant both in the volume of the liquid monomer and in the interface layer is fully determined by the nature of the chemical interaction of the radical active center with the functional group of a monomer.

From equation (4.49) we obtain $k_t = 16\pi r_i D_i N_A$ for the bimolecular chain termination rate constant. By estimation of the diffusion coefficient of a macroradical in a range of three orders of magnitude (from the diffusion coefficient in the liquid phase to the diffusion coefficient in the solid phase) $D_i = 10^{-13}$–10^{-10} m^2 s^{-1} we find: $k_t = 5\times(10^2 - 10^5)$ m^3 mol^{-1} s^{-1}. In the liquid monomeric phase the value k_{tv}, in accordance with the literature, is estimated to be in the range $k_{tv} = 10^2$–10^4 m^3 mol^{-1} s^{-1}. Comparison of the obtained estimations does not confirm this, but at the same time does not exclude the possibility of the diffusion control on the rate of bimolecular chain termination in the liquid monomeric phase. Simultaneously, the values k_{t1} and k_{t2} from Table 4.3 are 6–7 orders of magnitude smaller than k_t and convincingly confirm that they do not present the bimolecular chain termination in the interface layer. That is why the question about the diffusion controls on k_{t1} and k_{t2} should be discussed in another way.

According to the conception of chain propagation rate control on the termination rate [64], which is realized at full freezing of translational and segmental mobilities of macroradicals, the connection (or relation) between k_p and k_t based on the Smolukhovkij diffusion equation (4.49) is as follows:

$$k_t = {}^{32}\!/_{3}\,\pi\, r_i\, r_j^2\, k_p [\mathrm{M}] N_A \qquad (4.50)$$

Again, by assuming sufficiently low estimations of $r_i = r_j = 10^{-10}$ m and $[\mathrm{M_m}] = [\mathrm{M_0}] = 2.5\times10^3$ mol m^{-3}, we find that $k_p/k_t = 10$. Comparing this value with the experimentally determined $k_{pv}/k_{tv} \approx 10^{-1}/(10^2$–$10^4) \approx 10^{-3}$–$10^{-5}$, and $k_{p1}/k_{t1} \approx 10^4$ and $k_{p2}/k_{t2} \approx 10^5$, we conclude that the estimation of k_p/k_t calculated in accordance with equation (4.50) is 4–6 orders of magnitude higher than the experimentally obtained values in the liquid monomeric phase and 3–4 orders of magnitude lower than in the case of the interface layer. First, this confirms the fact that in the liquid monomeric phase control of the chain propagation rate on the rate of its termination is scarcely probable, even in the late states of the conversion, that is, in the ranges of $D_i=10^{-13}$ m^2 s^{-1}. Second, it confirms the fact that the chain termination in the interface layer is neither a fully bimolecular process nor a fully diffusion-controlled one.

Although in the presented kinetic scheme of a process ((4.23)–(4.25) in the interface layer the conception of the control of the chain propagation rate on the rate of its termination is used, here the chain termination is a monomolecular one and represents the radical self-burial act, but not the radical trapping. Such a mechanism is well-known and is discussed in the self-avoiding walks model, which describes the statistics of polymers well. However, direct comparison of the values of obtained constants k_{t1} and k_{t2} with the self-avoiding walks model is complicated: according to this model the probability of a chain surviving is determined by the coordinative number of grating, on which the probability of a free unit location in the nearest environment of occupied with an active center of radical will depend.

So, the proposed kinetic models of non-stationary processes in the form of equations (4.42) and (4.45) are based on hypotheses on the microheterogeneity of the polymerization system and on the special role of the interface layer on the "solid polymer–liquid oligomer" boundary. All characteristics of the bulk polymerization up to the high conversion state are explained, including the S-like character of the stationary kinetics curves, the number-average molecular weight of a polymer *via* the post-polymerization process and the presence in it of at least two characteristic sections determined by the life times of active primary and secondary radicals in the interface layer.

Furthermore, this model gives an explanation of the increase in radical concentration in the non-stationary light process coming into the solid polymeric matrix at the expense of the frozen radicals of the R_z and R_z' type.

Let us also note that, although a kinetic scheme of the polymerization process in the interface layer is essentially idealized and formal, it can be simply explained physically. Such an explanation is based on the mutuality of kinetic characteristics of the polymerization in the interface layer and, for example, matrix photopolymerization [65], a polymerization in the presence of porous fillers [66], or at the adsorptive immobilization on the carrier surface of the polymeric initiator [67]. This mutuality indicates that the solid phase forms a specially ordered structure of the nearest reactive space in which the translation and segmental mobilities of the macroradicals are sharply decreased, as, for example, in a system of microreactors weakly interacting with each other [65, 67]. In every microreactor in such a system, the chain termination is determined by the rate of its propagation and the radicals' death represents their self-burial act.

The proposed kinetic model of the postpolymerization describes the multiple experimental data well and is in good agreement with all the characteristics of the postpolymerization kinetics listed above. However, the introduction of two types of radicals sharply differing by characteristic life times into a kinetic scheme is an inevitable simplification of a real set of characteristic life times of active radicals. Furthermore, it cannot be indirectly re-passed on the kinetics of monofunctional monomer postpolymerization which, the same as stationary kinetics, can be characterized by differences from the kinetics of bifunctional monomer postpolymerization. The term "monomolecular chain termination", introduced in Refs. [55, 56] as an active center of the radical self-burial act in the act of chain propagation, did not have a theoretical basis *via* the relation of k_t with k_p.

These problems will be discussed in detail in Chapter 7.

Acknowledgements

We would like to express our appreciation of the discussions on sections 4.1. and 4.2. of this monograph with colleagues, in particular with Professor Ewa Andrzejewska at the Poznań University of Technology, Institute of Technology and Engineering (Poznań, Poland).

REFERENCES

[1] Anseth K., Decker C., Bowman C. *Macromolecules,* **1995**, 28 (8), p. 4040.
[2] Mateo J., Calvo M., Serrano J., Bosch P. *Macromolecules*, **1999**, 32, p. 5243.
[3] Young J., Bowman C. *Macromolecules*, **1999**, 32, p. 6073.
[4] Schulz G. *Z. Phys. Chem.* (*Munich*), **1956**, 8, p. 290.
[5] Buback M., Huckestein B., Russell G. *Macromol. Chem. Phys.*, **1994**, 195, p. 539.

[6] Anseth S., Kline L., Walker T., Anderson K., Bowman C. *Macro*mol*ecules,* **1995**, 28, p. 2491.

[7] Mateo J., Serrano J.; Bosch P. *Macro*mol*ecules,* **1997**, 30. p. 1285.

[8] Stickler M. *Macro*mol. *Chem.,* **1983**, 184, p. 2563.

[9] Anseth K., Anderson K., Bowman C. *Macro*mol. *Chem. Phys.,* **1996**, 197, p. 833.

[10] Kloosterboer J., Lijten G. *Cross-Linking Photopolymerization of Tetraethyleneglycol Diacrylate.* In *Cross-Linked Polymers*, American Chemical Society: Washington, DC, **1988**.

[11] Russell G., Napper D., Gilbert R. *Macro*mol*ecules,* **1988**, 21, p. 2133.

[12] Russell G., Gilbert R., Napper D. *Macro*mol*ecules,* **1993**, 26, p. 3538.

[13] Scheren P., Russell G., Sangster D., Gilbert R., German A. *Macro*mol*ecules,* **1995**, 28, p. 3637.

[14] Smoluchovskij M. *Z. Phys. Chem.,* **1918**, 92, p. 129.

[15] Einstein A. *Ann. Phys.,* **1905**, 17, p. 549.

[16] Flory P. *Statistical Mechanics of Chain* Mol*ecules*, Wiley-Interscience: New York, NY, **1969**.

[17] Anseth K., Wang C., Bowman C. *Macro*mol*ecules,* **1994**, 27, p. 650.

[18] Tryson G., Shulz A. *J. Polym. Sci.*, **1979**, B17, p. 2059.

[19] Timpe H., Strehmel B. *Macro*mol. *Chem.*, **1991**, 192, p. 771.

[20] Anseth KS, Wang CM, Bowman CN. *Polymer*, **1994**, 35, p. 3243.

[21] Anseth KS, Bowman **CN,** Peppas NA. *J. Polym. Sci.*,**1994**, 32, p. 139.

[22] Nwabunma D**,** Kim KJ, Lin Y, Chien LC**,** Kyu T. *Macro*mol*ecules*, **1998**, 31, p. 6806.

[23] Lecamp L., Youssef B., Bunel C., Lebaudy P. *Polymer*, **1999**, 40. p. 1403.

[24] Lecamp L., Youssef B., Bunel C., Lebaudy P. *Nucl. Instr. Methods Phys. Res.*, **1999**, B151, p. 285.

[25] Lovell LG., Stansbury JW., Sypres DC., Bowman CN. *Macro*mol*ecules*, **1999**, 32, p. 3913.

[26] Miller U. J. *Pure Appl. Chem.*, **1994**, A31, p. 1905.

[27] Miller U., Jockusch S., Timpe H. *J. Polym. Sci., A: Polym. Chem.*, **1992**, 30. p. 2755.

[28] Andrzejewska E., Bogacki M. *Macro*mol. *Chem. Phys.*, **1997**, 198, p. 1649.

[29] Andrzejewska E., Bogacki M., Andrzejewski M., Tyminska B. *Polimery*, **2000**, 45, p. 502.

[30] Andrzejewska E. *Prog. Polym. Sci.*, **2001**, 26, p.605.

[31] Andrzejewska E., Bogacki M. *Papers and posters presented at Eighth Conference on Radiation Curing, RadTech. Europe '97, Lyon*, 17–18 June **1997**. p. 151.

[32] Andrzejewska E., Bogacki M., Andrzejewski M. *Conference Proceedings, RadTech Europe* 99, 8–10 November **1999**, Berlin, Germany, p. 779.

[33] Decker C., Elzaouk B. *Eur. Polym. J.*, **1995**, 31, p. 1155.

[34] Decker C., Elzaouk B., Decker D. *Macro*mol. *Sci., Pure Appl. Chem.*, **1996**, A33, p. 173.

[35] Kurdikar D., Peppas N. *Macro*mol*ecules*, **1994**, 27, p. 4084.

[36] Peppas S. *Radiation curing — a personal perspective. In: Radiation curing science and technology*. Plenum Press, New York, NY, **1992**, p. 1-20 (Chapter 1).

[37] Batch G., Macosko C. *J. Appl. Polym. Sci.*, **1992**, 44, p.1711.

[38] Kloosterboer J., Touwslager F. *Chain crosslinking photopolymerization of diacrylates and some applications in electronics. RadTech Asia UV/EB Conference Proceedings*, **1993**. p. 409.

[39] Kloosterboer J., Lijten G. *Biol. Synth. Polym. Networks*, **1988**, 345, p. 55.

[40] Kloosterboer J., Lijten G. *Polymer*, **1990**, 31, p. 95.

[41] Brosse J., Couvret D. *Macro*mol. *Chem., Rapid Commun.*, **1990**, 1, p. 123.

[42] Decker C., Moussa K. *Eur. Polym. J.*, **1991**, 27, p. 403.

[43] Decker C., Moussa K. *Macro*mol. *Chem., Rapid Commun.*, **1990**, 11, p. 159.

[44] Decker C., Moussa K. *Macro*mol. *Chem.*, **1991**, 192, p. 507.

[45] Moussa K., Decker C. J. *Polym. Sci., Part A: Polym. Chem.*, **1993**, 31, p. 2197.

[46] Bosch P., Serrano J., Mateo J., Calle P., Sieiro C. *J. Polym. Sci., Part A: Polym. Chem.*, **1998**, 36, p. 1783.

[47] Bosch P., Serrano J., Mateo J., Guzman J., Calle P., Sieiro C. *J. Polym. Sci., Part A: Polym. Chem.*, **1998**, 36, p. 2785.

[48] Lovell L., Stansbury J., Sypres D., Bowman C. *RadTech Europe'99, Conference Proceedings*, 8–10 November **1999**, p. 773.

[49] Berchtold K., Lovell L., Nie J., Hacioglu B., Bowman C. *World Polymer Congress, IUPAC Macro* 2000. Warsaw, Poland, 9–14 July **2000**, Book of Abstracts, p. 173.

[50] Odian G. *Principles of polymerization*. 2nd edn. Wiley-Interscience, New York, NY, **1981**.

[51] Russel G. *Macro*mol. *Theor. Simul.*, **1994**, 3, p. 439.

[52] Kloosterboer J., Lijten G. *Polym. Commun.*, **1987**, 28, p. 2.

[53] O' Shaughnessy B., Yu J*Macro*mol*ecules*, **1998**, 31, p. 5240.

[54] O' Shaughnessy B., Yu.J*Phys. Rev. Lett.*, **1998**, 80. p. 2957.

[55] Zaglad'ko O. *Kinetic model of the* 3 *D radical polymerization in the bulk*, Ph. D Thesis in chemical sciences, Lviv, **1998**, 104 p.
[56] Zaglad'ko O., Turovskij A., Medvedevskikh Yu., Skorobagatyj Yu. 3 *D radical polymerization modeling*, Lviv, **2000**, 87 p.
[57] Minko S., Sydorenko A., Voronov S. *Vysoko*mol. *soed.* (*A*), **1995**, 37 (8), p. 1403.
[58] Medvedevskikh Yu., Zaglad'ko O., Turovskij A. *Int. J. Polym. Matter.*, **1998**, 41, p.1.
[59] Medvedevskikh Yu., Zaglad'ko E., Turovskij A., Zaikov G. *Russ. Polym. News*, **1999**, 4 (3), p. 33.
[60] Zaglad'ko O., Medvedevskikh Yu., Turovskij A. *Int. J. Polym. Mater.,* **1999**, 38, p. 227.
[61] Stepanian A., Zaremskij M., Olenin A. *Doklady AN USSR*, **1984**, 274 (3), p. 655.
[62] Liklema Y. *Fraktaly v phisikie*; M.: Myr, **1988**, 672 p.
[63] Treushnikov V., Zelentsova N., Olejnik A. *Zhurnal nauchn. i prikl. photo- i kinematographii*, **1988**, 33 (2), p. 146.
[64] Litvinenko G., Lachinov M., Sarkisova E. *Vysoko*mol. *soed.* (*A*), **1994**, 36 (2), p. 327.
[65] Ivanchev S., Dmitrenko A., Krupnik A. *Vysoko*mol. *soed.* (*A*), **1988**, 30 (9), p. 1951.
[66] Ivanov V., Romaniuk A., Shybanov V. *Vysoko*mol. *soed.* (*A*), **1994**, 35 (2), p. 119.
[67] Treushnikov V., Esin S., Zujeva T. *Vysoko*mol. *soed.* (*A*), **1995**, 37 (12), p. 1973.

Chapter 5. Stationary kinetics of 3-D polymerization up to the high conversion state

5.1. Conception of two reactive zones

The microheterogeneous conception [1–6] considers a polymerization process as a total combination of chemical transformations and structural-phase changes, such as phase distribution, aggregation of chains, local glass transition and appearance of supermolecular structures at the evolution of a fluctuation of the density of the polymerization system. As a result, the above-mentioned processes lead to the microheterogeneity of a polymerization system. Micronon-uniformity can be caused by the following factors: firstly, by thermodynamical non-compatibility of highly crosslinked polymers with the liquid phase of the reactive medium. In accordance with the Flory–Renner ratio, the separation (or syneresis) of the system with the elimination of the excess of the liquid phase from the crosslinked polymer is started under the condition that in a micro-volume of the reactive mix (in a micro-gel) the density of polymer network exceeds the value for equilibrium swelling. In highly crosslinked polymers, as a result of strong diffusion complications, separation does not lead to macroscopic distribution; instead, microseparation (or microsyneresis) takes place.

Accumulation of the polymer network junction points into highly crosslinked micro-volumes causes the degeneration of the segmental moving of macroradicals in polymeric chains and, at a determined stage of the conversion, glass transition of the micro-gel particles is started (the so-called local glass transition, with the formation of the glass-like polymeric grains). At the beginning of glass transition the micronon-uniform, because of the density of the polymeric network polymerizate, becomes microheterogeneous.

Structural-physical transformations depend on the chain termination rate and viscosity of reactive medium. Increasing the local viscosity in macromolecular balls, formed as a result of inter- and intramolecular aggregations of the polymeric chains, leads to a decrease of the chain termination rate and an increase of the polymerization rate, respectively, causing the increase of the local viscosity. The process in such micro-volumes proceeds by auto-catalysis as a local gel-effect.

For multifunctional monomers the micronon-uniformity of the polymerizing system (i.e., the availability in it the micrograins of the solid polymer with limited conversion distributed in the liquid monomeric phase or (after phase inversion) microdrops of the liquid monomer (which are distributed in the solid polymeric matrix) is a factor involved, as proven both by direct and indirect experimental methods. For example, it was determined using sol–gel analysis and IR-spectroscopy [4, 7] in the investigation of the polymerization of some oligoesteracrylates, that at all percentages of the conversion the sol represents the starting oligomer with the conversion equal to zero and the gel is the non-soluble highly crosslinked polymer with limited conversion. On the basis of these data it was assumed that the polymerization proceeds on the surface of the polymeric grains.

At some stage of the polymerization, when the volumetric part of the propagating polymeric particles becomes essential for their mutual contact, these particles are joined (also called monolytization of the solid). The joining into the indivisible framework structure in the monolytization stage proceeds at the expense of polymerization transformations in the contact zones.

The monolytization and joining of polymeric grains (supermolecular level) can be considered as the second gel-point at the supermolecular level (by analogy to the usual gel-point, corresponding to the formation of the overall 3-D network of polymeric links at the molecular level). A spring-like change of the polymer properties should be observed at the stage of the monolytization in a rather narrow range of conversion states. The final polymer consists of the highly crosslinked polymeric grains and liquid strata between them [8].

Thus, the polymerization process can be presented as follows: origination of polymeric grains from fluctuations in the polymer concentration → microdistribution → polymeric chain propagation → monolytization. Moreover, as was mentioned earlier, even at the initial stages of the process, the system consists of a polymer with a practical limited state of the conversion and starting monomer and conversion degree equal to zero. So, two reactive zones appear, namely: first, the stratum between grains, or the so-called liquid homogeneous medium, and the second is the superficial layer at the "solid polymer–liquid monomer" phase division, contributing the main share to the polymerization process [4, 7].

5.2. Kinetic model of the thermoinitiated polymerization

We propose the following kinetic equation, which allows us a quantitative description of the 3-D polymerization of the polyfunctional oligomers in bulk in the entire interval of transformation level [8–10]:

$$dP/dt=k_1\,(1-P)\,v_i^{1/2} + k_2P\,(1-P)\,v_i \qquad (5.1)$$

here $P=\Gamma/\Gamma_0$ is relative degree of polymerization determined by the relation of current conversion $\Gamma=([M]_0-[M])/[M]_0$ at time t to limit $\Gamma_0=([M]_0-[M]_{00})/[M]_0$ at $t\rightarrow\infty$; $[M]_0$. $[M]$ and $[M]_{00}$ are starting, current and final concentrations of monomer expressed relative to the volume of the whole system; k_1 and k_2 are constants; v_i is initiation rate.

Kinetic equation (5.1) shows a S-shaped dependence of P on t with ordinate P_0 being the point of inflection, which is determined from the condition $d^2P/d^2t=0$

$$k_2v_i-k_1v_i^{1/2}-2k_2Pv_i + [k_1(1-P)\,v_i^{1/2}/2+ k_2P\,(1-P)\,v_i]\,dv_i/v_idP=0 \qquad (5.2)$$

The estimates [8] of the maximum rate of the process $W_0=(dP/dt)_{max}$ corresponding to ordinate P_0. point of inflection of the kinetic curve $P=P(t)$, show that the principal contribution to the kinetics of the process is made by the second term in equation (5.1). Therefore, without a significant error, the expression in square brackets in equation (5.2) may be considered equal to dP/dt. From the expression $v_i=-dc/dt=k_ic$, in which c is concentration of the initiator and k_i is the rate constant of its decomposition, follows $dv_i/v_idt=-k_i$. Thus, the solution of equation (5.2) can be written as

$$P_0=(a-b)/2a \qquad (5.3)$$

where $a=k_2v_i^{1/2}/k_1$, $\qquad b=1 + k_i/k_1v_i^{1/2}$.

According to equation (5.3) in the area of positive P, the ordinate P_0. representing the point of inflection of the kinetic curve $P=P(t)$, may have a value in the range $0\leq P\leq 0.5$. This agrees with the experimental data presented in Figures 5.1–5.8.

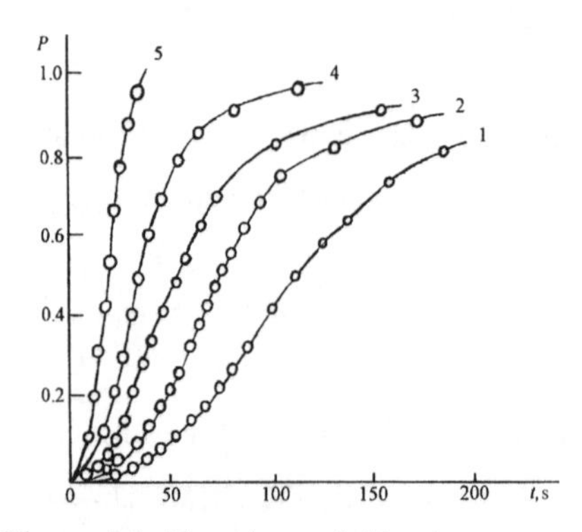

Figure 5.1. Experimental kinetic curves of MDF-2 photopolymerization at different thicknesses of layer ℓ: 7×10^{-4} (1), 4.8×10^{-4} (2), 2.9×10^{-4} (3), 1.5×10^{-4} (4), 0.3×10^{-4} m (5). c_0=0.021 mol/l, E_0=40.6 W/m^2.

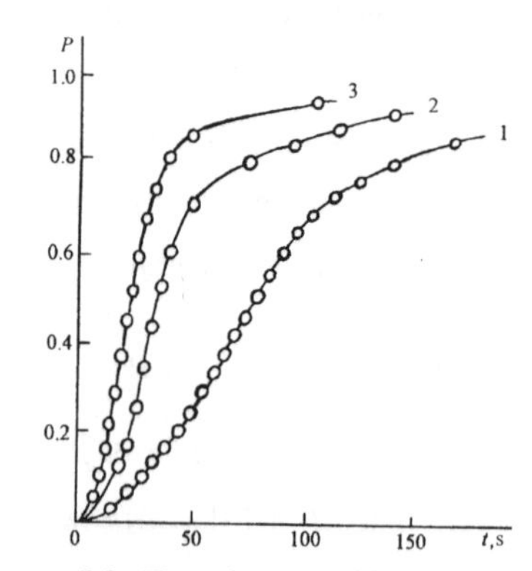

Figure 5.2. Experimental kinetic curves of MDF-2 photopolymerization at different concentrations of photoinitiator c_0: 0.021 (1), 0.040 (2), 0.105 mol/l (3). E_0=40.6 W/m^2, $\ell=5\times10^{-4}$ m.

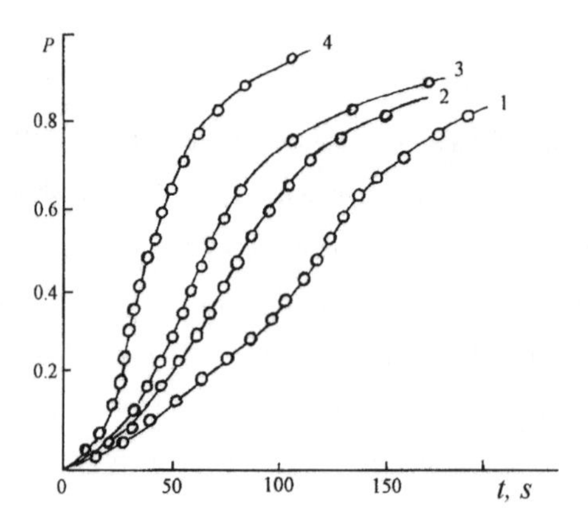

Figure 5.3. Experimental kinetic curves of MDF-2 photopolymerization at different UV-illumination powers E_0: 36.2 (1), 40.6 (2), 59.03 (3) and 79.04 W/m^2 (4). c_0=0.086 mol/l, $\ell=7\times10^{-5}$ m.

For the maximum rate W_0 of the process we have the expression

$$W_0=k_1 v_i^{1/2} (a + b)(a-b + 2) \tag{5.4}$$

From this equation follows, that at $a >> b$ and $P \approx ½$, then $W \approx k_2 v_i/4$. This allows us to estimate $k_2 v_i$ on W_0 directly. In accepted terms of a normalized rate of the process we have $-d(\ln(1-P))/dt$. Its maximum corresponds to the point of inflection of the kinetic curve of the form $\ln(1-P)=f(t)$ with the ordinate P^0 and rate W^0 of the process connected according to the analysis of equation (5.1) by the relation

$$P^0=W^0/k_i-1/2a \qquad (5.5)$$

Solution of the equation (5.5) does not limit the values of P^0 in the area of positive $P \geq 0$; this also agrees with the experimental data [1].

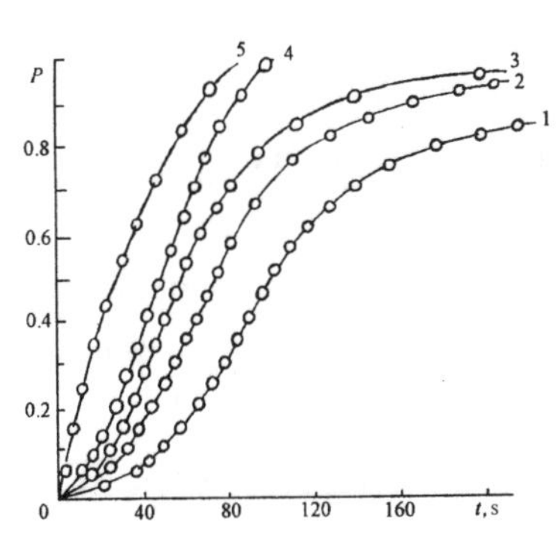

Figure 5.4. Experimental kinetic curves of OCM-2 photopolymerization at different thicknesses of layer ℓ: 4.7×10^{-4} (1), 4.1×10^{-4} (2), 3.5×10^{-4} m (3). c_0=0.021 mol/l, E_0=40,6 W/m^2.

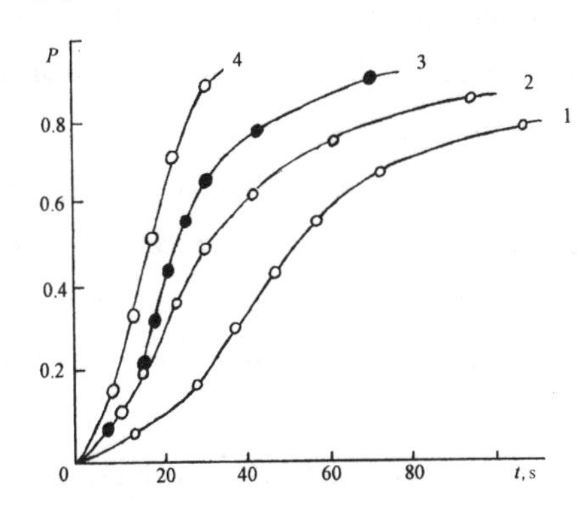

Figure 5.5. Experimental kinetic curves of OCM-2 photopolymerization at different concentrations of photoinitiator c_0: 0,016 (1), 0.02 (2), 0.04 (3), 0.06 mol/l (4). E_0=40.6 W/m^2; $\ell=6\times10^{-4}$ m.

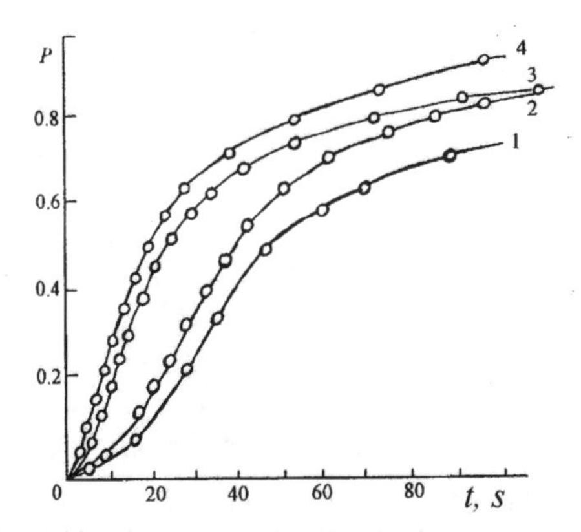

Figure 5.6. Experimental kinetic curves of OCM-2 photopolymerization at different UV-illumination powers E_0: 36.2 (1), 40.6 (2), 59.03 (3) and 79.04 W/m^2 (4). c_0=0.016 mol/l, $\ell=5\times10^{-4}$ m.

As we can see, equation (5.1) is in good agreement with the main peculiarities of kinetic curves of 3-D polymerization [8–10]. At this stage, the form of equation (5.1) means that the microheterogeneous model can be used as a base for derivation of this equation. The main positions of the above-mentioned model have been formulated by us as follows [8]:

(a) the rate of 3-D polymerization in volume may be represented by the sum of the rates of the homophase process, running in the volume of liquid monomer *via* the classic scheme with bimolecular chain termination and heterophase, which runs in the interlayer volume on "solid polymer–liquid oligomer" boundary in the regime of gel effect;

(b) clusters of solid polymer in the liquid monomeric phase and clusters of liquid monomer in the solid polymeric matrix are characterized by a fractal structure;

(c) the "catalytic" role of the interface layer, i.e., the gel effect, leads to a decreasing rate of radical trapping and to control transition from chain termination rate to rate of its propagation.

It follows from (a) above that the observed rate of process can be represented as

$$w = \varphi_v w_v + \varphi_m w_m \tag{5.6}$$

Here φ_v and φ_m are volumetric parts of liquid monomeric phase and interface layer; w_v and w_m are respective specific rates of polymerization.

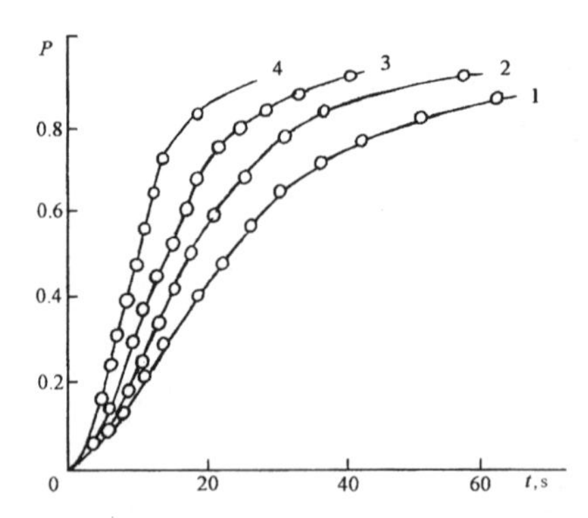

Figure 5.7. Experimental kinetic curves of DEGDA photopolymerization at different thicknesses of layer ℓ: 1.8×10^{-4} (1), 0.9×10^{-4} (2), 3.9×10^{-4} (3), 0.9×10^{-4} m (4). c_0=0.08 mol/l; E_0=40.6 W/m^2.

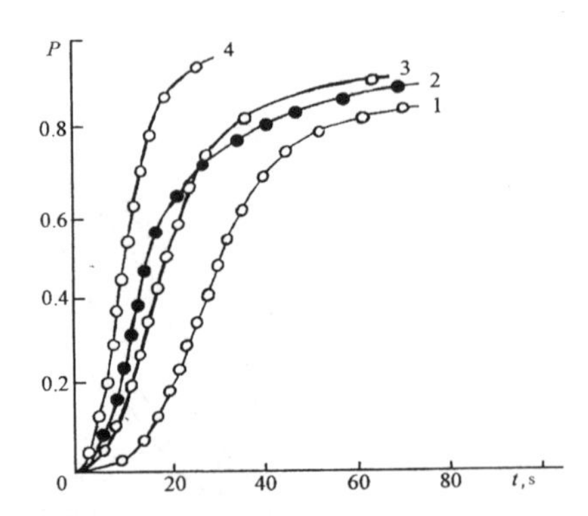

Figure 5.8. Experimental kinetic curves of DEGDA photopolymerization at different concentrations of photoinitiator c_0: 0.08 (1), 0.06 (2), 0.03 (3), 0.02 mol/l (4). E_0=40.6 W/m^2; ℓ=2×10^{-4} m.

The relationship between w=d[M]/dt and dP/dt is determined by the ratio dP/dt=w/[M]$_0\Gamma_0$.

So, primarily two terms in equation (5.1) and equation (5.6) characterize the polymerization rate in the liquid monomeric phase and the secondary ones at the interface. At this time, w_v and w_m are described as rates of chain propagation

$$w_v=k_{pv}\,[M_v]\,[R_v],$$

$$w_m=k_{pm}\,[M_m]\,[R_m] \qquad (5.7)$$

where $[M_v]$, $[M_m]$, $[R_v]$ and $[R_m]$ are concentrations of monomer and radicals, respectively, in the liquid phase and at the interface; k_{pv} and k_{pm} are the respective chain propagation rate constants.

It has been taken into account in the derivation of equation (5.1) from equation (5.6), according to experimental data, that at the earlier stage of bulk polymerization (at P=0.01) the monomeric phase becomes saturated as a relatively new polymeric phase, but the saturation level is low. That is why a concentration of monomer $[M_v]$ in a saturated monomer/polymer solution is constant and, practically, equal to the concentration at the start of the polymerization, i.e., $[M_v]=[M_0]$. In contrast, the volumetric part of the liquid phase (φ_v) has a variable value and is changed *via* polymerization conversion as $\varphi_v=1-P$.

The question is: how is the initiator distributed between the solid and liquid phase and also between them and interface layer? There are two probable alternative variants: we will consider both variants, since direct proof for a preference for one of each has not yet been found.

The first variant [8] is based on the assumption that the solubility of initiator in the polymer is rather small and that is why it is displaced from the solid phase into the liquid phase *via* polymerization, in which its concentration is increased as a result of the decrease in its volumetric part φ_v. Under this situation the relationship $v_i=v_{iv}\varphi_v$ or $v_{iv}=v_i/(1-P)$ should exist between the specific rate of initiation v_i, determined by assuming that the initiator is uniformly distributed over the volume of the whole system, and the specific rate of initiation v_{iv}, under the assumption that all of the initiator is located in the liquid phase volume. Analysis shows, that in the presence of such a relationship the final equation, that will be equal to starting equation (5.1), cannot be obtained from equations (5.6) and (5.7).

This is why the second variant is accepted, which assumes an uniform distribution of initiator between phases and interface layer and leads to $v_i = v_{iv}$.

We adopt that the polymerization process in the liquid monomer volume proceeds in accordance with the classic kinetic scheme with bimolecular chain termination:

$$M_v + R_v \xrightarrow{k_{pv}} R_v$$
$$R_v + R_v \xrightarrow{k_{tv}} \text{product of termination} \tag{5.8}$$

where k_{pv} is the chain propagation rate constant; k_{tv} is the bimolecular chain termination rate constant.

In accordance with kinetic scheme (5.8) the concentration of radicals into liquid phase is equal to:

$$[R_v] = (v_i / k_{tv})^{1/2} \tag{5.9}$$

We find the following expression by combining the obtained ratio (5.9) with equation (5.7), taking into account the volumetric part φ_v of the liquid phase:

$$\varphi_v w_v = (k_{pv} / k_{tv}^{1/2})\ [M]_0\ (1-P)\ v_i^{1/2} \tag{5.10}$$

A fractal approach [11] instead of a geometric one [8] has been used by us for functional determination of the volumetric part φ_m of the interface layer. In accordance with this fractal approach we suppose that propagating grains or clusters of solid polymeric phase are characterized by their fractal structure, the dimensionality of which generally does not coincide with the dimensionality of the Euclidean space. For example, the volume V of cluster and its radius r are connected by ratio $V \sim r^{d_v}$ in which $d_v \leq 3$. We can also write $S \sim r^{d_s}$, where $2 \leq d_s \leq 3$ [**11, 12**].

Using the fractal dimensionalities we can express the volumes of solid polymeric phase V_t and interface layer $V_m(0)$ at the starting stage by:

$$V_t = nF'_v\ r^{d_v} = V_0 P$$

$$V_m(0) = nhS = nh\ F'_s\ r^{d_s} \tag{5.11}$$

here V_0 is the general volume of system; n is the number of clusters or grains in the polymeric phase; h is the thickness of the interface layer; F'_v and F'_s are fractal characteristics of volume and surface of polymeric clusters not depending on r.

We obtained the next equation by solving equation (5.11) and r:

$$V_m(0)=nhF'_s(PV_0/nF'_v)^{d_s/d_v} \quad (5.12)$$

It follows from estimations of d_v and d_s [13] that the cluster that propagates according to mechanism of diffusion-limited aggregation is characterized by the properties of the mass fractal, and by d_v=d_s [12].

We obtained $\varphi_m(0)=h(F'_v/F'_s)P$, starting from the assumption about equality (d_v=d_s) for the initial stage of the process of equation (5.12), taking into account the ratio $\varphi_m=V_m/V_0$. For the final stage of the process, namely after phase inversion, analogous assumptions lead to the expression $\varphi_m(1)=h(F''_s/F''_v)(1-P)$, in which F''_s and F''_v are fractal characteristics of surface and volume of liquid oligomeric clusters in polymeric pattern.

Evidently, grains propagation and their aggregation proceed simultaneously and uninterrupted [3]; according to previous analyses the first process increases the volume of the interface layer proportional to P, and the second one decreases it proportional to 1–P. Since the general view of the dependence of φ_m on P is unknown, we approximate this dependence by the function:

$$\varphi_m=h(F_s/F_v)\,P\,(1-P) \quad (5.13)$$

here F_s/F_v is average fractal characteristic of surface and volume of clusters.

Function (5.13) satisfies the condition of non-interruption of φ_m upon P change and, as partial cases, contains only expressions for φ_m (0) and φ_m (1).

The "catalytic" role of interface layer in 3 -D polymerization is certainly connected with the decreasing chain termination rate. Both rate of radical trapping and also the kinetic order of the reaction are decreased from the second to the first. This can be explained by several possible causes for this effect. In accordance with Ref. [7] chain termination represents the diffusion-control reaction and its rate depends on the mobility of macroradicals. If this mobility is decreased, so that condition $t_d/t_p >> 1$ takes place (t_d and t_p are characteristic moving times of the active center of macroradicals as a result of translational or segmental mobilities and chain propagation reaction, respectively), then the radical termination rate is determined by the chain propagation rate of the general polymerization process. Here the effective chain termination rate constant is proportional to k_p[M] [14].

Acceleration of radical polymerization in the presence of porous fillers [15] indicates that porous structures which are formed *via* aggregation of ultradispersed particles of filler can be considered as so-called microreactors which are not easy available for macroradicals. That is why the rate of radical trapping in these isolated microreactors is determined mainly by the rate of their propagation. The same is role of isolated microreactors in pattern photopolymerization [16, 17].

Finally, let us notify that the first order of dependence of polymerization rate upon initiation rate at the adsorptive immobilization on the surface of filler of polymeric initiator can also be explained by the essential contribution of reactive diffusion to chain termination, i.e., supplying the reactive centers of macroradicals to one another at the expense of their propagation [18].

The above-mentioned factors take place simultaneously at the interface, finally leading to the control of the chain propagation rate, taking into account the chain termination rate. In this case two acts, namely chain propagation and chain termination, represent two results of interaction of a radical with a functional group of the oligomer. These results can be separated after introducing the notion of so-called "living" and "frozen" radicals [8].

$$R_m + M_m \longrightarrow \begin{cases} \xrightarrow{k_{p1}} R_m \\ \xrightarrow{k_{t1}} R_z \end{cases} \qquad (5.14)$$

Thus, in accordance with scheme (5.14), chain propagation with the probability $k_{p1}/(k_{p1} + k_{t1})$ forms the "living" radical and with the probability $k_{t1}/(k_{p1} + k_{t1})$ forms the "frozen" one. Since the chain termination in scheme (5.14) is a singl e act of chain propagation, we should assume in equation (5.7) that $k_{pm}=k_{p1} + k_{t1}$; howver, we have $k_{pm}=k_{p1}$, because $k_{t1} << k_{p1}$.

There are contrasting points of view about influence of the gel effect on the chain propagation rate and on the initiator decomposition rate [1, 19–21]. The most probable is that the gel-effect in the chain propagation should be regarded as a second-order reaction, but less than for the termination reaction. This means that k_{p1} cannot be equal to k_{pv} at the interface, due to the gel effect. For decomposition of initiator the gel-effect is observed *via* a "cell" effect, e.g., *via* initiation efficiency f, but not *via* constant rate of decomposition k_d. Under this assumption the initiation rate at the interface and in the liquid oligomeric phase is described by expressions $v_{im}=2k_d f_m c$, $v_{iv}=2k_d f_v c$. The next relation follows from these expressions, namely $v_{im}=(f_m/f_v)v_i$, since it was assumed that $v_{iv}=v_i$.

The condition of stationarity of "living" radicals, according to equation (5.7), we can now write as $(f_m/f_v)v_i=k_{t1}$ $[M_m][R_m]$. With the use of this ratio in equation (5.7) we obtained:

$$w_m=(f_m/f_v)\ (k_{p1}/k_{t1})\ v_i \qquad (5.15)$$

The contribution of the heterophaseous process of the polymerization with the specific rate w_m to the total will be written as follows, taking into account the approximate function (equation 5.11) for the volumetric part of the interface layer:

$$\varphi_m w_m=(f_m/f_f)\ (k_{p1}/k_{t1})\ h\ (F_s/F_v)\ P\ (1-P)\ v_i \qquad (5.16)$$

Using equations (5.10) and (5.16) in the integrated equation (5.6), and taking into account equation (5.12), we obtained the starting kinetic equation (5.1), in which physical sense of kinetic constants k_1 and k_2 are determined by the following expressions:

$$k_1=k_{pv}/k_{tv}^{1/2}\,\Gamma_0 \qquad (5.17)$$

$$k_2=h\ (F_s/F_v)\ (f_m/f_v)\ k_{p1}/k_{t1}\ \Gamma_0\ [M]_0 \qquad (5.18)$$

So, taking into account the factor Γ_0, which means the fullness of the 3-D polymerization proceeding, k_1 is the usual constant of homophaseous polymerization rate in the ranges of long chains with bimolecular termination. In contrast, k_2 is not the polymerization constant rate in the rigid sense of the word. Since the value of the polymerization constant rate depends upon the fractal characteristics of solid polymeric clusters in the liquid monomeric phase and liquid monomeric clusters in the solid polymeric pattern, k_2 is a fluctuation-depending parameter of the process determining the bed reproducing the kinetic measurements [22] and "whims" [23] of 3-D polymerization.

5.3. Peculiarities of the photoinitiated polymerization in a layer with an illumination gradient

The next classical kinetic equation for the initial stage of the process is typically used for quantitative description of the photoinitiated polymerization, namely [24]

$$-\frac{d[\mathrm{M}]}{dt}=\frac{k_{\mathrm{p}}}{k_{\mathrm{t}}^{1/2}}[\mathrm{M}]\left\{\gamma_{\mathrm{i}}I_{0}(1-e^{-\varepsilon[\mathrm{In}]\ell})\right\}^{1/2}, \tag{5.19}$$

in which γ_i is the quantum yield of a photoinitiation; I_0 is the intensity of the UV-illumination on the surface of the photocomposition; ε is the extinction coefficient; [In] is concentration of the photoinitiator; ℓ is the thickness of the photocomposition layer; k_{p} and k_{t} are rate constants of the chain propagation and bimolecular chain termination respectively, which are changed *via* the process proceeding.

Equation (5.19) has been derived on the assumption of bimolecular chain termination. Furthermore, this equation does not take into account the availability of concentration gradients of photoinitiator, monomer and stage of the conversion on the layer of the photoinitiator, which are caused by the illumination gradient. That is why equation (5.19) will be just for the description of a process proceeding under ideal mixing.

On the contrary, even the starting low photopolymerization compositions are characterized as a rule by high viscosity, increasing by some orders of magnitude *via* the polymerization process. Taking into account also a short time of the photo-curing we can assume, that the diffusion mass-transfer has a weak influence on the gradients of a photoinitiator and a monomer concentrations caused by the illumination gradient on the layer. By neglecting the equilibrant action of the mass-transfer, the kinetics of the photoinitiator decomposition with one-sided illumination of a plate layer to the normal of its surface is described by a system of non-linear differential equations as partial derivatives [25]:

$$\partial c/\partial t=-\gamma\varepsilon cJ \tag{5.20}$$

$$\partial J/\partial x=-\varepsilon cJ \tag{5.21}$$

where $I=I\ (x,\ t)$, $c=c\ (x,\ t)$ are light intensity and photoinitiator concentration in the layer with the coordinate x depending on the surface illuminated at time t; ε is the molar photoinitiator extinction coefficient; γ is quantum yield.

For solving of the system of equations (5.20) and (5.21) let us use the dimensionless values [26, 27]:

$$z_1 = c / c_0, \qquad z_2 = J / J_0$$

$$y = \varepsilon c_0 x, \qquad \tau = \gamma \varepsilon J_0 t \tag{5.22}$$

where $c_0=c\ (x,\ t=0)$ and $I_0=I\ (x=0.\ t)$ are starting concentration of the photoinitiator for all layers at $t=0$ and the intensity of the UV-illumination falling on the surface of the photocomposition.

With new variables (5.22) the system of equations (5.20) and (5.21) becomes:

$$\partial z_1 / \partial \tau = -z_1 z_2 \tag{5.23}$$

$$\partial z_2 / \partial y = -z_1 z_2 \tag{5.24}$$

with the limiting conditions:

$$z_1(y, \tau = 0) = 1, \qquad z_1(y = 0, \tau) = e^{-\tau}$$

$$z_2(y = 0, \tau) = 1,\ z_2(y, \tau = 0) = e^{-y} \tag{5.25}$$

Taking into account the limiting condition for z_1 at $\tau=0$ in equation (5.25), from equation (5.23) we will obtain the following equation:

$$\ln z_1 = -\int_0^\tau z_2 \mathrm{d}\tau \tag{5.26}$$

Using

$$\omega = \int_0^\tau z_2 \mathrm{d}\tau \tag{5.27}$$

we have

$$z_1 = e^{-\omega} \tag{5.28}$$

$$\partial\omega / \partial\tau = z_2 \tag{5.29}$$

at this $\omega(y, \tau=0)=0$.

With the following differentiation of equation (5.27) on y we obtain:

$$\partial\omega / \partial y = \int_0^{\tau} (\partial z_2 / \partial y) \mathrm{d}\tau \tag{5.30}$$

Taking into account equations (5.24), (5.28) and (5.29) we find that

$$\partial\omega / \partial y = -\int_0^{\tau} \exp\{-\omega\} \mathrm{d}\omega \tag{5.31}$$

It follows from this at the limited condition ω $(\tau=0)=0$. that

$$\partial\omega / \partial y = e^{-\omega} - 1 \tag{5.32}$$

We find, after integration of equation (5.32) with the limited condition ω $(y=0)=\tau$, that

$$y = -\ln\left(\frac{e^{\omega} - 1}{e^{\tau} - 1}\right) \tag{5.33}$$

By solving equation (5.33) relative to $z_1=e^{-\omega}$ we will obtain

$$z_1 = [1 + e^{-y}(e^{\tau} - 1)]^{-1} \tag{5.34}$$

In accordance with equations (5.24) and (5.34)

$$\partial \ln z_2 / \partial y = -[1 + e^{-y}(e^{\tau} - 1)]^{-1} \tag{5.35}$$

It follows from this, that

$$z_2 = e^{\tau - y}[1 + e^{-y}(e^{\tau} - 1)]^{-1} \tag{5.36}$$

Thus, the photoinitiator concentration and the light intensity are functions on the coordinate of layer (*via* $y=\varepsilon\, c_0\, x$) and time (*via* $\tau=\gamma\, \varepsilon\, I_0\, t$):

$$c = c_0[1+e^{-y}(e^{\tau}-1)]^{-1} \tag{5.37}$$

$$J = J_0 e^{\tau-y}[1+e^{-y}(e^{\tau}-1)]^{-1} \tag{5.38}$$

The photoinitiator decomposition rate as a function of the same parameters is determined *via* the following expression:

$$-\partial c/\partial t = \gamma\varepsilon c_0 J_0 e^{\tau-y}[1+e^{-y}(e^{\tau}-1)]^{-2} \tag{5.39}$$

Graphical interpretation of the dependencies (5.37) and (5.38) for the layers with rather high optical density $y=\varepsilon\, c_0\, x$ in the Figures 5.9 and 5.10 shows, that in this case the photoinitiator decomposition process can be concentrated in the small part of the layer. During the photoinitiator consumption in this part of a layer, the decomposition process is moved inside the layer and is characterized by the moving wave view.

The rate $(\partial y/\partial\tau)_c$ of a photoinitiator concentrated front moving can be found from the ratio

$$(\partial c/\partial\tau)(\partial\tau/\partial y)(\partial y/\partial c) = -1 \tag{5.40}$$

Using equations (5.37)–(5.39) in equation (5.40) we obtain

$$\partial y/\partial\tau)_c = e^{\tau}/(e^{\tau}-1) \tag{5.41}$$

As we can see, all points of the concentrated profile are moved with the same rate (wave moving), which quickly decreases exponentially with τ increasing until the stationary value $(\partial y/\partial\tau)_c=1$.

Analogously, from the ratio

$$(\partial J/\partial\tau)(\partial\tau/\partial y)(\partial y/\partial J) = -1 \tag{5.42}$$

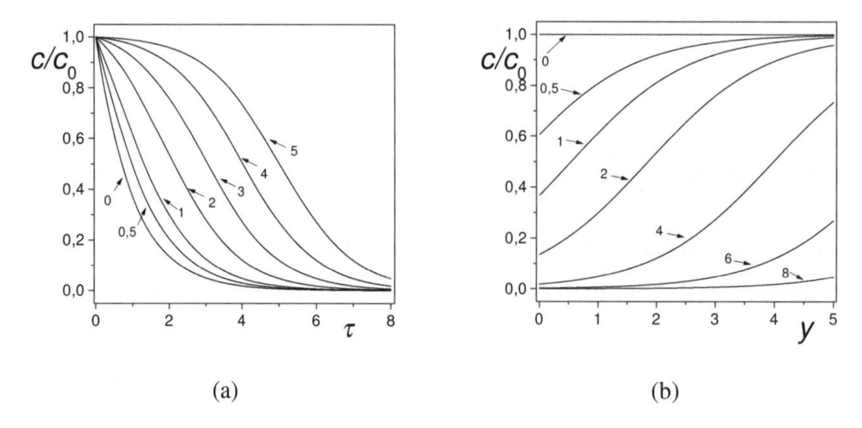

Figure 5.9. The change of relative concentration of photoinitiator with time (numbers near the curves are the values of y) (a) and with layer coordinate (numbers near the curves are the values of τ) (b).

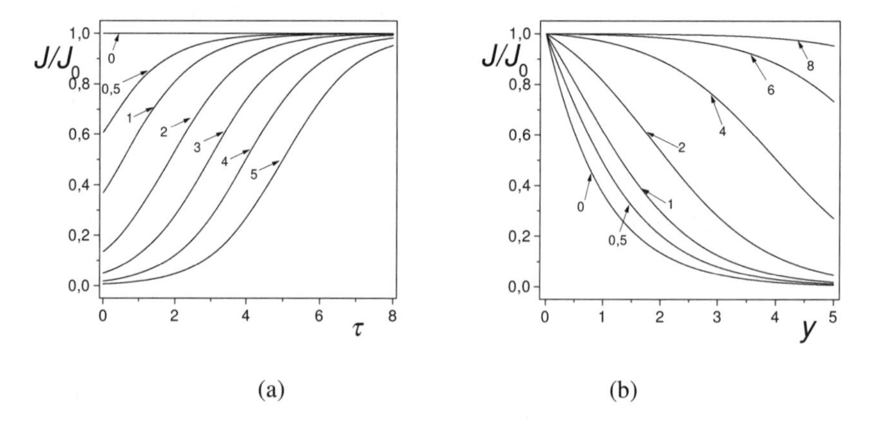

Figure 5.10. The change of relative light intensity with time (numbers near the curves are the values of y) (a) and via the layer coordinate (numbers near the curves are the values of τ) (b).

Taking into account equations (5.37) and (5.38) we find the expression for the rate $(\partial y/\partial\tau)_y$ of moving the illumination front on a layer

$$(\partial y/\partial \tau)_J = e^y/(e^y - 1) \tag{5.43}$$

Accordingly to equation (5.43) the rate of moving an illumination front is constant for all values I and is quickly decreases exponentially with increasing y until the stationary value equal to 1 is reached.

Practically the same can be said about the initiation rate represented by equation (5.39), which for the graphical illustration can be written more conveniently as

$$v_{\text{in}} / v_{\text{in,max}} = e^{\tau - y}[1 + e^{-y}(e^{\tau} - 1)]^{-2} \tag{5.44}$$

Here: $v_{\text{in}}=-\partial c/\partial t$, $v_{\text{in.max}}=\gamma\varepsilon c_0 I_0$ is the maximum initiation rate.

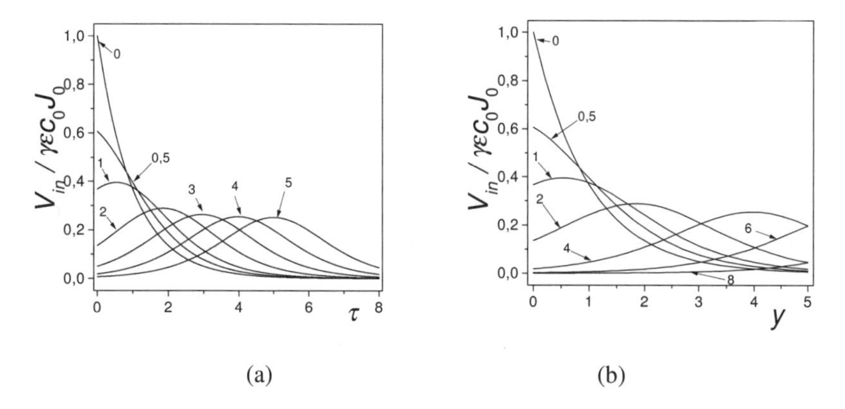

Figure 5.11. The dependence of the relative initiation rate on the characteristic time τ (a) and thickness of layer y (b) of the photocomposition. The calculation has been done in accordance with equation (5.44).

As we can see in Figure 5.11a,b when τ=1 a rapid decrease of the initiation rate takes place in a layer with thickness y=1 and after that the maximum of the ratio $v_{\text{in.}}/v_{\text{in,max}}$ is frontally displaced inside the photocomposition layer.

Let us analyze the kinetics of layer photoinitiated polymerization in its simplest classic variant, for which the process rate expressed in the units of the conversion is described by the following equation [28]:

$$\partial p / \partial t = k(1 - p)v_{\text{i}}^{1/2} \tag{5.45}$$

Here, $P=P\,(x, t)$ and $v_{\text{i}}=v_{\text{i}}\,(x, t)$ are conversion and initiation rate in the layer x+dx at time t, respectively; k is effective polymerization constant rate. Usually in the expression of the initiation rate not only the initiator decomposition rate is included, but also the initiation coefficient. In the case presented here it cannot be included, since the quantum yield γ plays approximately the same role. That is why we assume $v_i=-\partial c/\partial t$ according to equation (5.39). By substituting equation (5.39) in equation (5.44) we have

$$\partial p / \partial t = k(1-p)(\gamma \varepsilon c_0 J_0)^{1/2} e^{(\tau-y)/2}[1+e^{-y}(e^{\tau}-1)]^{-1} \quad (5.46)$$

This expression represents the differential rate of the polymerization in the layer y, $y + dy$ at time τ. A graphical interpretation of this expression in the form of the rate of a process $w=-\partial \ln(1-P)/\partial t$ relating to the maximal possible one $w_{max}=k(\gamma \varepsilon c_0 J_0)^{1/2}$ is as follows:

$$w / w_{max} = e^{(\tau-y)/2}[1+e^{-y}(e^{\tau}-1)]^{-1} \quad (5.47)$$

As we can see in Figure 5.11, the ratio w/w_{max} is quickly decreased in a layer of characteristic length $y=1$ for characteristic time $\tau=1$, as result of which the polymerization rate of a process is increased inside the layer; at this, the maximum of the ratio w/w_{max} is constant in time, frontally displacing on the depth of a layer. Exactly this variant describes the so-called step-by-step photo-curing. A front of the polymerization first is concentrated in a small part ($y \approx 1$) of a layer y of the illuminated surface, practically completely absorbing the light: as the photoinitiator is consumed, the front gradually moves inside the layer. At higher concentrations of photoinitiator, the layer in which the polymerization process is concentrated becomes smaller and the rate of front polymerization moving is lower. As a result, as it will be shown below, the observed order of the integral, averaged over all thickness of a photocomposition layer polymerization rate, can be negative.

Further, by integration of equation (5.46) on time at y=constant we obtain for the conversion as a function of the layer coordinate and time, that

$$p = 1 - \exp\left\{ -\frac{2k(c_0 / \gamma \varepsilon J_0)^{1/2}}{(1-e^{-y})^{1/2}} \arctan \frac{(1-e^{-y})^{1/2}(e^{\tau/2}-1)}{e^{y/2}+e^{-y/2}(e^{\tau/2}-1)} \right\} \quad (5.48)$$

In accordance with equation (5.48), a limited (at $\tau \to \infty$) conversion of a monomer *via* the photoinitiated polymerization process formally does not equal 1 and is determined by the expression:

$$P_\infty = 1 - \exp\left\{ -\frac{2A}{(1-e^{-y})^{1/2}} \arctan \frac{(1-e^{-y})^{1/2}}{e^{y/2}} \right\} \quad (5.49)$$

where $A=k(c_0/\gamma\varepsilon J_0)^{1/2}$, P_∞ $(y=0)=1-\exp\{-2A\}$ and P_∞ $(y >> 1)=1-\exp\{-\pi A\}$, respectively.

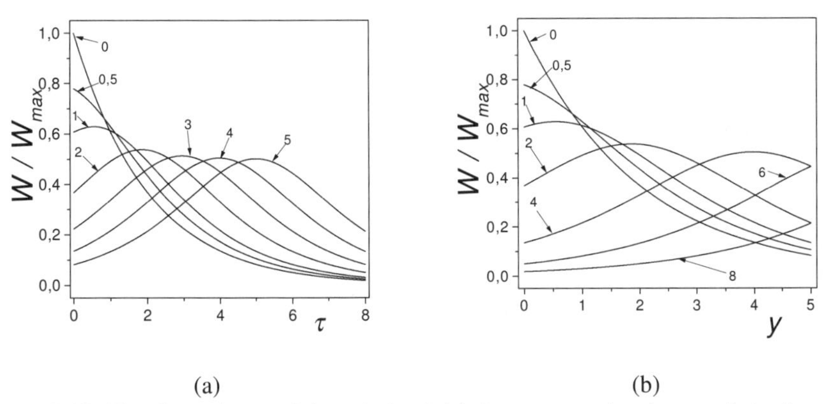

(a) (b)

Figure 5.12. The dependence of the relative initiating rate on the characteristic time τ (a) and thickness of the photocomposition layer y (b). The calculation has been done in accordance with equation (5.47).

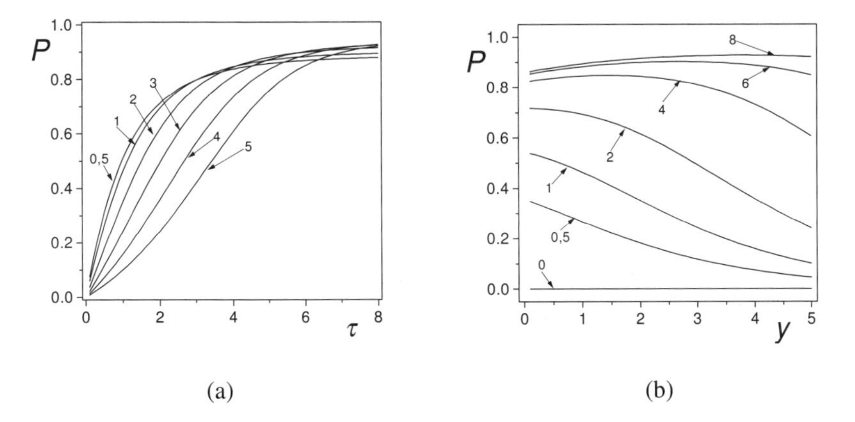

(a) (b)

Figure 5.13. The conversion of a monomer by photoinitiated polymerization as a function of the characteristic time τ (a) and thickness of layer y (b). The calculation has been done in accordance with equation (5.49) at A=1.

Thus, limited conversion inside a layer is somewhat higher than near the illuminated surface. This is connected with different lengths of the polymerization chains on the layer and depends on the value A (see Figures 5.12 and 5.13). At $A >> 1$ $P_\infty \approx 1$ for all layers. Usually the same is observed on a practice. At this, however, the number-averaged molecular mass of the forming polymer proportional to the length of a chain will increase from the illuminated surface to the inside of the layer, thus forming a gradient of the properties of a polymer on the layer. In some cases it can be obtained and be desirable.

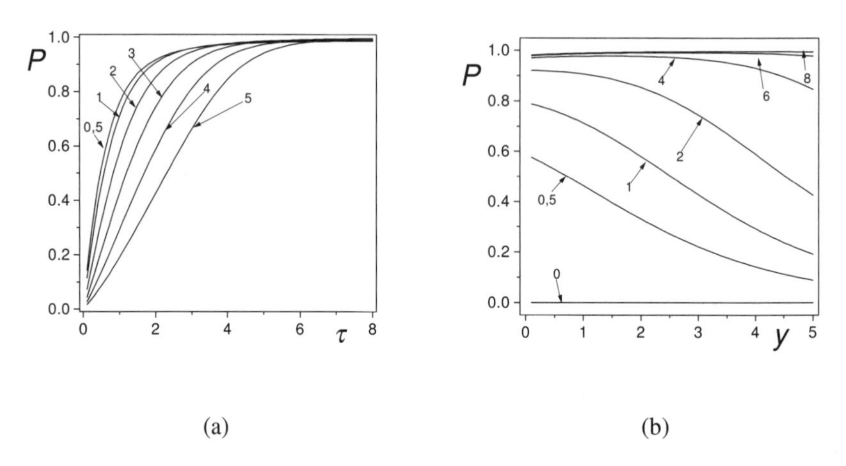

Figure 5.14. The conversion of a monomer at photoinitiated polymerization as a function of the characteristic time τ (a) and thickness of a layer y (b). The calculation has been done in accordance with the equation (5.49) a*t* A=2.

However, in practice only those methods are used which do not to allow the measurement of the differential characteristics of the process $P\ (x,\ t)$ or $\partial P\ (x,\ t)/\partial t$. Differential characteristics of the photoinitiated process, i.e., $P\ (y,\ \tau)$, $w\ (y,\ \tau)$, $v_{\text{in.}}\ (y,\ \tau)$, etc., are very interesting from the theoretical point of view, but their experimental proof is complicated. As a rule, the photoinitiated polymerization process in the photocomposition layer is studied taking into account the general contraction of a layer which represents an integral characteristic of the process that is averaged over all layers of the photocomposition. In practice, we do not measure the values $P\ (x,\ t)$ or $\mathrm{d}P\ (x,\ t)/\mathrm{d}t$ but their integral analogs, which can be obtained *via* the following transformations

$$P(t) = \frac{1}{\ell}\int_0^\ell p(x,t)\mathrm{d}x \qquad (5.50)$$

$$dP(t)/\mathrm{d}t = \frac{1}{\ell}\int_0^{\ell}\left(\partial p(x,t)/\partial t\right)\mathrm{d}t \tag{5.51}$$

Here l is the thickness of the photocomposition layer.

Thus, in order to compare the theoretical dependencies of the (5.46) and (5.48) type with experimental data, it is necessary to additionally transform them into integral forms (5.50) and (5.51), respectively.

We could obtain only the expression for the integral polymerization rate; putting equation (5.46) into equation (5.50) we found

$$-\frac{\mathrm{d}\ln(1-P)}{\mathrm{d}t} = \frac{2}{\ell}(\gamma\varepsilon c_0 J_0)^{1/2}\frac{e^{\tau}}{(e^{\tau}-1)^{1/2}}\arctan\left[\frac{e^{y_0/2}-1}{(e^{\tau}-1)^{1/2}}\right] \tag{5.52}$$

Here $y_0=\varepsilon c_0 l$.

Formally, integrating equation (5.52) over time, we found an expression for the integral conversion as function of time. However, the same analytical expressions cannot always be regarded in this simple view. That is why in the case of complicated kinetics we should make simplifications allowing us to have less restricted kinetic equations, but comparable with the experimental data. For simplification it is evident that we need: an approximation of the initial stage of the process in accordance with the condition $\tau \Rightarrow 0$ and an approximation of the endless thin layer according to condition $l \rightarrow 0$. These and other approximations will be discussed in the analysis of concrete experimental data.

An approximation of the initial stage of a process

The initial stage of a process [29] satisfies the condition $\tau \rightarrow 0$ or $\tau < \delta$, where $\tau=\gamma\varepsilon J_0 t$ is the characteristic photoinitiator decomposition time and δ is a small value. That is why we can determine the initial stage of the process as an interval of the characteristic time $\Delta\tau$ in which we can neglect the photoinitiator concentration Δc in comparison with its starting value c_0: $\Delta c << c_0$. For the long-chain polymerization process this time interval $\Delta\tau$ can correspond to final conversion changes.

The main differential characteristics of a process in the initial stage in a layer of end thickness ℓ we can obtain from general equations, for example, equations (5.46) and (5.48), directing $\tau \rightarrow 0$. More conveniently, however, it is to consider the starting equations (5.20), (5.21) and (5.44), assigning $c=c_0$=constant. We have

$$\partial J / \partial x = -\varepsilon c_0 J \tag{5.53}$$

$$\partial c / \partial t = -\gamma\varepsilon c_0 J \tag{5.54}$$

$$\partial p / \partial t = k(1-p)v_i^{1/2} \tag{5.55}$$

where *I*, *c* and *P* are functions only of the coordinate of a layer. It follows from equations (5.53)–(5.55) that:

$$J = J_0 \exp\{-\varepsilon c_0 x\} \tag{5.56}$$

$$v_i = -\partial c / \partial t = \gamma\varepsilon c_0 J_0 \exp\{-\varepsilon c_0 x\} \tag{5.57}$$

$$\partial p / \partial t = k(1-p)(\gamma\varepsilon c_0 J_0)^{1/2} \exp\{-\varepsilon c_0 x / 2\} \tag{5.58}$$

These expressions are illustrated by the curves in Figures 5.10b, 5.11b and 5.12b for which τ=0.

Using equations (5.57) and (5.58) after integral transformations on equation (5.51) we obtain the equation for the integral initiation rate $V_{int.}$ and integral rate $W_{int.p.}$ of the polymerization:

$$V_{int} = \frac{\gamma J_0}{\ell}(1-\exp\{-\varepsilon c_0 \ell\}) \tag{5.59}$$

$$W_{int.p} = -\frac{d\ln(1-p)}{dt} = \frac{2k}{\ell}\left(\frac{\gamma J_0}{\varepsilon c_0}\right)^{1/2}(1-\exp\{-\varepsilon c_0 \ell / 2\}) \tag{5.60}$$

It follows from equation (5.59) that for the optically transparent layers ($\varepsilon c_0 \ell << 1$) the maximal integral initiation rate is $V_{max}=\gamma\ell J_0 c_0$ and does not depend on ℓ. For dense optical layers ($\varepsilon c_0 \ell >> 1$) $V_{max}=\gamma J_0/\ell$. This circumstance is essentially reflected in the dependence of integral rate of the polymerization on the initiator concentration. For the optically transparent layers ($\varepsilon c_0 \ell << 1$), it follows from equation (5.60) that

$$-\frac{\mathrm{d}\ln(1-p)}{\mathrm{d}t}=k\left(\gamma J_0 \varepsilon c_0\right)^{1/2} \qquad (5.61)$$

A typical order on the photoinitiator is equal to ½. For dense optical layers ($\varepsilon c_0 \ell >> 1$) we have

$$-\frac{\mathrm{d}\ln(1-p)}{\mathrm{d}t}=\frac{2k}{\ell}\left(\frac{\gamma J_0}{\varepsilon c_0}\right)^{1/2} \qquad (5.62)$$

Thus, the integral rate of the polymerization varies inversely with layer thickness and its formal order on the initiator is equal to ½.

Equation (5.60) is easy integrated and this allows us to obtain the expression for integral conversion as a function of time:

$$P=1-\exp\left\{-\frac{2k}{\ell}\left(\frac{\gamma J_0}{\varepsilon c_0}\right)^{1/2}\left(1-\exp\{-\varepsilon c_0 \ell/2\}\right)\, t\right\} \qquad (5.63)$$

An approximation of the endless thin layer

The endless thin layer is determined starting from the condition that $y=\varepsilon c_0 l << 1$, that is, as a endless optically transparent layer. That is why under the condition of the endless layer ℓ can correspond to it at limited low photoinitiator concentration $c_0 \to 0$ or at the final value c_0, a limited low thickness of layer $l \to 0$.

In this case we can neglect by the illumination gradient ($I=I_0$) and differential rates of the initiation and the process proceeding are the same as the integral ones. That is why [30]

$$\frac{\mathrm{d}c}{\mathrm{d}t}=-\gamma\varepsilon c J_0 \qquad (5.64)$$

$$\partial p/\partial t=k(1-p)v_i^{1/2} \qquad (5.65)$$

It follows from equation (5.64), that

$$c = c_0 \exp\{-\gamma\varepsilon J_0 t\}, \tag{5.66}$$

representing only the function of time.

That is why

$$\frac{dP}{dt} = k(1-P)(\gamma\varepsilon c_0 J_0)^{1/2} \exp\{-\gamma\varepsilon J_0 t\} \tag{5.67}$$

Here the dependence of polymerization rate on time is determined not only by monomer consumption, that is, by conversion P increasing, but also by the decrease with time of the photoinitiator concentration.

By integrating equation (5.67) we obtain

$$P = 1 - \exp\left\{-2k(c_0 / \gamma\varepsilon J_0)^{1/2}\left(1 - \exp\{-\gamma\varepsilon J_0 t / 2\}\right)\right\} \tag{5.68}$$

As we can see, expressions (5.63) and (5.68) are essentially different.

thus, we find that the kinetics of a photoinitiated polymerization in a photocomposition layer is characterized by characteristics, which can essentially distort, for example, the observed orders on photoinitiator and illumination intensity, the form of the kinetic curve depending on the thickness of the layer, etc. That is why taking into account these characteristics is necessary for the analysis of photoinitiated polymerization.

5.4. Technique of the experiment

Photoinitiated polymerization has been carried out using the laser interferometry method (one of the dilatometer methods [24]), the principal and geometric schemes of which are presented in Figures 5.15 and 5.16, respectively.

The composition to be studied (**3**) is located on the thermostatting lining (**1**) under a 0.15-mm-thick covering glass (**4**). A monochromatic beam with the wavelength λ=633,1 nm, emitted by a laser (**5**), is falling on the surface of quartz glass.

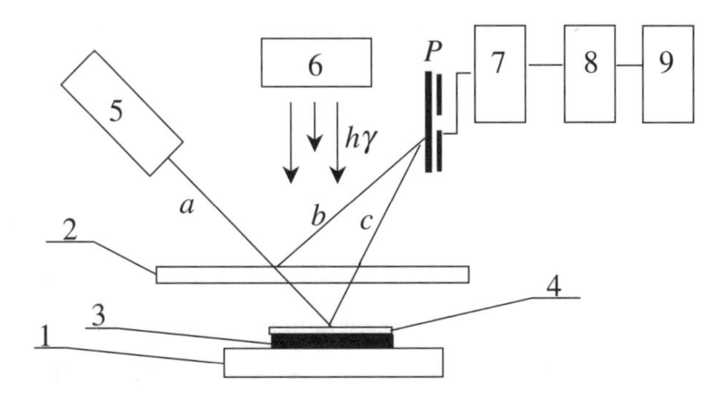

Figure 5.15. Principal scheme of a set-up for laser interferometry. Here, **1** is the thermostating metallic lining; **2** is a quartz glass; **3** is the monomeric composition; **4** is a covering medical glass; **5** is the laser LGN-207A; **6** is a source of the UV-illumination mercury-quartz lamp DRT-400; **7** is a photo-receiver; **8** is an amplifier; **9** is the automatic potentiometer KSP-4-014; **a** is the falling laser beam; **b** and **c** are reflected from the surface of a quartz glass and the monomeric polymerizing composition laser beams

The angle of slope β (see Figure 5.16) between the plains of the composition (**3**) and the quartz glass (**2**) is controlled by screws and, in this case, reflected from the surface of quartz glass and surface of the monomeric photocomposition laser beams **b** and **c** form the interferential picture in the plain P (see Figure 5.15). The interferential picture thus formed is received by the photo-receiver (**7**). The signal from the photo-receiver is intensified by means of the amplifier (**8**) and after that is controlled by the automatic potentiometer (**9**).

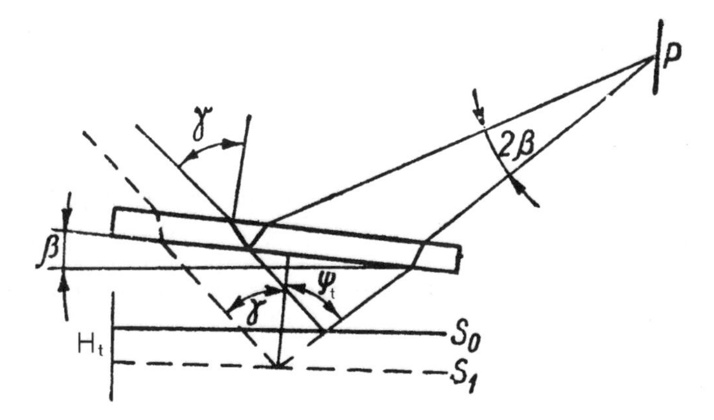

Figure 5.16. Geometric scheme of plant for laser interferometry

Contraction (H_t) of the photocomposition layer at time t was determined according to the formula

$$H_t=\lambda\,\varphi_t/(4\,\pi\cos\gamma) \qquad (5.69)$$

where λ=633,1 nm is the wavelength of the laser beam; φ_t is the difference between the phases of beams reflected from the auxiliary quartz glass and from the surface of photocomposition; $\gamma=\pi/4$ is the angle of a beam falling on a surface of photocomposition.

A change of the interferential picture *via* contraction of the polymerization composition was registered step-by-step over time by the photo-receiver and was written on the diagram band with the use of potentiometer as interferogram. A typical experimental spectrum of the photoinitiated polymerization is presented in Figure 5.17.

The light intensity I under the interference of the beams ***b*** and ***c*** is described by the following ratio:

$$I=I_1+I_2+(I_1\,I_2)^{1/2}\cos\varphi_t \qquad (5.70)$$

Here I is the general intensity in the fixed point of the interferential plain; I_1 is the intensity of laser beam ***b***; I_2 is the intensity of laser beam ***c***.

The difference $\varphi_t=n_t\,\pi$ between the phases corresponds to maxima of the intensity on the interferograms; n_t is the peak number of at time t.

Contraction of the layer in the moment of time t for the respective maximum of a peak was determined by the ratio

$$H_t=\lambda\,n_t/4\cos(\pi/4)=2.24\times10^{-7}\,n_t \qquad (5.71)$$

The value 2.24×10^{-7} m determines also the error of measurement. Limited achieved contraction H_0 was calculated starting from the general number of peaks n_0 on the interferogram $H_0=2.24\times10^{-7}\,n_0$.

The relative integral degree of the monomer transformation (or conversion P) was calculated with the use of the following ratio

$$P=H_t/H_0 \qquad (5.72)$$

On the basis of the H_0 value the thickness of the photocomposition layer ℓ was calculated as follows

$$\ell = H_0/\alpha \quad (5.73)$$

where α is the contraction coefficient determined according to the formula

$$\alpha = (\rho_p - \rho_M)/\rho_p \quad (5.74)$$

Here ρ_p and ρ_M are densities of the polymeric and monomeric phase, respectively.

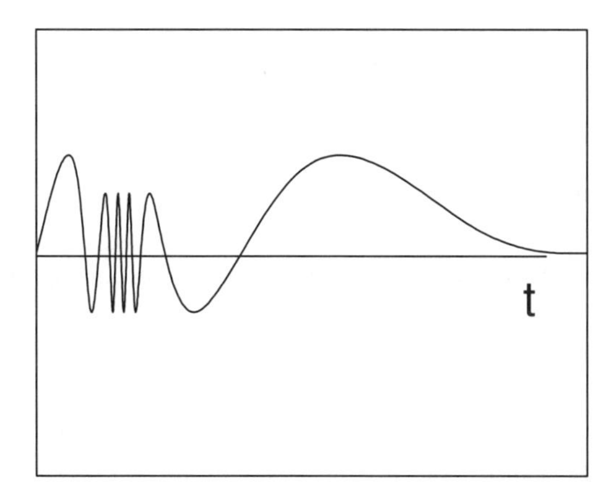

Figure 5.17. Typical interferogram of the photopolymerization.

Photopolymerization of the monomeric composition has been carried out under the covering medical glass (0.10–0.15 mm thick) with the aim of preventing inhibition by oxygen in the surface layer of the photocomposition and in its local (at the point of a laser beam reflection) deformation caused by tensions at the phase boundary (solid and liquid) division at the frontal character of a process.

Spectroscopic investigations showed that such covering medical glass does not significantly distort the integral spectrum of an UV-illumination using a DRT–400 mercury-quartz lamp for the photoinitiator decomposition (see Figure 5.18).

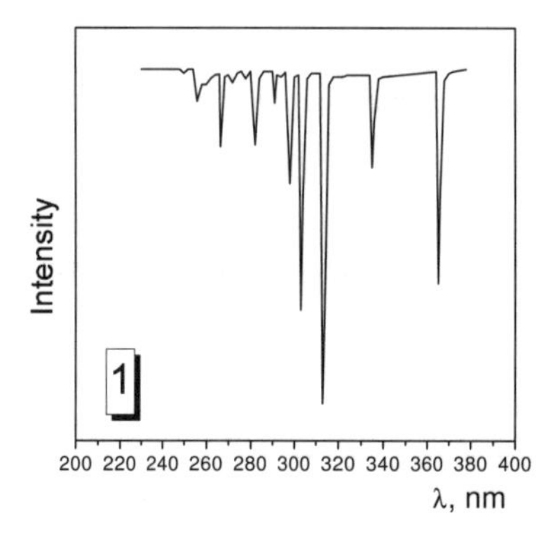

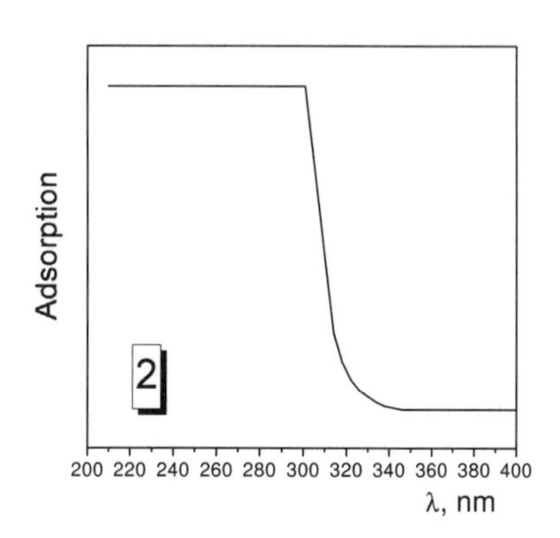

Figure 5.18. Spectrum of the DRT-400 lamp illumination (1) and spectrum of the covering medical glass adsorption (2).

5.5. Kinetic model of the photoinitiated polymerization and its comparison with experimental data

The proposed microheterogeneous model (see equation 5.1) for thermal polymerization needs modification. It is necessary to take into account the presence of a light gradient in the layer of the photopolymerization composition, gradients in the concentrations of the photoinitiator of the oligomer and the depth of the photopolymerization. That is why in this variant of the photoinitiated polymerization equation (5.1) should be written as a partial derivative [8]:

$$\partial p / \partial t = k_1(1-p)v_{\mathrm{i}}^{1/2} + k_2 p(1-p)v_{\mathrm{i}} \tag{5.75}$$

Here, $p\ (x,\ t)=\Gamma\ (x,\ t)/\Gamma_0$ is relative depth of polymerization in the layer x, $x+\mathrm{d}x$ at time t determined by the ratio of current depth of polymerization $\Gamma\ (x,\ t)$ and the limited achieved one $\Gamma_0\ (t \rightarrow \infty)$.

However, as a rule, the experimentally measured parameters of the photoinitiated polymerization are not the differential rate of the process $\partial p/\partial t$, or relative depth of the polymerization P, but their integral, that is, the average of the layer analogs. They are determined by the expression:

$$W = \mathrm{d}P / \mathrm{d}t = \ell^{-1} \int_0^{\ell} (\partial P / \partial t) \mathrm{d}x \qquad P = \ell^{-1} \int_0^{\ell} p \mathrm{d}x \tag{5.76}$$

where ℓ is thickness of the photopolymerization composition.

Using equations (5.39) and (5.43) in the integral conversion (see equation 5.76), and taking into account the theorem about the mean, the expression for the polyfunctional (meth)acrylates will be [8–10]

$$W = \mathrm{d}P / \mathrm{d}t = k_1 (1 - P) \mathrm{Int}(1) + k_2 P (1 - P) \mathrm{Int}(2) \tag{5.77}$$

$$\mathrm{Int}(1) = 2\ell^{-1} \left(\frac{\gamma I_0}{\varepsilon c_0} \right)^{1/2} \left(\frac{e^{\tau}}{e^{\tau} - 1} \right)^{1/2} \arctan \frac{(e^{\tau} - 1)^{1/2} (1 - e^{-y_0/2})}{1 + e^{-y_0/2} (e^{\tau} - 1)} \tag{5.78}$$

$$\mathrm{Int}(2) = \gamma I_0 \ell^{-1} \frac{1 - e^{-y_0}}{1 + e^{-y_0} (e^{\tau} - 1)} \tag{5.79}$$

where $y_0 = \varepsilon c_0 \ell$ is initial optical density of the layer.

Indirect comparison of experimental data with general kinetic model (see equations (5.77)–(5.79)) is complicated. Therefore, we consider two approximations. the first is based on the assumption that $c = c_0$. v_i=constant, which can be considered as a model of the initial stage of the process. Using condition $\tau << 1$ in equations (5.77)–(5.79) we obtain:

$$\mathrm{d}P / \mathrm{d}t = \ell^{-1} [2k_1 (\gamma I_0 / \varepsilon c_0)^{1/2} (1 - e^{-y_0/2})(1 - P) + k_2 \gamma I_0 (1 - e^{-y_0}) P (1 - P)] \tag{5.80}$$

Analyzing equation (5.80) for the extremes of the function, we find that the relation should determine the ordinate P_0 of the point of the inflection of the kinetic curve

$$P_0 = (A - 1) / 2A \tag{5.81}$$

where

$$A = \frac{k_2}{2k_1} (\gamma \varepsilon c_0 I_0)^{1/2} \frac{1 - e^{-y_0}}{1 - e^{-y_0/2}} \tag{5.82}$$

According to the experimental data $P_0 \approx$ ½, and using equation (5.81) it follows that $A >> 1$. This leads to the expression for the maximum rate of the photopolymerization $W_0 = (\mathrm{d}P/\mathrm{d}t)_0$, which corresponds to the point of the inflection of the kinetic curve

$$W_0 = (k_2\gamma I_0)/4\ell)\cdot(1-e^{-y_0}) \tag{5.83}$$

As it follows from equation (5.83), at given concentration of the photoinitiator c_0 and small thickness of the layer of the composition ($y_0=\varepsilon c_0\ell << 1$), the maximum rate of the photopolymerization must be first-order on initiator: $W_0 \approx k_2\ \gamma\ \varepsilon\ c_0\ I_0/4$; however, at large thickness of the layer ($y_0=\varepsilon c_0\ell >> 1$) it must be zero-order: $W_0 \approx k_2\ \gamma I_0/4\ell$.

Experimental data shown in Figure 5.19 confirm this dependence: at $\ell \to 0$ the values W_0 depend on the concentration of the photoinitiator, but at $\ell \to \infty$ they approach, showing the independence of W_0 on c_0. At the same time, the maximum rate of the process for all ℓ is a linear function of UV-illumination power falling on the surface of the composition (see Figure 5.20).

The calculated dependences of W_0 from ℓ are presented in Figure 5.19 as solid lines. These data were obtained by entering the following set of parameters in equation (3.38): $k_2\gamma I_0=2.2\times10^{-5}$ s (at $E_0=40.6$ J m^{-2} s^{-1}) and $\varepsilon=87$ m^3 mol^{-1} m^{-1}.

As it is impossible to indicate the wavelength of the UV-radiation, which leads to decomposition of the photoinitiator, we arbitrarily selected an average wavelength between 280 and 300 nm, which led to $I_0=2.5\times10^{-6}E_0$ mol×quantum m^{-2} s^{-1}. Using the values $k_2\gamma I_0=2.2\times10^{-5}$ s at $E_0=40.6$ J m^{-2} s^{-1}, this allowed us to estimate $k_2\ \gamma=0.22$ m^3 mol^{-1}. Equation (5.80) at $A >> 1$ after integration has the form:

$$\ln[(1+AP)/(1-P)] = k_2\gamma I_0(1-e^{-y})t/\ell = 4W_0t \tag{5.84}$$

This offers the possibility of estimating the parameter A from experimental data by the relation

$$A = \exp(4W_0t_{1/2}) - 2 \tag{5.85}$$

where $t_{1/2}$ is time of attainment of the ordinate $P=1/2$ on the kinetic curve.

Since the errors in determining W_0 and $t_{1/2}$ are included in the exponent, the estimation of A using equation (5.85) has a significant experimental error.

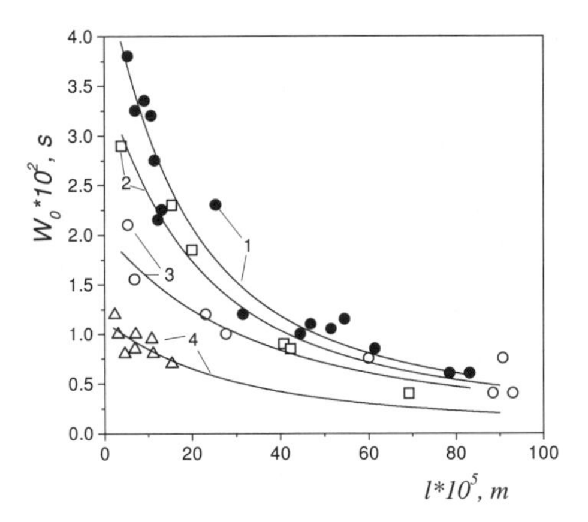

Figure 5.19. Experimental dependencies (dots) and dependencies calculated using equation (5.83) (lines) of the maximal rate of MDF-2 photopolymerization on the thickness of layer ℓ at different concentrations of photoinitiator c_0: 0.105 (1), 0.064 (2), 0.040 (3), 0.021 mol/l (4). E_0=40.6 W/m^2, ℓ=2×10^{-4} m.

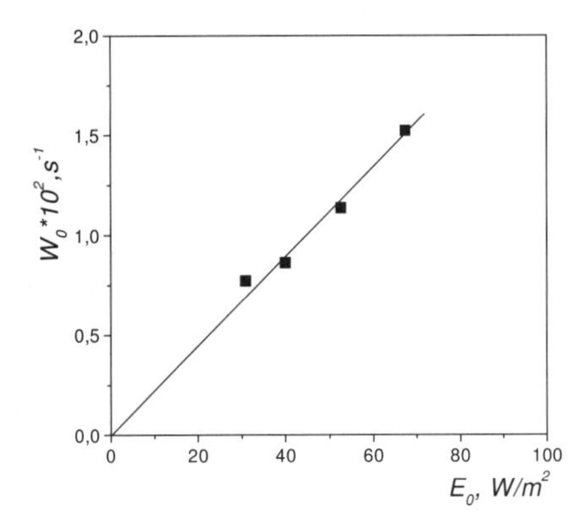

Figure 5.20. Dependence of maximal rate W_0 for MDPh-2 photopolymerization on UV-illumination power. c_0=0.021 mol/l, ℓ=1×10^{-4} m.

Figure 5.21 shows the dependence of A on ℓ, which agrees with expression (5.82), according to which the increase of the thickness of the layer parameter A must coincide with a decrease of value $A_0=k_2\,(\gamma\,\varepsilon\,c_0\,I_0)^{1/2}/k_1$ at $\ell=0$ to value $A=A_0/2$ by a factor 2.

The calculated dependence of A on ℓ (see Figure 5.21), was obtained at value $A_0=25$, that led to the estimation $k_1\gamma^{1/2}=8.4\times10^{-3}$ m^3(mol^{-1} s^{-1})$^{1/2}$. The calculated values of parameters ε, $k_1\gamma^{1/2}$ and $k_1\gamma$ allow us to interpret the experimental data in coordinates of equation (5.84). Parameter A is calculated using equation (5.82).

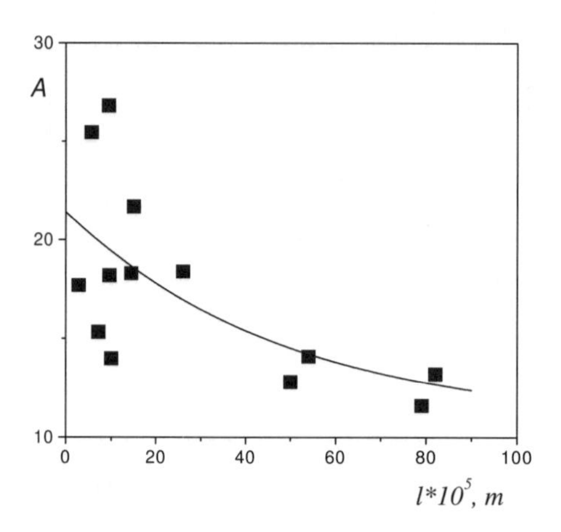

Figure 5.21. Experimental dependencies (dots) and dependencies calculated using equation (5.82) (lines) of parameter A on the thickness of layer ℓ for MDF-2 photopolymerization at c_0=0.021 mol/l and E_0=40.6 W/m^2.

As shown in Figure 5.22 the initial plots of the kinetic curves are well linearized in coordinates of equation (5.84). The values of the slope tangent of the straight-line plots are within the limits of the error measured for values of $4W_0$. The deviation of the experimental data from the straight-line relation in the coordinates of equation (5.84) begins at values P_k=0.7–0.9. Note that higher values of ℓ correspond to smaller values of P_k. This allows us to conclude that the observed deviation is caused by neglecting the denpendency of the initial stage of the photoinitiation rate on time in this model. The estimations of the quantum yield γ from experimental data were made using the second approximation, which corresponds to the model of the endless thin layer [30].

Using the condition $y_0=\varepsilon c_0 \ell << 1$ for the layer 0,0+dx accurate determination of the endless thin layer differential and integral characteristics of the process coincide. Therefore, the photopolymerization rate may be described by the initial equation (5.75), in which $v_i=\gamma\varepsilon c_0 I_0 \exp(-\gamma\varepsilon I_0 t)$ and is a function of time.

Thus, in the model of the endless thin layer

$$\mathrm{d}P/\mathrm{d}t = k_1(\gamma\varepsilon c_0 I_0)^{1/2}(1-P)e^{-\tau/2} + k_2\gamma\varepsilon c_0 I_0 P(1-P)e^{-\tau} \qquad (5.86)$$

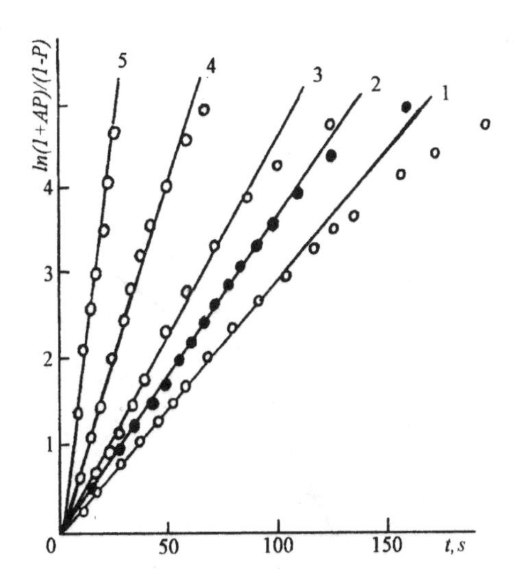

Figure 5.22. Interpretation of experimental kinetic curves of MDF-2 photopolymerization in coordinates of equation (5.84) at different concentrations of photoinitiator c_0 and layer thickness ℓ: c_0=0.105 mol/l, $\ell=3\times10^{-4}$ m (1); c_0=0.105 mol/l, $\ell=1.5\times10^{-4}$ m (2); c_0=0.105 mole/l, $\ell=5.5\times10^{-4}$ m (3); c_0=0.021 mol/l, $\ell=1.2\times10^{-4}$ m (4); c_0=0.105 mol/l, $\ell=8.6\times10^{-4}$ m (5). E_0=40.6 W/m^2.

The analysis of equation (5.86) on maximum normalized rate of photopolymerization which in terms of the accepted value $dP/(1-P)dt = -d\ln(1-P)/dt$ establishes the relationship between the ordinate P^0 and rate $W^0=(dP/dt)$, which corresponds to the point of inflection of the kinetic curve of the shape $ln(1-P)=f(t)$ and the maximum polymerization rate $(-d(\ln(1-P))/dt)_{max}$

$$P^0 = W^0 / \gamma\varepsilon I_0 - 1/2A_0 \qquad (5.87)$$

Figure 5.23 shows the interpretation of experimental data in the form of the kinetic curves $\ln(1-P)=f(t)$, which satisfy the approximation of endless thin layer $y_0=\varepsilon\, c_0 l \ll 1$. From the ordinate of inflection and slope tangents the values $\ln(1-P^0)$ and $(dP/(1-P)dt)_{max}$ were found. This allowed us to calculate P^0 and W^0. These and the calculated value of A_0 allowed us to calculate $\gamma\varepsilon I_0$.

The average of two calculations (see Figure 5.23) gave the value $\gamma\, \varepsilon\, I_0 \approx 8*10^{-3}$ at $I_0=10^{-4}$ molxquantum m^{-2} s^{-1}, thus γ=0.9 is found. With the already known values $k_1\gamma^{1/2}$

and $k_2\gamma$ we can estimate the numerical values of the two principal parameters of the initial kinetic equation: k_1=8.8×10^{-3} (m^3 mol^{-1} s^{-1})$^{1/2}$ and k_2=0.24 m^3 mol^{-1}.

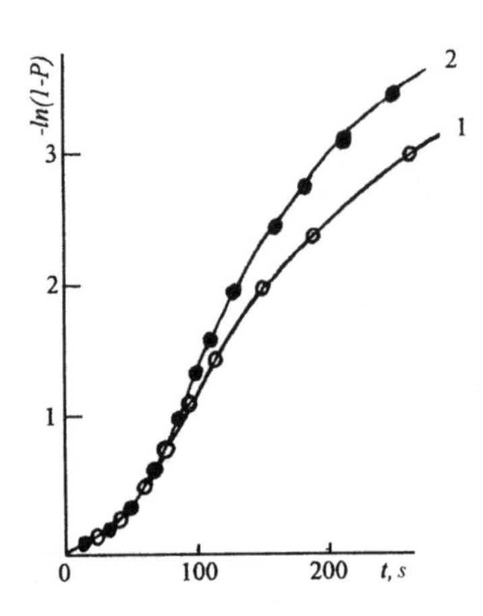

Figure 5.23. Interpretation of experimental kinetic curves in coordinates indicating the presence of maximal reduced rate of MDF-2 photopolymerization: c_0=0.021 mol/l, ℓ=1.4×10^{-4} m (1); c_0=0.021 mol/l, ℓ=9×10^{-4} m (2). E_0=40.6 W/m^2.

Analysis of kinetic curves of polymerization of diacrylate OCM-2 and acrylate DEGDA showed, that the ordinate of inflection, corresponding to maximum of reaction rate, does not equal P_0=0.5 and is lower. Because of this fact we cannot use the condition $A >> 1$ for the analysis of experimental data. Therefore, we rewrite the equation (5.77) as

$$\mathrm{d}P/\mathrm{d}t = \ell^{-1}2k_1(\gamma I_0/\varepsilon c_0)^{1/2}(1-e^{-y_0/2})(1-P)(1+AP)\,, \quad (5.88)$$

where A is described by equation (5.82).

As was shown earlier, the ordinate of inflection P_0 of the kinetic curve corresponding to the maximum rate of the process is determined by ratio (5.81). By transforming equation (5.88) with the use of equation (5.82) we obtained the relation between the maximal rate of polymerization process W_0 and the ordinate of inflection of the kinetic curve P_0:

$$W_0/(1-P_0)^2 = (k_2\gamma I_0/\ell)\cdot[1-e^{-y_0}] \quad (5.89)$$

Figures 5.24 and 5.25 show the dependencies of experimentally found values $W_0/(1-P_0)^2$ on thickness of composition layer (ℓ) at different concentrations of photoinitiator. According to equation (5.89) in the case that $y_0=\varepsilon c_0\ell << 1$, the ratio $W_0/(1-P_0)^2$ should be a linear function of photoinitiator concentration: $W_0/(1-P_0)^2 \equiv k_2\gamma\varepsilon c_0 I_0$; in the case that $y_0=\varepsilon c_0\ell >> 1$, the ratio $W_0/(1-P_0)^2$ does not depend upon thickness: $W_0/(1-P_0)^2 \equiv k_2\gamma I_0/\ell$.

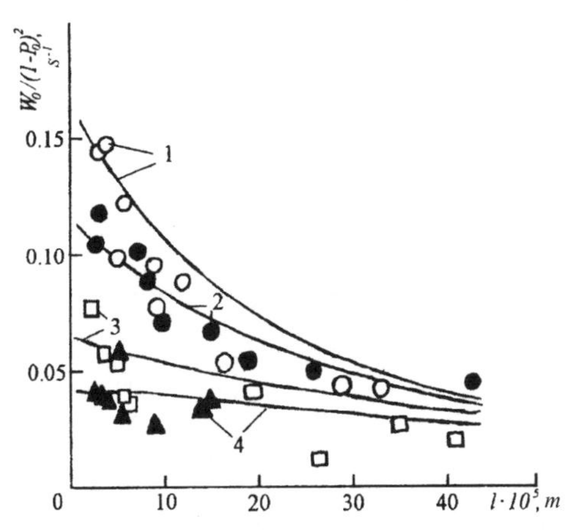

Figure 5.24. Experimental dependencies (symbols) and dependencies calculated using equaiton (5.86) (lines) of maximal photopolymerization rate for OCM-2 on thickness of layer ℓ at different concentrations of photoinitiator c_0: 0.06 (1), 0.04 (2), 0.02 (3), 0.016 mol/l (4). E_0=40.6 W/m^2, ℓ=2×10^{-4} m.

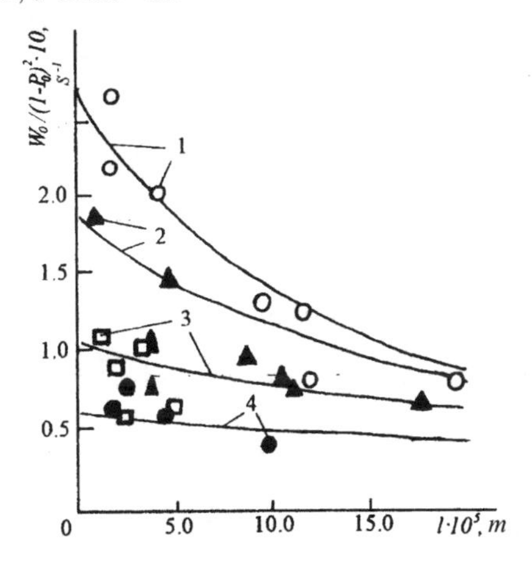

Figure 5.25. Experimental dependencies (symbols) and dependencies calculated using equation (5.86) (lines) of maximal photopolymerization rate for DEGDA on thickness of layer ℓ at different concentrations of photoinitiator c_0: 0.08 (1), 0.06 (2), 0.03 (3), 0.02 mol/l (4); E_0=40.6 W/m^2; ℓ=2×10^{-4} m.

Experimental data from Figures 5.24 and 5.25 confirm this fact: at $\ell \to 0$ values of $W_0/(1-P_0)^2$ are divergent proportionally to photoinitiator concentration and at $\ell \to \infty$ these are coinciding, proving the independence of $W_0/(1-P_0)^2$ from c_0. At the same time, according to equation (5.89) and experimental data, for all ℓ the ratio $W_0/(1-P_0)^2$ is a linear function on power of UV-illumination falling upon the surface of the composition (see Figure 5.26). Calculated dependencies of $W_0/(1-P_0)^2$ on ℓ are presented in Figures 5.24 and 5.25 as full lines. Such dependencies were obtained with the following parameters of equation (5.89): E_0=40.6 J m^{-2} s^{-1}, $k_2\gamma I_0$=1.74×10^{-5} s for OCM-2 and 1.9×10^{-5} s for DEGDA, ε=160 m^{3}mol^{-1} m^{-1}.

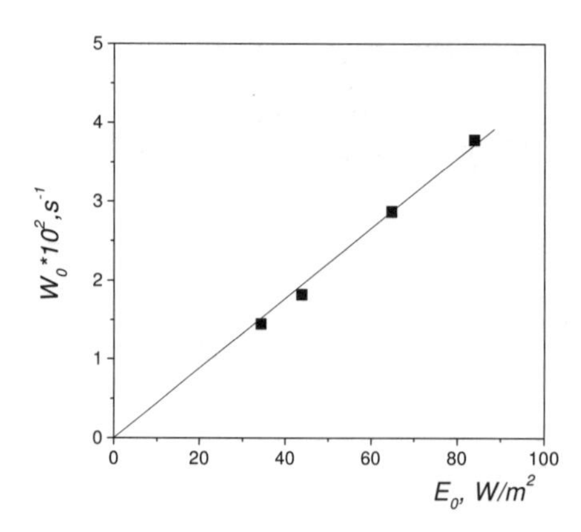

Figure 5.26. Dependence of maximal rate W_0 for OCM-2 photopolymerization on UV-illumination power. c_0=0.02 mol/l, ℓ=2*10^{-5} m.

By using the relation I_0=2.5×10$^{-6}E_0$ mol×quantum m^{-2} s^{-1} we obtain $k_2\gamma$=0.174 m^3 mol^{-1} for OCM-2 and 0,19 m^3 mol^{-1} for DEGDA.

Comparison of the experimentally found values for parameter A with the calculated dependence of A on ℓ using equation (5.82) is shown in Figures 5.27 and 5.28.

As we can see, the dependencies of the experimentally found values for parameter A on ℓ are in good agreement with equation (5.82), according to which with increasing thickness of layer the parameter A is decreased from the value $A_0=k_2(\gamma\varepsilon\, c_0 I_0)^{1/2}/k_1$ at ℓ=0 to $A=A_0/2$ at $\ell \to \infty$. At the same time, calculated dependencies of A from ℓ were obtained at values $k_1\gamma^{1/2}$=0.04 (m^3 mol^{-1} s^{-1})$^{1/2}$ for OCM-2 and 0.06 (m^3 mol^{-1} s^{-1})$^{1/2}$ for DEGDA.

Estimation of the quantum yield accordingly to experimental data has been done in the second approximation, i.e., according to the model of think layer, which is based on the condition $\varepsilon c_0 \ell << 1$ in accordance with equation (5.86).

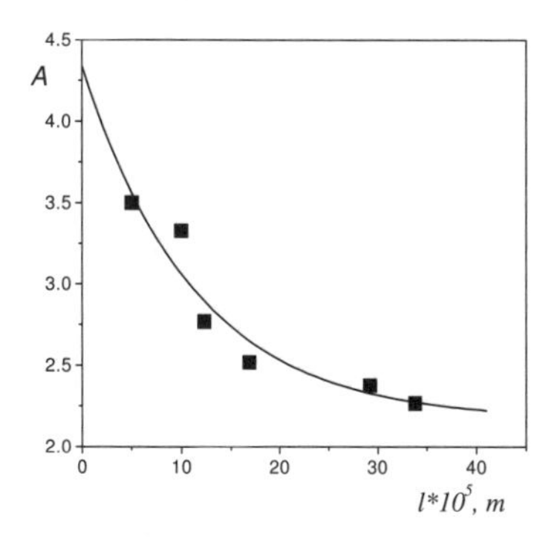

Figure 5.27. Experimental dependencies (squares) and dependencies calculated using equation (5.82) (lines) of parameter *A* on the thickness of layer ℓ for OCM-2 photopolymerization at c_0=0.06 mol/l and E_0=40.6 W/m^2.

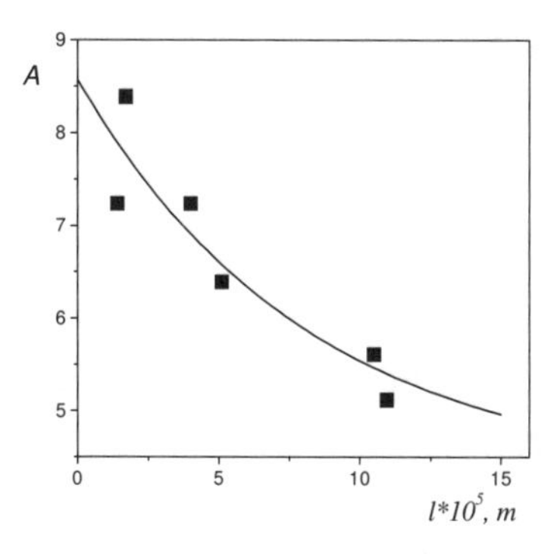

Figure 5.28. Experimental dependencies (squares) and dependencies calculated using equation (5.82) (lines) of parameter *A* on the thickness of layer ℓ for DEGDA photopolymerization at c_0=0.08 mol/l and E_0=40.6 W/m^2.

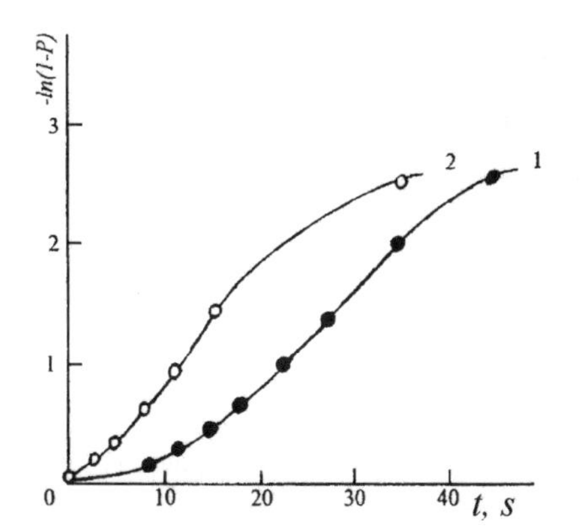

Figure 5.29. Interpretation of experimental kinetic curves in plots indicating the presence of a maximum specific rate of OCM-2 photopolymerization: c_0=0.06 mol/l, ℓ=2×10^{-5} m (1); c_0=0.04 mol/l, ℓ=2.5×10^{-5} m (2). E_0=40.6 W/m^2.

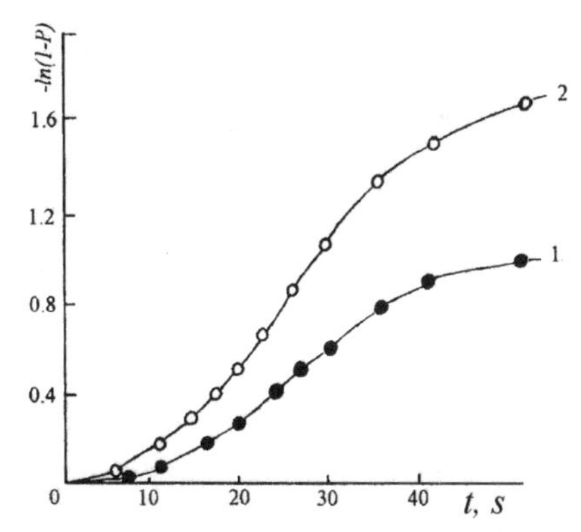

Figure 5.30. Interpretation of experimental kinetic curves in plots indicating the presence of a maximum specific rate of DEGDA photopolymerization: c_0=0.06 mol/l, ℓ=1.4×10^{-4} m (1); c_0=0.08 mol/l, ℓ=1×10^{-4} m (2). E_0=40.6 W/m^2.

Interpretation of experimental data, which satisfy the condition of approximation for endless thin layer $\varepsilon c_0\ell << 1$, is presented in Figures 5.29 and 5.30 as kinetic curves $\ln(1-P)=f(t)$. The calculated values of A_0 allowed to estimate the quantum yield of initiation γ and, on its basis, to estimate the constants of homophaseous (k_1) and heterophaseous (k_2) polymerization processes.

The estimations obtained for the constants of the kinetic model for 3-D polymerization of the di(meth)acrylates investigated are presented in Table 5.1.

The values we found for the molar coefficient of extinction were nearly 14-times higher than the value determined from the UV-spectrum at wavelength $\lambda_{max} \cong 340$ nm, corresponding to the maximum of the adsorption band for 2,2-diizopropyloxyacetophenone (DiPAPh).

Table 5.1. Kinetic parameters of the photoinitiated 3-D polymerization of di(meth)acrylates in the microheterogeneous model

Di(meth)acrylate	$k_1 \times 10^2$ $(m^3\ mol^{-1}\ s^{-1})^{1/2}$	$k_2 \times 10^2$ $(m^3\ mol^{-1}\ s^{-1})^{1/2}$	ε $(m^3\ mol^{-1}\ m^{-1})$	γ
MDF-2	0.8	2.4	87	0.9
OCM-2	4.3	2.1	160	0.85
DEGDA	6.4	2.3	160	0.84

As we can see, in accordance with obtained estimations (Table 5.1) the constant rate of the heterophaseous process k_2 practically does not depend on nature of the di(meth)acrylate, whereas the constant rate of homophaseous process k_1 increases with molecular mass and decreasing monomer viscosity.

Values k_1, k_2 and also ε and γ (Table 5.1) for 3-D photopolymerization of MDPh-2 were introduced into the approximate kinetic equation (5.86) of the endless thin layer model and also into the general kinetic equation (5.77)–(5.79) for direct calculation of kinetic curves $P=f(t)$ by numerical integration of equation (5.86) and equations (5.77)–(5.79).

Parameters of the photopolymerization model for OCM-2 and DEGDA were entered into equation (5.86) of the endless thin layer model and with the use of numerical integration the theoretical curves $P=f(t)$ have been calculated. As we can see from Figure 5.32, the calculated dependences $P=f(t)$ are in good agreement with the experimental kinetic curves.

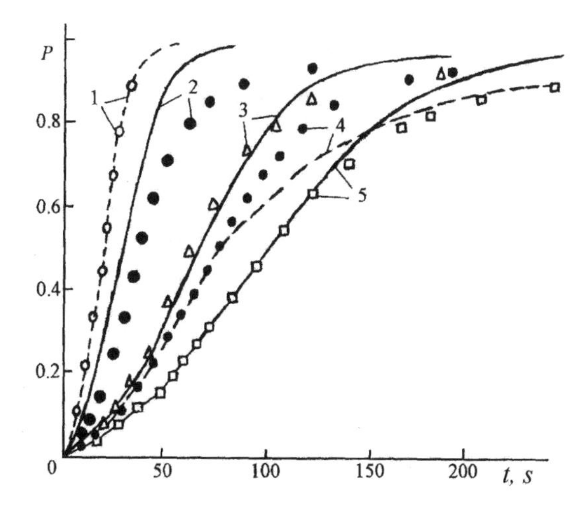

Figure 5.31. Experimental kinetic curves (symbols) and kinetic curves calculated using equations (5.80) and (5.86) (continuous and dashed lines, respectively) of MDPh-2 photopolymerization: c_0=0.105 mol/l, $\ell=3\times10^{-4}$ m (1); c_0=0.105 mol/l, $\ell=1.5\times10^{-4}$ m (2); c_0=0.105 mol/l, $\ell=5.5\times10^{-4}$ m (3); c_0=0.021 mol/l, $\ell=1.2\times10^{-4}$ m (4); c_0=0.105 mol/l, $\ell=8.6\times10^{-4}$ m (5). E_0=40.6 W/m^2.

The calculated dependencies $P=f(t)$ were compared with the experimental kinetic curves (Figures 5.31 and 5.32). From Figures 5.31 and 5.32 we can see that, although the calculated dependences $P=f(t)$ do not fully agree with the experimental kinetic curves, the quantitative and qualitative agreement between them is satisfactory. This fact and also the proximity of the quantum yield of the photoinitiator decomposition to 1 are essential arguments for the benefit of the accordance of the proposed kinetic models (equations (5.77)–(5.79)) with the experimental data. Another argument supporting this conclusion is found in the correlation of the numerical value k_1, found by us, with those reported in the literature. If we transform the dimensionality k_1 into a more usable one, then we obtain $k_1=0.28$ (l mol^{-1} s^{-1})$^{1/2}$. According to equation (5.17) $k_1=k_{pv}/\Gamma_0 k_{tv}^{1/2}$, where Γ_0 is limited achieved conversion and k_{pv} and k_{tv} are rate constants of chain propagation and chain termination in the volume of liquid oligomeric phase, respectively. By assuming that $\Gamma_0=0.8$ we find $k_{pv}/k_{tv}^{1/2}=0.22$ (l mol^{-1} s^{-1})$^{1/2}$. This value coincides, for example, with $k_p/k_t^{1/2}$ for methylmethacrylate [31].

Divergences between calculated and experimental kinetic curves, shown in Figures 5.31 and 5.32, can be explained by approximate estimation of kinetic constants k_1, k_2, ε and γ on the basis of a complete set of experimental material.

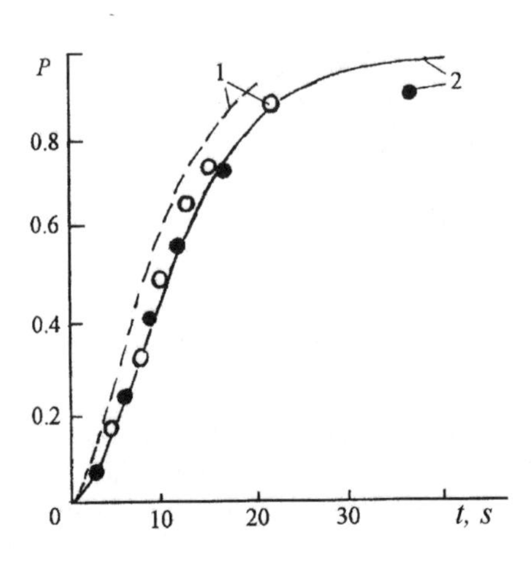

Figure 5.32. Experimental kinetic curves (dots) and kinetic curves calculated using equation (5.86) for OCM-2 and DEGDA (continuous and dashed lines, repsectively) of photopolymerization at E_0=40.6 W/m^2: OCM-2, c_0=0.060 mol/l, ℓ=1×10^{-5} m (1); DEGDA, c_0=0.060 mol/l, ℓ=2×10^{-5} m (2).

A more detailed analysis of the kinetic curves showed that some of them can be described with a little error with the use of individual set of constants k_1, k_2, ε at γ=1, which is near to 1. That is why, from our point of view, the difference between calculated and experimental dependencies $P=P(t)$ is caused not only by approximate estimation of kinetic constant set, but also by other physical factors.

Note that the instrumental error in the construction of the individual kinetic curve in separate experiments is rather small: we can observe this from the location of the points on the kinetic curve (see Figures 5.1–5.8). At the same time, comparison of the individual kinetic curves give the parameter scattering, calculated with the use of these kinetic curves (for instance, values W_0 in Figures 5.19, 5.24, 5.25), that exceeds the error of individual experiments manifold. This phenomenon is well-known for 3-D polymerization and is qualified as a bad reproduction of the kinetic measurements or "whims" of the process. From our point of view, the above-mentioned characteristic of the 3-D polymerization is direct proof of the microheterogeneity of the process, its active role in it the "liquid oligomer–solid polymer" interface layer and fluctuating mechanism of its formation both at the stage of solid polymeric phase initiation and at the stage of monolytization, i.e., phase inversion. It is reflected in the kinetic constant k_2, the numerical value of which depends on the ratio of fractal characteristics of the surface and volume of clusters of the solid polymeric phase into liquid oligomeric matrix, and liquid oligomer into solid polymeric matrix. Since the contribution of the polymerization process at the interface is predominant, its fluctual dependence is transformed in the total process.

In conclusion, reviewing the literature and experimental material obtained from stationary photoinitiated radical polymerization of bifunctional (meth)acrylates in bulk allowed us to obtain the kinetic equation describing the process at all depths of conversion. A mechanism of 3-D polymerization combining the kinetic schemes of homophaseous and heterophaseous processes has been proposed on the basis of the microheterogeneous model and also effective rate constants of these processes have been estimated.

REFERENCES

[1] Berlin A. A., Korolev G. V., Kepheli T. Ja. *Acrylic oligomers and materials on their basis.* M.: Khimija, **1983**, 232 p.

[2] Vasil'yev `D. K., Belgovsky I. M. *Vysokomol. soed.*, **1990**, 32 B (9), p. 678.

[3] Vasil'yev D. K., Belgovsky I. M. *Vysokomol. soed.*, **1989**, 31 A (6), p. 1233.

[4] Korolev G. V., Mogilevich M. M., Golikov N. V. *Sietchastyje poliakrylaty. Mikrogeterogennyje structury, phizicheskije sietki, dieformacionno-prochnostnyje svojstva.* M.: Khimija, **1995**, p. 48.

[5] Volkova M. I., Bel'govskij I. M., Golikov N. V. *Vysokomol. soed.*, **1987**, 28 A (6), p. 435–440.

[6] Chiferri A., Word I. *Svierkhvysokomodul'nyje polimery.* L.: Khimija, **1983**, 272 p.

[7] Kloosterboer I.G., Lijten G.F. *Crosslinking polymers: chemistry, properties & application*, ACS, **1988**, 367, p. 409.

[8] Medvedevskikh Yu., Zaglad'ko O., Turovskij A., Scorobogatyi Ya., Zaikov G. *Int. J. Polym. Mater.*, **2000**, 48, p. 237.

[9] Zaglad'ko O., Medvedevskikh Yu., Turovskij A., Zaikov G. *Int. J. Polym. Mater.*, **1998**, 39, p. 227.

[10] Medvedevskikh Yu., Zaglad'ko O., Turovskij A., Zaikov G. *Chem. Phys. Rep.*, **2000**, 18 (9), p. 1653.

[11] Pfejpher P. *Fraktaly v phisikie, Book of Abstracts of symposium on fractals in physics,* M.: Myr, **1988**, p. 72.

[12] Shefer D., Kefer K. *Fraktaly v phisikie, Book of Abstracts of symposium on fractals in physics,* M.: Myr, **1988**, p. 62.

[13] Smirnov B. M. *Uspiekhi physicheskikh nauk,* **1986**, 149 (2), p. 178.

[14] Kaminskiy V. A., Ivanov V. A., Brun E. B. *Vysokomol. soed.*, **1992**, 34 A (9), p. 14.

[15] Minko S. S., Luzinov I. A., Smirnov B. R. *Vysokomol. soed.,* **1990**, 32 A (10), p. 750.

[16] Ivanov V. B., Romaniuk A. P., Shybanov V. V. *Vysokomol. soed.,* **1994**, 35 A (2), p. 119.

[17] Ivanchiov S. S., Dmitriyenko A. V., Krupnik A. A. *Vysokomol. soed.,* **1988**, 30 A (9), p. 1951.

[18] Minko S. S., Sidorenko A. A., Voronov S. A. *Vysokomol. soed.,* **1995**, 30 A (8), p. 1403.

[19] Kurdikar D. L., Peppas N. A. *Macromolecules,* **1994**, 27, p. 733.

[20] Budtov V. P., Podosenova N. G., Zotikov E. G. *Vysokomol. soed.,* **1983**, 25 A (4), p. 237.

[21] Revnov B. V., Podosenova N. G., Ivanchiov S. S. *Vysokomol. soed.,* **1988**, 30 A (3), p. 184.

[22] Kolegov V. I. *Vysokomol. soed.,* **1995**, 37 A, (1), p. 16.

[23] Treushnikov V. M., Zelentsova N. V., Olejnik A. V. *Zhurnal nauchn. i prikl. photo- i kinematographii,* **1988**, 33 (2), p. 146.

[24] Grishchenko V. K., Masluik A. F., Gudzera S. S. *Zhidkije photopolimeryzujushchijesya kompoziciji*, K.: Naukova dumka, **1985**, p. 90.

[25] Ivanov V. V. and Smirnov B. R. *Vysokomol. soed.,* **1991**, 33 (B), p. 807.

[26] Ovchinnikov V., Kharchijev V. *Zhurnal ekspier. i tieor. phisiki*, **1965**, 49, № 1 (7), p. 315.

[27] Borovich B., Zuyev V., Krokhin O. *Zhurnal ekspier. i tieor. phisiki*, **1973**, 64 (4), p. 1184.

[28] Medvedevskikh Yu. G., Berlin A. A. *Tieoret. i ekspier. khimija,* **1992**, 28 (5/6), p. 420.

[29] Medvedevskikh Yu. G., Simonenko V. V. *J. Phys. Chem.,* **1992**, 66, p. 1432.

[30] Medvedevskikh Yu. G., Simonenko V. V. *J. Phys. Chem.,* **1992**, 66, p. 1656.

[31] Kargin V. A. *Encyklopedia polimerov,* M.: Myr, **1972**, v. 2, p. 185.

Chapter 6. Stationary kinetics of the linear polymerization up to the high conversion state

6.1. Conception of the three reactive zones

The kinetics of mono- and multifunctional polymerization up to the high conversion state are characterized both by general and differential characteristics of the process. The main general ones are: (i) the S-like shape of the kinetic curves, indicating the presence of autoacceleration and autodeceleration processes; (ii) a great post-effect, that is, in the dark after UV-illumination, stopping the postpolymerization process observed from the autoacceleration stage; and (iii) a high (to 10^{-1} mol/m^3) concentration of radicals discovered by EPR-spectroscopy *in situ* until the end of the polymerization process [1–8].

Two main conceptions have been formulated with the aim of explaning such characteristics for the polymerization process up to the high conversion state. The first is based on the diffusion-controlled character of elementary reactions assigned to the classic kinetic scheme. Here, the kinetic equation of the initial stage of the process is the starting point in the diffusion-controlled reaction conception. Parameters of the above-mentioned equation are functions of the current state of the monomer–polymeric solution [9]. The second conception or microheterogeneous model [1, 10–14] is based upon the principle that, in the process kinetics, in its initial stage, there is no homophaseous polymerization in the liquid monomer-solution, but a heterophaseous one, proceeding at the boundary of the "liquid monomer–solid polymer" microgranules under gel-effect conditions. The latter conception for 3-D polymerization is described in detail in Chapter 5.

The main deference between the linear and the 3-D polymerizations is the higher value of the conversion (P) in the start point of the autoacceleration process (P=0.2–0.4 for the linear polymerization and P << 0.1 for the 3-D polymerization). The main reason of the above-mentioned fact consists in the difference in solubility of the polymers. Thus, it is well-known that the linear polymer is characterized by better solubility in monomer and elimination of the polymeric phase at the higher conversion P. But, however, we cannot talk about the solid polymeric phase as the final product with limited conversion P (as is the case with 3-D polymerization). In other words, the clear separation of polymer and monomer (two phases) will not be observed *via* the polymerization process. On the basis of the above, we can conclude that the linear polymerization cannot be described starting from the positions of the conception about two reactive zones and be a simple copy of the kinetic model for the 3-D polymerization. That is why, taking into account the similarities and differences of the polymerization we imagine the polymerization process to proceed as follows [15, 16]:

1. To the monomer concentration $[M_v^0]$, corresponding to conversion $P_v^0=([M_0]-[M_v^0])/[M_0]$, where $[M_0]$ is the initial concentration of the monomer in the block, the polymerization system is monophaseous and represents the solution of the polymer in the monomer; we call this solution the monomer–polymeric phase (MPPh). MPPh polymerization process proceeds in accordance with the classic kinetic scheme with bimolecular chain termination:

$$\begin{aligned} &M + R \xrightarrow{k_{pv}} R \\ &R + R \xrightarrow{k_{tv}} \text{products of reaction} \end{aligned} \qquad (6.1)$$

The linear dependence of conversion on time in MPPh is explained by partial diffusion control at the linear chain termination.

2. At the achievement of conversion P_v^0, the monomer–polymeric solution becomes saturated, relative to polymer, and at some super-saturation the new polymer–monomer phase (PMP) is selected from it. It represents, by itself, the saturated solution of a monomer in a polymer with the concentration of the monomer $[M_s^0]$, which corresponds to conversion P_s^0: $P_s^0=([M_0]-[M_s^0])/[M_0]$.

Concentrations of monomer $[M_v^0]$ in MPPh and $[M_s^0]$ in PMPh are functions of the nature of the monomer–polymeric system and temperature. The super-saturation of MPPh disappears after the spontaneous origin of PMPh embryos, needed to initiate the new phase. Therefore, the polymerization process is further accompanied only by the propagation of embryos or micrograins of PMPh.

3. The photoinitiator is distributed non-uniformly between MPPh and PMPh, but in equilibrium, that is, in accordance with the law of substance distribution between two phases. Since PMPh is rather viscous, the effects of mixing inside the micrograins can be neglected; therefore, the photoinitiator concentration will be a variable of the radius of micrograins in accordance with the time of a given micrograins-layer selection.

4. In the case of a microheterogeneous system, polymerization proceeds in three reactionary zones; namely, in the saturated monomer–polymeric solution, at the interface layer on the boundary of MPPh and PMPh micrograins and in the "solid" (conditionally) polymer–monomeric solution.

5. In MPPh polymerization proceeds in accordance with the same classic kinetic scheme (6.1), albeit at a constant concentration $[M_v^0]$ of the monomer and viscosity of the solution. Only the concentration of the photoinitiator and volumetric parts (φ_v) of MPPh are varied.

6. In the case of an interface layer at the boundary of MPPh and PMPh polymerization proceeds according to scheme (6.1), but the monomolecular chain termination rate constant k_{tvs} is different. It is connected with this fact, that "solid" PMPh creates an especially ordered structure of the nearest reactionary space, in which there is sharply reduced transmission and segmental mobility of macroradicals. We assume that the concentration of the monomer and photoinitiator in the interface layer are equal or proportional to their concentration in MPPh.

The volume and the volumetric part of the interface layer are a complete function of the propagation and aggregation process (stage of monolytization [1]) of PMPh micrograins. The volumetric part of the interface layer will be proportional to the volumetric part φ_s of the PMPh, namely $\varphi_{vs} \sim \varphi_s$ only while $\varphi_v >> \varphi_s$ is on the stage of micrograins propagation; at the end of the monolytization stage, when $\varphi_v << \varphi_s$, it is opposite: $\varphi_{vs} \sim \varphi_v$. Supposing, that micrograins of the PMPh are the mass fractals for which the fractal dimensions of surface and volume coincide [17, 18] and using the probability of contact in the model of black and white balls in accordance with Ref. [19], we approximate the relationship of φ_{vs}, φ_v and φ_s using the approximate function

$$\varphi_{vs} \cong F_{vs}\varphi_v\varphi_s \qquad (6.2)$$

where the coefficient of proportionality F_{vs} depends upon the fractal characteristics of the micrograins of the PMPh in the MPPh and the microdrops of the MPPh in the PMPh, their number and the thickness of the interface layer.

In an approximation when the volume of the interface layer is small and the relationships $\varphi_{vs} << \varphi_v + \varphi_s$, $\varphi_v + \varphi_s \cong 1$ are fulfilled, we obtained equation (6.3) instead of equation (6.2):

$$\varphi_{vs} \cong F_{vs}\varphi_s(1-\varphi_s) \tag{6.3}$$

7. Polymerization in the polymer–monomeric phase is characterized by two main conditions. First, it proceeds under a gel-effect condition, at which, by virtue of sizeable loss of transmission and segmental mobility of macroradical control on the rate of chain termination passes to the rate of its propagation. This means, that the acts of chain propagation and its termination take place as two different outcomes of the interaction of the active radical R_s (propagation of chain) or a frozen one, the so-called self-burial, in accordance with the terminology of Refs. [20, 21], that is, an inactive radical R_z (monomolecular chain termination). This can be represented by scheme (6.4):

$$R_s + M \begin{cases} \xrightarrow{k_{ps}} R_s \\ \xrightarrow{k_{ts}} R_z \end{cases} \tag{6.4}$$

From a stationary condition upon active radicals R_s it follows, that the specific rate of polymerization in PMPh is equal to:

$$w_s=(k_{ps}/k_{ts})v_{is} \tag{6.5}$$

where v_{is} is the specific rate of initiation in PMPh, and k_{ps}/k_{tss} is the length of the chain.

The second, *via* polymerization proceeding in the new polymeric phase of PMPh, that is, a polymer with limited conversion near 1, cannot be allocated as an independent phase, since PMPh is very viscous. Therefore, polymerization inside the PMPh micrograins can be considered as a glass transition process, leading to reduction of the reactive volume of PMPh. The glass transition process of micrograins develops non-uniformly upon the radius of the micrograins in accordance with the different time of a given layer selection. That is why the summarized rate of polymerization in the PMPh will be characterized with an integral character.

6.2. Deduction of the kinetic model

Homogeneous system

Polymerization proceeds in a homophaseous system in accordance with the classic kinetic scheme (6.1) to the monomer concentration $[M] \geq [M_v^0]$ or conversion $P \leq P_v^0$ with a specific rate

$$-d[M]/dt = k_{pv}[M]v_{iv}^{1/2}/k_{tv}^{1/2} \tag{6.6}$$

where v_{iv} is the initiation rate. *Via* conversion P the polymerization rate is described by the expression

$$dP/dt = k_{pv}(1-P)v_{iv}^{1/2}/k_{tv}^{1/2} \tag{6.7}$$

The weak gel-effect, which is exhibited as a linear section of the kinetic curve up to the autoacceleration stage, is caused by partial diffusion control on the rate of bimolecular chain termination. Thus, we present the chain termination rate constant k_{tv} in the following form:

$$1/k_{tv} = 1/k_c + 1/k_d \tag{6.8}$$

where k_c and k_d are constant rates of termination, controlled by the chemical act of radical interaction and, accordingly, delivery by diffusion to the point of interaction.

The viscosity of MPPh (η_v) is described by a linear function in the interval $0 \leq P \leq P_v^0$:

$$\eta_v = \eta_0\left(1 + \frac{\eta_v^0 - \eta_0}{\eta_0 P_v^0} P\right), \tag{6.9}$$

where η_0 and η_v^0 are viscosities of pure monomer in the bulk and monomer–polymeric solution, respectively, at $P = P_v^0$.

Accepting k_d as inversely proportional to the viscosity of the solution ($k_d\eta$ =constant), we can rewrite

$$\frac{1}{k_{\mathrm{d}}}=\frac{\left(1+\frac{\eta_{\mathrm{v}}^{0}-\eta_{0}}{\eta_{0}P_{\mathrm{v}}^{0}}P\right)}{k_{\mathrm{d}}^{0}},\tag{6.10}$$

where k_{d}^{0} is rate constant of radical diffusion in the monomer.

By combining equations (6.8) and (6.10), we have

$$1/k_{\mathrm{tv}}=(1+aP)/k_{\mathrm{tv}}^{0}\tag{6.11}$$

where

$$1/k_{\mathrm{tv}}^{0}=1/k_{c}+1/k_{\mathrm{d}}^{0}\tag{6.12}$$

$$a=k_{\mathrm{tv}}^{0}(\eta_{\mathrm{v}}^{0}-\eta_{0})/k_{\mathrm{d}}^{0}\eta_{0}P_{\mathrm{v}}^{0}\tag{6.13}$$

Thus, k_{tv}^{0} is a constant rate of chain termination in a pure monomer (P=0). With substitution of equation (6.11) into an equation (6.7), the rate of polymerization in the MPPh at $P \le P_{\mathrm{v}}^{0}$ will be described by the equation:

$$\mathrm{d}P/\mathrm{d}t=(k_{\mathrm{pv}}/k^{0\ 1/2}_{\mathrm{tv}})(1-P)(1+aP)^{1/2}v_{\mathrm{iv}}^{1/2},\ P\le P_{\mathrm{v}}^{0}\tag{6.14}$$

Microheterogeneous system

At $P \ge P_{\mathrm{v}}^{0}$ polymerization proceeds in three reactionary zones, and that is why its rate is described by the sum

$$-\mathrm{d}[\mathrm{M}]/\mathrm{d}t=w_{\mathrm{v}}\varphi_{\mathrm{v}}+w_{\mathrm{vs}}\varphi_{\mathrm{vs}}+\langle w_{\mathrm{s}}\rangle\varphi_{\mathrm{s}}\tag{6.15}$$

where w_{v} and w_{vs} are specific rates of the process in a saturated monomer–polymeric solution and the interface layer at the boundary of MPPh and PMPh, respectively; $\langle w_{\mathrm{s}}\rangle$ is the average specific rate of polymerization in PMPh per volume of micrograins. Instead of equation (6.15) we can write:

$$\mathrm{d}P/\mathrm{d}t=w_{\mathrm{v}}\varphi_{\mathrm{v}}/[\mathrm{M}_0]+w_{\mathrm{vs}}\varphi_{\mathrm{vs}}/[\mathrm{M}_0]+\langle w_{\mathrm{s}}\rangle\,\varphi_{\mathrm{s}}/[\mathrm{M}_0]\tag{6.16}$$

The contribution $(\mathrm{d}P/\mathrm{d}t)_{\mathrm{v}}=w_{\mathrm{v}}\varphi_{\mathrm{v}}/[\mathrm{M}_0]$ of the polymerization process proceeding in the MPPh to the summarized rate is described by the same equation (6.14), but at a constant value of $P=P_{\mathrm{v}}^{0}$ and variable $\varphi_{\mathrm{v}}=1-\varphi_{\mathrm{s}}$:

$$(\mathrm{d}P/\mathrm{d}t)_{\mathrm{v}}=(k_{\mathrm{pv}}/k^{0}{}_{\mathrm{tv}}{}^{1/2})(1-P_{\mathrm{v}}{}^{0})(1+aP_{\mathrm{v}}{}^{0})^{1/2}(1-\varphi_{\mathrm{s}})v_{\mathrm{iv}}{}^{1/2} \quad (6.17)$$

The specific rate of polymerization in the interface layer is described by the classic equation (6.6), but with the constant rate of termination k_{tvs}, in which the contribution of diffusion control is increased considerably, we obtain:

$$w_{\mathrm{vs}}=-\mathrm{d}[\mathrm{M}]/\mathrm{d}t=(k_{\mathrm{pv}}/k_{\mathrm{tvs}}{}^{1/2})[\mathrm{M}_{\mathrm{vs}}]v_{\mathrm{ivs}}{}^{1/2} \quad (6.18)$$

When phases coexist, k_{tvs}=constant. Accepting that in the interface layer the concentration of the monomer $[\mathrm{M}_{\mathrm{vs}}]$ is proportional to $[\mathrm{M}_{\mathrm{v}}{}^{0}]([\mathrm{M}_{\mathrm{vs}}]=k_{\mathrm{m}}[\mathrm{M}_{\mathrm{v}}{}^{0}])$, and that the initiation rate v_{ivs} is proportional to v_{iv} ($v_{\mathrm{ivs}}=k_{\mathrm{v}}v_{\mathrm{iv}}$), let us describe the contribution of polymerization rates in the interface layer $(\mathrm{d}P/\mathrm{d}t)_{\mathrm{vs}}=w_{\mathrm{vs}}\varphi_{\mathrm{vs}}/[\mathrm{M}_0]$ into the summarized kinetics, taking into account equation (6.3), by the next expression

$$(\mathrm{d}P/\mathrm{d}t)_{\mathrm{vs}}=k_2(1-P_{\mathrm{v}}{}^{0})\varphi_{\mathrm{s}}(1-\varphi_{\mathrm{s}})v_{\mathrm{iv}}{}^{1/2} \quad (6.19)$$

where

$$k_2=(k_{\mathrm{p}}/k_{\mathrm{tvs}}{}^{1/2})F_{\mathrm{vs}}k_{m}k_{\mathrm{v}} \quad (6.20)$$

k_2 is an effective constant rate of polymerization in the interface layer.

Summarizing equation (6.17) and equation (6.19), namely $(\mathrm{d}P/\mathrm{d}t)_{\mathrm{v}}$ + $(\mathrm{d}P/\mathrm{d}t)_{\mathrm{vs}}=(\mathrm{d}P/\mathrm{d}t)_{\mathrm{v+vs}}$ we obtain

$$(\mathrm{d}P/\mathrm{d}t)_{\mathrm{v+vs}}=k_1(1-P_{\mathrm{v}}{}^{0})[(1+aP_{\mathrm{v}}{}^{0})^{1/2}+k_2\varphi_{\mathrm{s}}/k_1](1-\varphi_{\mathrm{s}})v_{\mathrm{iv}}{}^{1/2} \quad (6.21)$$

where

$$k_1=k_{\mathrm{p}}/k^{0}{}_{\mathrm{tv}}{}^{1/2} \quad (6.22)$$

The magnitude $(\mathrm{d}P/\mathrm{d}t)_{\mathrm{v+vs}}$ determines the rate of the new PMPh selection from MPPh. Next, we will determine the connection between them. At times t and $t+\mathrm{d}t$, the quantity of monomer in MPPh is equal to $[\mathrm{M}_{\mathrm{v}}{}^{0}](1-\varphi_{\mathrm{s}})$ and $[\mathrm{M}_{\mathrm{v}}{}^{0}](1-\varphi_{\mathrm{s}}-\mathrm{d}\varphi_{\mathrm{s}})$, respectively. The quantity of monomer passed into PMPh for time dt is equal to $[\mathrm{M}_{\mathrm{s}}{}^{0}]\mathrm{d}\varphi_{\mathrm{s}}$. At the expense of co-polymerization into MPPh and the interface layer $\mathrm{d}[\mathrm{M}]_{\mathrm{v+vs}}$ was reacted. From the balance, it follows that $-\mathrm{d}[\mathrm{M}]_{\mathrm{v+vs}}=([\mathrm{M}_{\mathrm{v}}{}^{0}]-[\mathrm{M}_{\mathrm{s}}{}^{0}])\mathrm{d}\varphi_{\mathrm{s}}$.

Thus, we have

$$(\mathrm{d}P/\mathrm{d}t)_{\mathrm{v+vs}}=(P_{\mathrm{s}}^{0}-P_{\mathrm{v}}^{0})\mathrm{d}\varphi_{\mathrm{s}}/\mathrm{d}t \tag{6.23}$$

Comparing equation (6.21) and equation (6.23), we can rewrite the latter relative to $\mathrm{d}\varphi_{\mathrm{s}}/\mathrm{d}t$:

$$\frac{\mathrm{d}\varphi_{\mathrm{s}}}{\mathrm{d}t}=\frac{1-P_{\mathrm{v}}^{0}}{P_{\mathrm{s}}^{0}-P_{\mathrm{v}}^{0}}k_1\left[(1+aP_{\mathrm{v}}^{0})^{1/2}+\frac{k_2}{k_1}\varphi_{\mathrm{s}}\right](1-\varphi_{\mathrm{s}})v_{\mathrm{iv}}^{1/2} \tag{6.24}$$

In surnmary, we can describe the summarized rate of the process in a microheterogeneous system as

$$\mathrm{d}P/\mathrm{d}t=(P_{\mathrm{s}}^{0}-P_{\mathrm{v}}^{0})\mathrm{d}\varphi_{\mathrm{s}}/\mathrm{d}t+\langle w_{\mathrm{s}}\rangle\,\varphi_{\mathrm{s}}/[\mathrm{M}_0] \tag{6.25}$$

Let us return to $\langle w_{\mathrm{s}}\rangle$, which is the average specific rate of polymerization in the volume of PMPh micrograins, even though t_{v}^{0} is the time needed for the achievement of conversion P_{v}^{0} in MPPh and for the appearance of the PMPh nucleus. Let us take this time as the emanating point, assuming that $t_{\mathrm{v}}^{0}=0$ is the beginning of the new phase separation. Let us consider the element $\mathrm{d}\varphi_{\mathrm{s}}$, which was allocated to MPPh in the moment of time $t \geq 0$. At this moment, the concentration of the monomer in it was equal to $[\mathrm{M}_{\mathrm{s}}^{0}]$, and the part $1-\beta$ of the glass-transitioned, so-called inactive, part of the polymer with limited conversion, is equal to zero. At time $\tau \geq t$ the part of the non-glass-transitioned matter at the expense of polymerization is equal to $\beta \geq 0$. So, at time τ the specific rate of polymerization in the element volume $\mathrm{d}\varphi_{\mathrm{s}}$, separated at time t, is equal to

$$-\mathrm{d}[\mathrm{M}]_{\mathrm{s}}/\mathrm{d}\tau=w_{\mathrm{s}}\beta \tag{6.26}$$

where β is a function of $\tau-t$ and the specific rate w_{s} of the process.

In this case, at time t, by removing the unit volume of PMPh–$\beta=1$, the amount of monomer is equal to $[\mathrm{M}_{\mathrm{s}}^{0}]$. At time $\tau > t$ we obtained $\beta < 1$, and the amount of monomer in a unit volume is equal to $[\mathrm{M}_{\mathrm{s}}^{0}]\beta$. *Via* time $\mathrm{d}\tau$ an amount $\mathrm{d}[\mathrm{M}_{\mathrm{s}}]$ of monomer will be reacted and remain in the non-glass-transitioned part of the unit volume $[\mathrm{M}_{\mathrm{s}}^{0}](\beta+\mathrm{d}\beta)$. It follows from this that $\mathrm{d}[\mathrm{M}_{\mathrm{s}}]=[\mathrm{M}_{\mathrm{s}}^{0}]\mathrm{d}\beta$ or

$$\mathrm{d}[\mathrm{M}_{\mathrm{s}}]/\mathrm{d}\tau=[\mathrm{M}_{\mathrm{s}}^{0}]\mathrm{d}\beta/\mathrm{d}\tau \tag{6.27}$$

Comparing equation (6.26) and equation (6.27), we obtain

$$d\beta/\beta = -w_s d\tau/[M_s^0] \tag{6.28}$$

With the photoinitiator concentration variations being constant as a result of its photodecomposition, we accept that v_{is} and w_s in accordance with equation (6.5), do not depend upon the time interval τ–t after removing the given volume of PMPh. Therefore, by integrating equation (6.28) in accordance with condition $\beta=1$ at time t for the PMPh removal, we have

$$\beta = \exp\{-w_s(\tau-t)/[M_s^0]\} \tag{6.29}$$

The summarized rate of polymerization in the removed unit $d\varphi_s$ of the PMPh volume is equal to

$$-\frac{d[M_s]}{d\tau} d\varphi_s = w_s \beta d\varphi_s \tag{6.30}$$

Let us substitute equation (6.29) into equation (6.30) with the replacement $d\varphi_s = \frac{d\varphi_s}{dt} dt$ in its right part, and integrate equation (6.30) on φ_s on the left and on t on the right, taking into account the fact that in accordance with the mean-value theorem we have

$$-\int_0^{\varphi_s} \frac{d[M_s]}{d\tau} d\varphi_s = -< \frac{d[M_s]}{d\tau} > \varphi_s = < w_s > \varphi_s \tag{6.31}$$

we obtain

$$< w_s > = \frac{1}{\varphi_s} \int_0^{\tau} w_s \exp\{-\frac{w_s}{[M_s^0]}(\tau - t)\} \frac{d\varphi_s}{dt} dt \tag{6.32}$$

It follows from this that the rate contribution $(dP/dt)_s = <w_s> \varphi_s/[M_0]$ of the polymerization process to PMPh in a microheterogeneous system can be determined by the expression

$$\left(\frac{dP}{d\tau}\right)_S = \int_0^\tau \frac{w_s}{[M_0]} \exp\{-\frac{w_s}{[M_s^0]}(\tau - t)\} \frac{d\varphi_s}{dt} dt \qquad (6.33)$$

The general kinetic equation of polymerization in a microheterogeneous system at $P \geq P_v^0$ can be written as follows:

$$\frac{dP}{d\tau} = (P_s^0 - P_v^0)\frac{d\varphi_s}{d\tau} + \int_0^\tau \frac{w_s}{[M_0]} \exp\{-\frac{w_s}{[M_s^0]}(\tau - t)\} \frac{d\varphi_s}{dt} dt \qquad (6.34)$$

At this time, we can write, in accordance with equation (6.5)

$$w_s/[M_0]=k_3 v_{is},\ w_s/[M_s^0]=k_3 v_{is}/(1-P_s^0) \qquad (6.35)$$

where

$$k_3=k_{ps}/k_{ts}\ [M_0] \qquad (6.36)$$

Equations (6.24) and (6.34) (taking into account that $dt=d\tau$) represent the kinetic model of polymerization in a microheterogeneous system in the integral-differential form. We obtain the integrated form of the kinetic model when we rewrite

$$P=P_v^0(1-\varphi_s) + \langle P_s \rangle \varphi_s \qquad (6.37)$$

where $\langle P_s \rangle$ is the average conversion in PMPh in at the present time τ.

As it was shown earlier, the amount of monomer in the unit volume of PMPh, taking into account the share of β of its non-glass-transitioned part is equal to $[M_s]=[M_s^0]\beta$ or

$$[M_s] = [M_s^0]\exp\left\{-\frac{w_s}{[M_s^0]}(\tau - t)\right\} \qquad (6.38)$$

By multiplying the left part of equation (6.38) by $d\varphi_s$, and the right one by $\frac{d\varphi_s}{dt}dt$, and by integrating it, taking into account the mean-value theorem, we obtain

$$\langle [M_s] \rangle = \frac{[M_s^0]}{\varphi_s} \int_0^\tau \exp\{-\frac{w_s}{[M_s^0]}(\tau - t)\} \frac{d\varphi_s}{dt} dt \qquad (6.39)$$

Its follows from this that

$$<P_s>= 1-\frac{1-P_s^0}{\varphi_s}\int_0^{\tau}\exp\{-\frac{w_s}{[M_s^0]}(\tau-t)\}\frac{d\varphi_s}{dt}dt \qquad (6.40)$$

Thus, the integral kinetic model of polymerization in a microheterogeneous system can be represented, in accordance with equation (6.37) and equation (6.40), as follows:

$$P=P_v^0 + (1-P_v^0)\varphi_s-(1-P_s^0)\int_0^{\tau}\exp\left\{-\frac{w_s}{[M_s^0]}(\tau-t)\right\}\frac{d\varphi_s}{dt}dt \qquad (6.41)$$

The difference between the rates of initiation v_{iv} in MPPh and v_{is} in PMPh is determined by the initiator distribution character between the two phases at the moment of PMPh removal from MPPh and also by the absence of mixing of the inside grains of the PMPh. We obtain, at the equilibrium distribution of the initiator, the following ratio:

$$c_s/c_v=L \qquad (6.42)$$

where L is the distribution coefficient, and c_v and c_s the molar-volumetric concentrations of the initiator in MPPh and removal of PMPh at the present time, respectively.

Let us assume, that the initiator concentration in MPPh at time τ is equal to c_v $(1-\varphi_s)$. At time τ+dτ we shall obtain, respectively, $(c_v + dc_v)(1-\varphi_s-d\varphi_s)$; at this time, the initiator concentration transported in PMPh in time dτ is equal to $c_s d\varphi_s$. It follows from this that $-c_v d\varphi_s+(1-\varphi_s)dc_v+c_s d\varphi_s=0$. We obtain equation (6.43) by the replacement of $c_s=c_vL$. So,

$$dc_v/c_v=(1-L)d\varphi_s/(1 - \varphi_s) \qquad (6.43)$$

Let us integrate equation (6.43) according to the condition that at the start of removal of PMPh (φ_s=0), the initiator concentration in the MPPh is equal to c_v^0. Then, we have

$$c_v=c_v^0/(1-\varphi_s)^{1-L} \qquad (6.44)$$

$$c_s=Lc_v^0/(1-\varphi_s)^{1-L} \qquad (6.45)$$

Thus, the rates of initiation at the thermal decomposition of the initiator in a microheterogeneous system can be as follows

$$v_{iv}=f_{iv}k_d c_v^0/(1-\varphi_s)^{1-L} \qquad (6.46)$$

$$v_{is}=f_{is}k_d L\, c_v^0/(1-\varphi_s)^{1-L} \qquad (6.47)$$

where k_d is the rate constant of the initiator decomposition and f_{iv} and f_{is} are initiation coefficients of MPPh and PMPh.

Taken together, equations (6.24), (6.34)–(6.36), (6.46) and (6.47) constitute the kinetic model of thermoinitiated polymerization in a microheterogeneous system at $P \geq P_v^0$.

The variant of the photoinitiated polymerization is considerably complicated, since, in this case, it is necessary to take into account the presence of a light exposure gradient on the layer of the photopolymerization composition, which reduces, accordingly, the rates of polymerization and conversion. These rates not only are functions, but also of the coordinates of a layer x of the illuminated surface (x=0), and appear, thus, in differential performances of the process in a layer x, x+dx. The transition from the differential characteristics $P(x, t)$ and $\partial P(x, t)/\partial t$ to the experimentally determined, that is, averaged on layer of photopolymerizing composition, $P(t)$ and d$P(t)$/dt, is carried out *via* integrated transformations

$$P(t)=\frac{1}{\ell}\int_0^{\ell} P(x,t)\mathrm{d}x, \qquad \frac{\mathrm{d}P(t)}{\mathrm{d}t}=\frac{1}{\ell}\int_0^{\ell}\frac{\partial P(x,t)}{\partial t}\mathrm{d}x \qquad (6.48)$$

where ℓ is the layer thickness.

With the dynamics of the initiator decomposition in the homogeneous phase taken into account, the light illumination gradient is described by a system of non-linear differential equations in partial derivatives:

$$\partial c/\partial t=-\gamma\varepsilon cJ\,,\ \partial J/\partial x=-\varepsilon cJ \qquad (6.49)$$

Here J=$J(x, t)$ and c=$c(x, t)$ are the light intensity and the photoinitiator concentration in layer x, x+dx is from the illuminated surface at time t, ε is the molar extinction coefficient of the photoinitiator and γ is the quantum yield of the initiator photodecomposition.

The solution of equation (6.49) we already know (see Chapter 5.3) and leads to the following expression of the differential rate of the initiator photodecomposition:

$$\partial c(x,t)/\partial t = \gamma\varepsilon J_0 \exp\{\gamma\varepsilon J_0 t - \varepsilon c_0 x\}/[1+\exp\{-\varepsilon c_0 x\}(\exp\{-\varepsilon J_0 t\}-1)] \qquad (6.50)$$

where J_0 is the intensity of the light falling on the surface of the polymerization composition and c_0 is the initial concentration of the photoinitiator.

However, equation (6.50) in the full form can be used only for the polymerization in MPPh until $P=P_v^0$. That is why in the analysis of the kinetic models obtained up to high conversions simplifications are used in the variant of the photoinitiated polymerization.

First, let us assume, that the characteristic time of the photoinitiator decomposition $t_d=(\gamma\varepsilon J_0)^{-1}$ is considerably longer than the time t of the polymerization; therefore, we can neglect the change of the photoinitiator concentration in time (typical approximation for long-chain processes), taking that $c=c_0$. Second, by neglecting the microheterogeneity of the system, let us accept that $\partial J/\partial x = -\varepsilon c_0 J$, and after that, we obtain $J=J_0\exp\{-\varepsilon c_0 x\}$ (see Chapter 5.3) and, respectively,

$$-[\partial c(x)/\partial t] = \gamma\varepsilon c_0 J_0 \exp\{-\varepsilon c_0 x\} \qquad (6.51)$$

Third, taking into account that the thickness of layer l of the photopolymerization composition and its optical density $\varepsilon c_0 l$ are small in our experiments, let us assume, that the obtained kinetic models will be approximately adequate in the variant of the photoinitiated polymerization in the case when we do not use the differential rate of the photoinitiator decomposition according to equation (6.51), but an average one upon the layer. Under the definition of average values in equation (6.51), we obtain

$$-\left\langle\left(\frac{\mathrm{d}c}{\mathrm{d}t}\right)^{1/2}\right\rangle = \frac{1}{\ell}(\gamma\varepsilon c_0 J_0)^{1/2}\int_0^{\ell}\exp\{-\varepsilon c_0 x/2\}\mathrm{d}x \qquad (6.52)$$

$$-\left\langle\frac{\mathrm{d}c}{\mathrm{d}t}\right\rangle = \frac{\gamma\varepsilon c_0 J_0}{\ell}\int_0^{\ell}\exp\{-\varepsilon c_0 x\}\mathrm{d}x \qquad (6.53)$$

therefore

$$-\left\langle \left(\frac{\mathrm{d}c}{\mathrm{d}t} \right)^{1/2} \right\rangle = \frac{2}{\ell} \left(\frac{\gamma J_0}{\varepsilon c_0} \right)^{1/2} (1 - \exp\{-\varepsilon c_0 \ell / 2\}) \tag{6.54}$$

$$-\left\langle \frac{\mathrm{d}c}{\mathrm{d}t} \right\rangle = \frac{\gamma J_0}{\ell} (1 - \exp\{-\varepsilon c_0 \ell\}) \tag{6.55}$$

It follows from equation (6.54) and equation (6.55), that, with the small relative density of the layer, we will obtain $\varepsilon c_0 l << 1$, $-<(\mathrm{d}c/\mathrm{d}t)^{1/2}>=(\gamma\varepsilon c_0 J_0)^{1/2}$, $-<\mathrm{d}c/\mathrm{d}t>=\gamma\varepsilon c_0 J_0$, and, at the large optical density $\varepsilon c_0 l >> 1$, $-<(\mathrm{d}c/\mathrm{d}t)^{1/2}>=(2/l)(\gamma\varepsilon c_0 J_0)^{1/2}$, $-<\mathrm{d}c/\mathrm{d}t>=\gamma J_0/l$. Thus, the observed order on photoinitiation in bimolecular chain termination can vary from 1/2 to –1/2, and 1 to 0 in monomolecular chain termination, depending on the thickness of layer ℓ or its optical density $\varepsilon c_0 l$. But, at the same time, the order on UV-illumination intensity is always equal to 1/2 or 1.

As we can see from the experimental data, the order on photoinitiation at the initial linear sections of the kinetic curves is similar, but less than 0.5, implying that the approximation of the infinitely think layer $\varepsilon c_0 l << 1$ in our experiments strictly is not enough. Therefore, taking into account that ℓ is small and varies in a narrow range according to the previous analysis, we use the average approximations instead of strict expressions (6.54) and (6.55):

$$-<(\mathrm{d}c/\mathrm{d}t)^{1/2}>=(\gamma J_0)^{1/2}(\varepsilon c_0)^m, \tag{6.56}$$

$$-<\mathrm{d}c/\mathrm{d}t>=\gamma J_0(\varepsilon c_0)^{2m}, \tag{6.57}$$

where $0 < m < 1/2$.

Introducing the factors f_v and f_s, we obtain the next equation for a homogeneous polymerization system at $P \leq P_\mathrm{v}^0$:

$$< v_{iv}^{1/2} >= (\gamma_v J_0)^{1/2} (\varepsilon c_0)^m ; \tag{6.58}$$

For a microheterogeneous polymerizing system at $P \geq P_\mathrm{v}^0$, taking into account expressions (6.44) and (6.45) at the replacement $c_\mathrm{v}^0 = c_0$, we obtain

$$<v_\mathrm{iv}^{1/2}>=(\gamma_\mathrm{v} J_0)^{1/2}(\varepsilon c_0)^m/(1-\varphi_\mathrm{s})^{m(1-L)} \tag{6.59}$$

$$<v_{is}^{1/2}> = \gamma_s J_0 (\varepsilon c_0 L)^{2m} / (1-\varphi_s)^{2m(1-L)} \qquad (6.60)$$

where $\gamma_v = f_v \gamma$ and $\gamma_s = f_s \gamma$ are quantum yields of photoinitiation in MPPh and PMPh, respectively.

By substituting $v_{iv}^{1/2}$ and v_{is} in the kinetic models (equations (6.14), (6.24), (6.34) and (6.35)) in their average analogs on the layer (equations (6.58)–(6.60)), we finally obtain the following equation for photoinitiated polymerization in a homogeneous system at $P \le P_v^0$, $t \le t_v^0$:

$$\frac{dP}{dt} = \bar{k}_1 J_0^{1/2} c_0^m (1-P)(1+aP)^{1/2} \qquad (6.61)$$

We obtained the following expressions at $P_v^0 \le P$, $t_v^0 \le t$ in the case of a microheterogeneous system:

$$\frac{d\varphi_s}{d\tau} = \frac{1-P_v^0}{P_s^0 - P_v^0} \bar{k}_1 (J_0)^{1/2} c_0^m \left[(1+aP_v^0)^{1/2} + \frac{\bar{k}_2}{\bar{k}_1} \varphi_s \right] (1-\varphi_s)^{1-m\alpha} \qquad (6.62)$$

$$\frac{dP}{d\tau} = (P_s^0 - P_v^0) \frac{d\varphi_s}{d\tau} + \bar{k}_3 J_0 (c_0 L)^{2m} U(\tau) \qquad (6.63)$$

$$U(\tau) = \int_0^\tau (1-\varphi_s)^{-2m\alpha} \exp\left\{ -\frac{\bar{k}_3 J_0}{1-P_s^0} \left(\frac{c_0 L}{(1-\varphi_s)^\alpha} \right)^{2m} (\tau - t) \right\} \frac{d\varphi_s}{dt} dt \qquad (6.64)$$

where $\alpha = 1-L$ and

$$\bar{k}_1 = k_1 \gamma_v^{1/2} \varepsilon^m = k_p \gamma_v^{1/2} \varepsilon^m / (k_{tv}^0)^{1/2} \qquad (6.65)$$

$$\bar{k}_2 = k_2 \gamma_v^{1/2} \varepsilon^m = k_p F_{vs} k_m k_v \gamma_v^{1/2} \varepsilon^m / (k_{tvs})^{1/2} \qquad (6.66)$$

$$\bar{k}_3 = k_3 \gamma_s \varepsilon^{2m} = k_p \gamma_s \varepsilon^{2m} / k_{ts} [M_0] \qquad (6.67)$$

Let us rewrite the integral equations (6.40) and (6.41) as follows:

$$< P_s > = 1 - [(1-P_s^0) V(\tau) / \varphi_s], \qquad (6.68)$$

$$P = P_v^0 + (1-P_v^0)\varphi_s - (1-P_s^0) V(\tau) \qquad (6.69)$$

$$V(\tau)=\int_0^\tau \exp\left\{-\frac{\bar{k}_3 J_0 (c_0 L)^{2m}}{(1-P_\text{s}^0)(1-\varphi_\text{s})^{2m\alpha}}(\tau-t)\right\}\frac{\text{d}\varphi_\text{s}}{\text{d}t}\text{d}t\,. \tag{6.70}$$

Equations (6.61)–(6.70) describe the kinetics of the linear photoinitiated polymerization of methacrylates in optically thin layers up to high degrees of conversion in the three reactive zones, namely the liquid monomer–polymeric phase, in the "solid" polymer–monomeric phase and at the boundary of the above-mentioned phase division, according to the conception about the microheterogeneity of system.

6.3. Comparison with experimental data

The kinetics of photoinitiated polymerization of glycidyl methacrylate (2.3-epoxypropylmethacrylate) was studied using a laser interpherometric set-up in thin layers of $(0.5–2)\times10^{-4}$ m in the presence of a photoinitiator, 2.2-dimethoxy-2-acetophenone (IRGACURE 651, $C_6H_5–C(OCH_3)_2–C(O)–C_6H_5$), under UV-illumination using a DRT-400 mercurial quartz lamp.

With the aim of initiating the photopolymerization process, 2.2-dimethoxy-1.2-diphenylethane-1-on was used as initiator, having the following formula

```
             O— CH3
             |
C6H5—C—C — C6H5
        ||   |
        O  O— CH3
```

The characteristics of this initiator are as follows: content of the main substance is ≥98%, molecular weight (M_n)=156.30, density (d_4^{20})=1.210 g cm^{-3}, melting temperature $(T_\text{melt.})$=67–71°C.

The UV-spectrum of the presented initiator adsorption is characterized by two maxima. The first is caused by adsorption of phenyl groups at wavelengths λ<300 nm (π–π* transition), and the second one is observed in the field of carbonyl group adsorption at λ=310–340 nm as a result of n–π transition (see Figure 6.1).

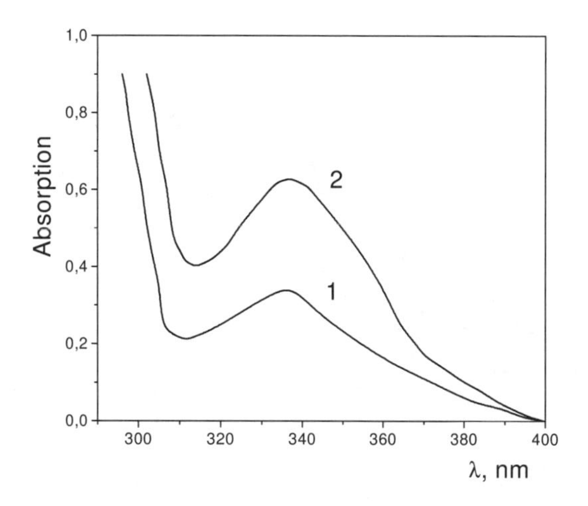

Figure 6.1. Adsorption spectrum of the initiator IRGACURE 651 at concentrations of 0.05% (1) and 0.10% (2) by mass. The solvent is chloroform.

It was determined with the use of NMR-spectroscopy (^{1}H and ^{13}C) that the photolysis of the photoinitiator proceeds *via* a α-chemical mechanism with the formation of benzoyl- and benzoylketal′ radicals (6.71), and also the formation of methyl radical from benzoylketal′ is observed.

$$
\begin{array}{ccccccccccc}
 & & & O\text{—}CH_3 & & & & & O\text{—}CH_3 & & \\
 & & & | & & & & & | & & \\
C_6H_5\text{—} & C & \text{—} & C & \text{—}C_6H_5 \Rightarrow C_6H_5\text{—} & C\bullet & + & \bullet & C & \text{—}C_6H_5 & \\
 & || & & | & & || & & & | & & \\
 & O & & O\text{—}CH_3 & & O & & & O\text{—}CH_3 & &
\end{array}
\qquad (6.71)
$$

The technique of the experiment is described in more detail in Ref. [19] and the typical kinetic curves of the glycidyl methacrylate polymerization are presented in Figure 6.2.

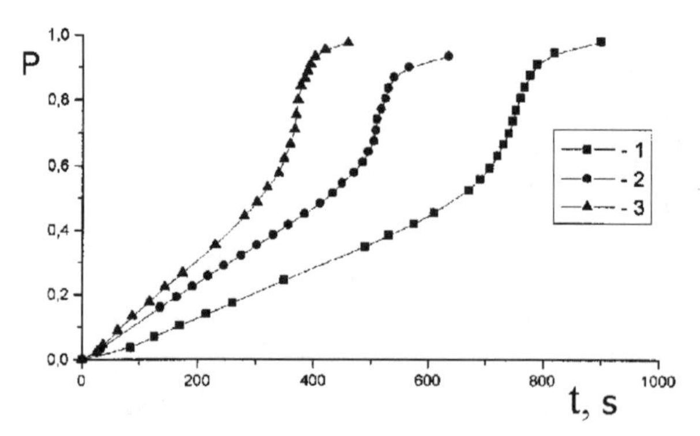

Figure 6.2. Typical kinetic curves of glycidyl methacrylate polymerization at photoinitiator concentrations of 0.5 (1), 1.5 (2) and 3.0% (3) by mass. T=283 K, $E_{0=}$37.4 W/m^2.

The concentration of the photoinitiator (0.5–3.0% by mass), temperature (10–30°C) and power E_0 of UV-illumination (37.4–65.0 W/m^2) on the surface of the photocomposition layer were varied.

The selected experimental kinetic curves of glycidyl methacrylate polymerization are presented in Figure 6.3 as a dependence of conversion on time. As we can see from the point's location, the instrumental error at the individual kinetic curve constructing is rather small. However, comparison the individual kinetic curves between each other (see Table 6.1) shows the scattering of the characteristic parameters (*e. g.*, the maximal rate W_0 of the process on the autoacceleration stage, conversion P_0 and time t_0 of achievement W_0, and also the rate W_1 of an initial linear section of the kinetic curve). Such scattering essentially exceeds the error of each individual experiment. At the same time, the scattering of the characteristics parameters represented in Table 6.1 fully shows the level of the influence of the layer thickness ℓ of the photopolymerizing composition *via* ranges of its changing. The poor reproducing of kinetic measurements or "whims" of the process are well known [22, 23] and they are a result of the fluctual sensitivity of the polymerization process, especially on an autoacceleration stage. That is why the results consisting of the five to eight kinetic curves represented in Figure 6.3 were obtained in the narrow limits of the layer thickness, changing for each set of the assigned parameters (starting concentration of the photoinitiator, temperature, and power of *UV*-illumination).

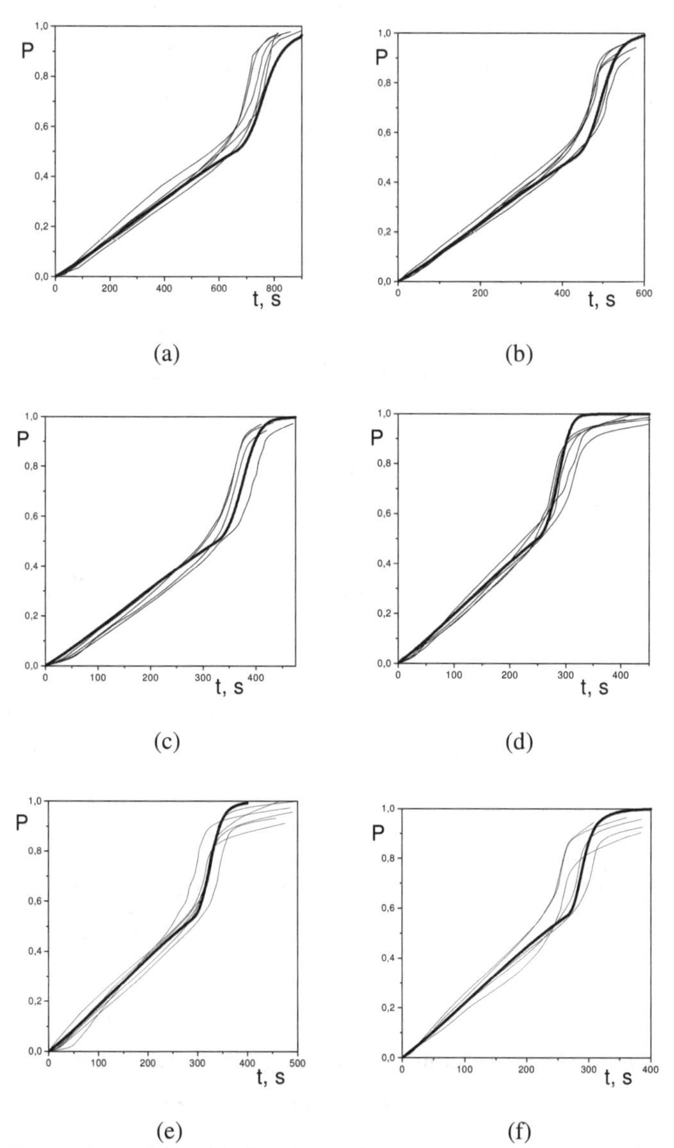

Figure 6.3. Comparison of the (thin lines) experimental data and (bold lines) calculated data in accordance with the equations (6.61)–(6.64) of the kinetic curves of glycidyl methacrylate photopolymerization in the following variants:

Variant	a	b	c	d	e	f
Photoinitiator concentration (%, by mass)	0.5	1.5	3.0	3.0	3.0	3.0
Power of UV-illumination E_0 (W/m^2)	37.4	37.4	37.4	65	37.4	37.4
Temperature (K)	283	283	283	283	293	303

Table 6.1. Characteristic parameters of glycidyl methacrylate polymerization kinetics.

No.	Thickness of layer $l \times 10^4$, m	Time of maximum rate reached (s)	Conversion at maximum rate, P_0	Maximum rate of process (s^{-1})	Rate of the linear site (s^{-1})
1	1.3	760	0.74	4.0	0.74
2	1.5	740	0.74	2.7	0.89
3	1.8	780	0.8	4.5	0.83
4	2.0	720	0.72	3.4	0.81
5	2.3	720	0.75	4.3	0.76

Photoinitiator concentration c_0=0.5% by mass, T=283 K, E_0=37.4 W/m^2.

The calculated kinetic curves were compared with the experimental data in Figure 6.3 and as we can see, they are in good agreement. The calculations were done with the use of the parameters of the constants of the model presented in Table 6.2.

Table 6.2. Parameters of proposed kinetic model for the different temperatures.

Temperature (K)	$\overline{k_1}$	$\overline{k_2}$	$\overline{k_3}$	P_v^0	P_s^0	a	L	m
283	0.020	0.50	2.40	0.50	0.80	6.5	0.50	0.4
293	0.024	0.6	2.87	0.53	0.76	6.5	0.55	0.4
303	0.029	0.72	3.45	0.57	0.73	6.5	0.60	0.4

Of these parameters, $\overline{k_1}$, a, P_v^0 and m were simply estimated from the initial sections of the kinetic curves. The others were selected manually with the aim of obtaining a satisfactory agreement with the experimental data in all ranges of the controlled parameters of process variation. As we can see from Table 6.2, with the exception of a=6.5 and m=0.4, the other parameters of this kinetic model change with temperature, but, at the same time, all of them do not depend upon the photoinitiator concentration and the intensity of the illumination. The values of the constants $\overline{k_1}$, $\overline{k_2}$ and $\overline{k_3}$ increased uniformly by 20% with an increase in temperature of 10°C, which corresponds to an effective activation energy of ≈ 12.6 kJ/mol. In accordance with the experimental data (see Figure 6.3c,e,f), the conversion P_v^0 in the saturated MPPh also increases with increasing temperature, which can be considered as polymer solubility propagation in the monomer with increasing temperature. Similarly, we can expect the monomer solubility in the polymer to increase, and that leads to P_s^0 diminution with increasing temperature. The factor L, the photoinitiator distribution between MPPh and PMPh, is increased at the expense of the above-mentioned effect.

The intensity of the UV-illumination (J_0) on the surface of the polymerization composition was calculated taking into account its power E_0 (W/m^2), in approximation, that it concentrated on the conditional wavelength λ=340 nm: $J_0=2.83\times10^{-6}$ E_0 mol×quant/m^2×sec. The starting concentrations of the photoinitiator c_0=(percent by mass)=ρ_m/M_{in} and the monomer in the bulk $[M_0]=\rho_m/M_m$ (where $\rho_m=1.04\times10^6$ g/m^3 is the density of the monomer, and M_{in}=256 and M_m=142 g/mol are the molecular masses of the initiator and the monomer, respectively) expressed in mol/m^3. For the given choice of the dimensions for J_0 and c_0, also $[M_0]$ determines the numerical values $\overline{k_1}$, $\overline{k_2}$ and $\overline{k_3}$ (Table 6.2).

From our point of view, taking into account the experimental data error, the accordance between them and calculated one can be estimated as satisfactory and it can be considered that the proposed kinetic model quantitatively explains the main characteristics of the photoinitiated linear polymerization up to the high conversion state in the layers with small optical densities.

The experimental data presented as a conversion dependence on time do not allow selection of the separate components of the linear polymerization process, to estimate their share into the summary rate of the process, or to underline its characterized peculiarities. The kinetic model gives this possibility.

Let us consider the most interesting stage of polymerization in the microheterogeneous system, that is, from the moment of polymer–monomer phase extraction.

The kinetic data for PMPh extraction from MPPh are presented in Figure 6.4 by the dependencies calculated according to equation (6.62), dependencies φ_s and $d\varphi_s/dt$, on time. The presented calculated results show that the maximum rate $(d\varphi_s/dt)_{max}$ of the PMPh from the MPPh extraction is observed practically at the same value $\varphi_{s(max)}$=0.525 at all concentrations of the photoinitiator. From the analysis of equation (6.62) on the function $d\varphi_s/d\tau$, it follows, that at the equilibrium distribution of the photoinitiator between MPPh and PMPh, that is, at L=1, the maximum value $d\varphi_s/d\tau$ should be observed at $\varphi_{s(max)} \leq 0.5$; at this time, the equality sign is fulfilled according to the condition $\overline{k_1} << \overline{k_2}$, at which the main share of $(d\varphi_s/d\tau)_{max}$ is contributed by the polymerization rate in the interface layer, whose maximum is determined by the function $\varphi_s(1-\varphi_s)$.

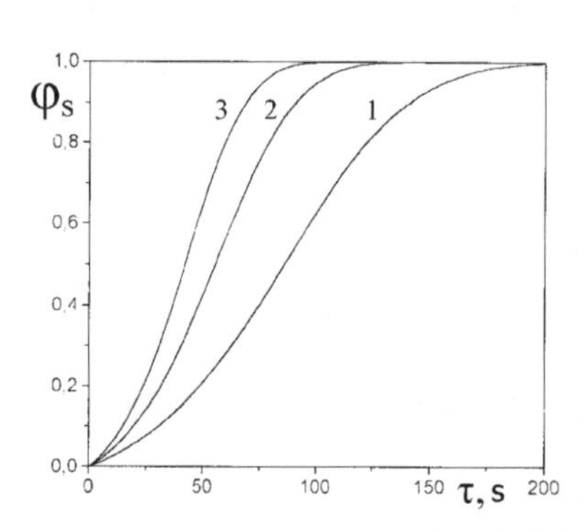

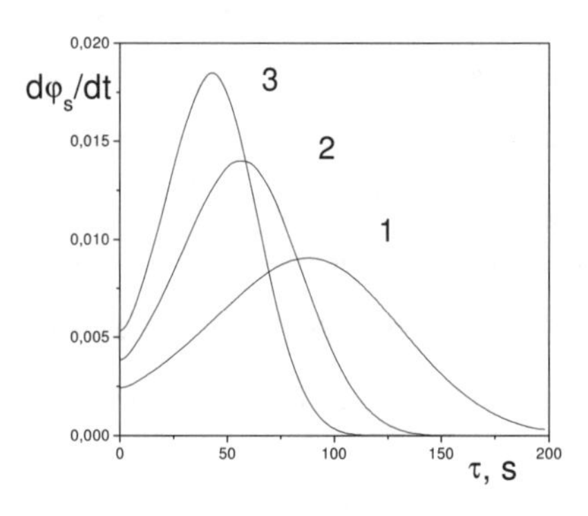

Figure 6.4. Dependencies calculated using equaiton (6.62) and at various concentration of the photoinitiator (percent by mass) in the microheterogeneous system: 0.5 (1), 1.5 (2), 3.0 (3) % by mass. E_0=37.4 W/m^2, T=283 K.

Under the same condition, i.e., $\overline{k_1} << \overline{k_2}$, but at a non-equilibrium photoinitiator distribution (L < 1), the maximum rate of polymerization in the interface layer, according to the function $\varphi_s(1-\varphi_s)^\beta$, where $\beta=1-m\alpha$, is achieved at the value $\varphi_{s(max)}=1/(1+\beta)$, that is, at $\varphi_{s(max)}=0.555$. The lesser value $\varphi_{s(max)}=0.525$ obtained in the numerical calculations indicates that the condition $\overline{k_1} << \overline{k_2}$ ($\overline{k_1}$ is 25 times less than $\overline{k_2}$) is not completely fulfilled and brings into the value $(d\varphi_s/d\tau)_{max}$ a small, but noticeable, share of polymerization in MPPh liquid.

Figure 6.5 compares the calculated summarized rate $dP/d\tau$ (equaiton (6.63)) and its constituents: $(dP/d\tau)_{v+vs}=(P_s^0-P_v^0)d\varphi_s/d\tau$ is the summarized rate of polymerization in MPPh and the interface layer on the boundary of MPPh and PMPh, and also $(dP/d\tau)_s = \overline{k}_3 J_0 (c_0 L)^{2m} U(\tau)$, the polymerization rate in PMPh. The maximum values $(dP/d\tau)_{v+vs}$ are observed at the same times as $d\varphi_s/d\tau$; the maximum values $(dP/d\tau)_s$ are observed at later times.

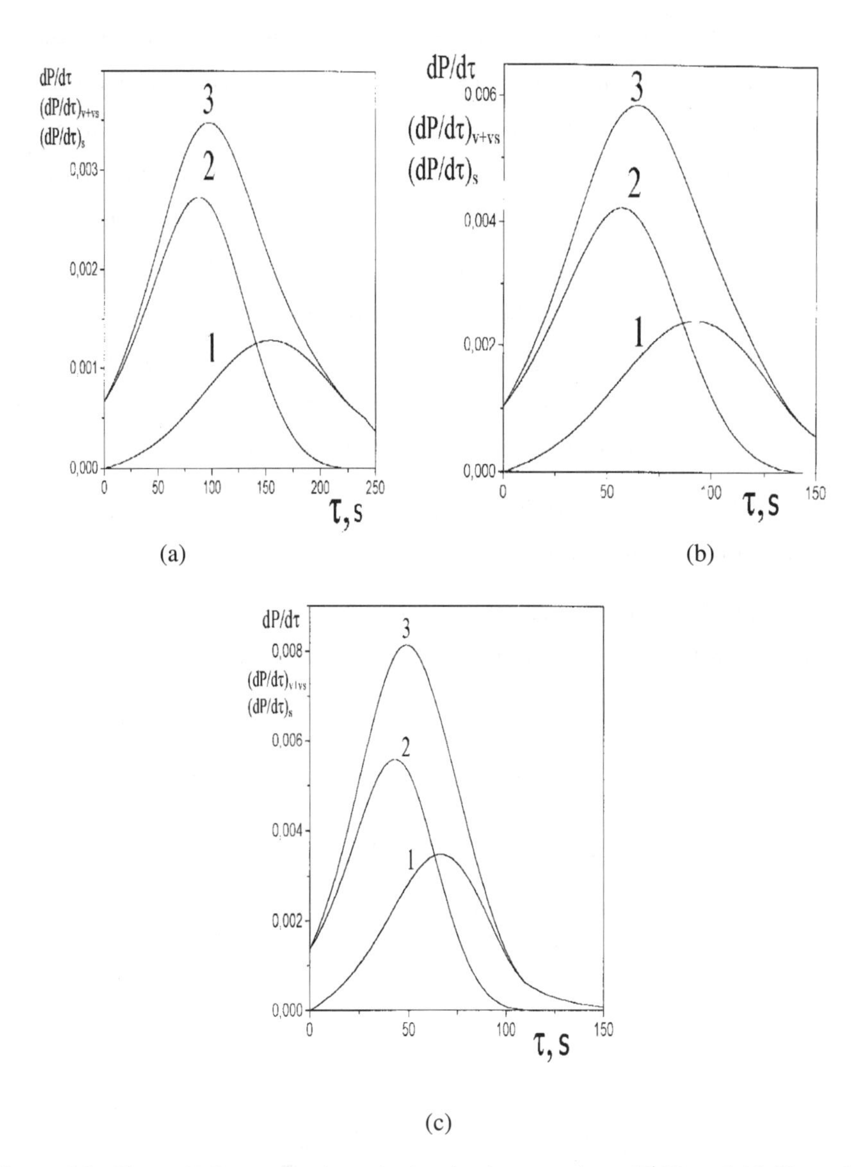

Figure 6.5. (Curve 3) Dependencies calculated using equations (6.62) and (6.63), and the summary polymerization rate $dP/d\tau$ in the microheterogeneous system and its components: namely, (curve 1) $(dP/d\tau)_s$, the polymerization rate in PMPh and (curve 2) $(dP/d\tau)_{v+vs}$, polymerization rate in MPPh and the interphase layer on the boundary of MPPh and PMPh.

The summarized rate of the process $dP/d\tau$ at short times t from the beginning of PMPh extraction is completely determined by $(dP/d\tau)_{v+vs}$, and at long τ, that is, in the finished stage of the polymerization process, by the rate $(dP/d\tau)_s$ of polymerization in PMPh. The extreme of the summarized rate $dP/d\tau$ of the polymerization *via* time is located between extremes $(dP/d\tau)_{v+vs}$ and $(dP/d\tau)_s$, but is considerably nearer to the extreme $(dP/d\tau)_{v+vs}$. In another case, the main share (up to 70%) of the maximum value $dP/d\tau$ gives $(dP/d\tau)_{v+vs}$. However, with increasing photoinitiator concentration this share decreased slowly and the share of $(dP/d\tau)_s$ is increased. When the extreme $dP/d\tau$ was slightly shifted relative to the extreme $d\varphi_s/d\tau$, the values $\varphi_s^0 > \varphi_{s(max)}$ correspond to maximum values $dP/d\tau$. The calculated data give a value $\varphi_s^0=0.625$, as well as $\varphi_{s(max)}$ practically not depending upon the photoinitiator concentration. Thus, the conversion P_0, corresponding to the maximal rate of polymerization, is determined, in our case, by the equation $P_0 \approx P_v^0(1-0.625) + \overline{P}_s^0\, 0.625$, where $\overline{P}_s^0$ is the average value of conversion to PMPh at the time corresponding to the maximum $dP/d\tau$.

The dependence of the values $\overline{P}_s$, calculated according to equation (6.68), on the time at various photoinitiator concentrations is presented in Figure 6.6.

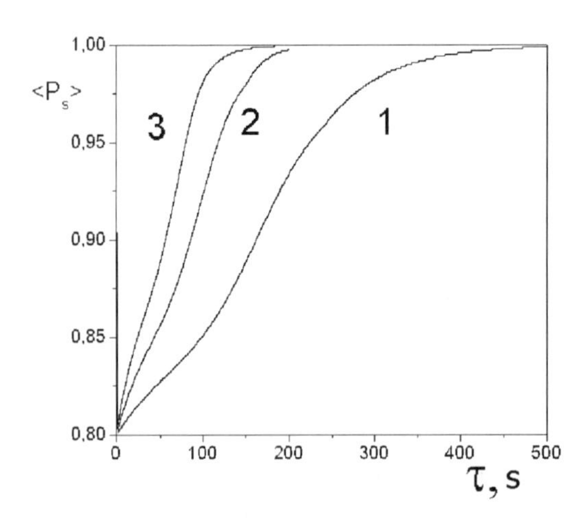

Figure 6.6. Average conversion to PMPh, calculated using equation (6.68), at different concentrations of the photoinitiator c_0 (% by mass): 0.5 (1), 1.5 (2), 3.0 (3). E_0=37.4 W/m^2, T=283 K.

The values $\overline{P}_s^0$, corresponding to the maximal rates of the summarized process, are very slightly increased (from 0.85 to 0.87) with increasing photoinitiator concentration.

From this it follows that the magnitudes P_0 should also be increased with the photoinitiator concentration increasing from ≈ 0.72 at c_0=0.5% by mass to 0.74 at c_0=3% by mass. The experimental data do not allow one to observe this weak effect.

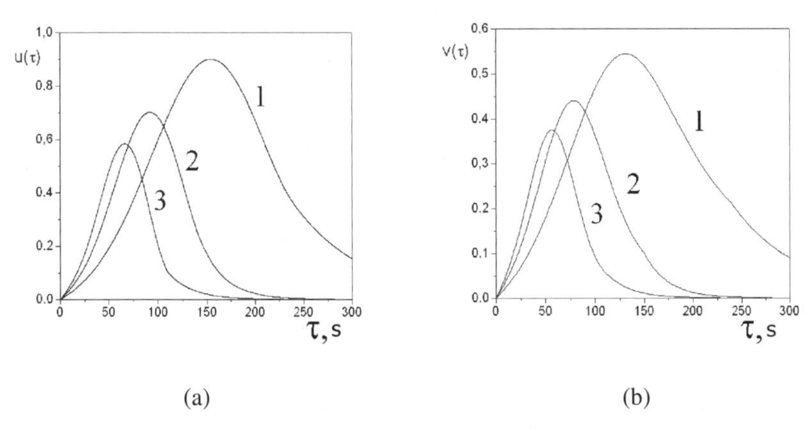

Figure 6.7. Behavior of integrals $U(\tau)$ and $V(\tau)$ in time, calculated in accordance with equations (6.64) and (6.70), at different concentrations of the photoinitiator c_0 (% by mass): 0.5 (1), 1.5 (2), 3.0 (3). E_0=37.4 W/m^2, T=283 K.

Even though the first order upon the photoinitiator, taking into account the gradient of illumination on the layer of the polymerization composition decreases to $2m$=0.8, is installed into the kinetic equation of the polymerization process in PMPh, the rate $(dP/d\tau)_s$ is increased 2.7-fold when increasing the photoinitiator concentration up to 6-fold (see Figure 6.5). It is connected with fact, that the co-multiplicant before the integral $U(\tau)$ is increased with increasing photoinitiator concentration in the rate expression $(dP/d\tau)_s$ for polymerization in the PMPh, but the integral $U(\tau)$ is decreased. The behavior of the integral $V(\tau)$ is similar. Their behavior in time at the different photoinitiator concentrations is shown in Figure 6.7.

REFERENCES

[1] Berlin A. A., Korolev G. V., Kepheli T. Ya. *Acrylic oligomers and materials on their basis;* M.: Khimija, **1983**; 232 p.

[2] Kurdikar D. L., Pepas N. A. *Polymer* **1994**, 35 (5), p. 1004.

[3] Shen I., Tian Y., Wang G. *Macromol. Chem.* **1991**, 192, 2669.

[4] Efimov A. L., Dyachkov A. I., Kuchanov S. I. *Vysokomol. Soed.* (*B*), **1982**, 24 (2), p. 83.

[5] Dyachkov A. I., Efimov A. L., Efimov L. I. *Vysokomol. Soed.* (*A*), **1983**, 25 (10), p. 2176.

[6] Zhu S., Tian Y., Hamilies A. *Macromolecules*, **1990**, 23, p. 1144.

[7] Efimov A. L., Bugrova T. A., Dyachkov A. I. *Vysokomol. Soed.* (*A*), **1990**, 32 (11), p. 2296.

[8] Doetschman D., Mechlenbacher R. *Macromolecules,* **1996**, 29, p. 1807.

[9] Broon E. B., Ivanov V. A., Kaminski V. A. *The Reports of the AS USSR*, **1986**, 291 (3), p. 618.

[10] Budtov V. P., Revnov B. V. *Vysokomol. Soed.* (*A*), **1994**, 36 (7), p. 1061.

[11] Vasylev D. K., Belgovsky I. M. *Vysokomol. Soed.* (*B*), **1990**, 32 (9), p. 678.

[12] Vasylev D. K., Belgovsky I. M. *Vysokomol. Soed.* (*A*), **1989**, 31 (6), p. 1233.

[13] Korolev G., Mogylevich I., Golikov I. Crosslinked polyacrylates. *Microheterogeneous structures, physical networks, deformation-strengthen properties;* M.: Khimija, **1995**; p. 48.

[14] Volkova M., Belgovsky I., Golikov I. *Vysokomol. Soed.* (*A*), **1987,** 28 (6), p. 435.

[15] Medvedevskikh Yu., Bratus A., Hafijchuk G., Zaichenko A., Kytsya A., Turovskij A., Zaikov G. *Journal of Applied Polymer Science*, **2002**, 86, p. 3556.

[16] Medvedevskikh Yu., Bratus A., Hafijchuk G., Zaichenko A., Kytsya A., Turovskij A., Zaikov G. *Polymer Yearbook* 18, **2003**, p. 101.

[17] Smirnov B. *Successes of the physical sciences*, **1986**, 149 (2), p. 178.

[18] Shaffer D., Kefer K. *Structure of random silicates: polymers, colloids and solids;* M.: Myr, **1988**; p. 62.

[19] Medvedevskikh Yu., Zaglad'ko O., Turovskyj A., Zaikov G. *Int. J. Polymer. Mater.*, **1999**, 43, p. 157.

[20] Zaglad'ko O., Medvedevskikh Yu., Turovskyj A., Zaikov G. *Int. J. Polymer. Mater.*, **1998**, 39, p. 227.

[21] Medvedevskikh Yu., Zaglad'ko O., Turovskyj A., Zaikov G. *Russian Polymer News*, **1999**, 4 (3), p. 33.

[22] Kolegov V. *Vysokomol. Soed.* (*A*), **1995.** 37 (1), p. 16.

[23] Treushnikov V., Zelentsova N., Olejnik A. *Journal of scientific & applied photo- and cinema*, **1988**, 33 (2), p. 146.

Chapter 7. Non-stationary kinetics (postpolymerization)

7.1. Experimental regularities of dimethacrylate postpolymerization

In Chapter 4 points of view on the postpolymerization process and mainly on the chemical mechanism of the monomolecular chain termination have been presented based on rather modest experimental material. In Chapter 7 more experimental data are presented, purposely obtained [1, 2] for the discovering of the nature of monomolecular chain termination.

Experiments have been carried out using the interferometric laser plant measuring current (H) and limited achieved (H_0) contractions of the photopolymerization composition layer. Taking taken into account the ratio of the above-mentioned contractions, the relative degree of polymerization (or conversion) was determined, $P=H/H_0$. Integral UV-illumination of a DRT-400 lamp with an intensity of 37.4 W/m^2 falling on the surface of the photopolymerization composition has been used in the experiments. UV-illumination was stopped at a defined stage of the luminous polymerization process, while continuing to register the contraction of composition layer in the dark; thereafter, the UV-illumination was switched back on, thus providing the finish of the polymerization process and the determination of the final contraction H_0. Taking into account this value H_0, the thickness of the layer has been determined. The thickness of the photocomposition layer was varied in the range of $(0.5–3.0)\times10^{-4}$ m. 2.2-Dimethoxy-1.2-diphenylethane-1-one was used as photoinitiator of the process at a starting concentration of 1% (by mass). Calculations of the monomolecular chain termination rate constants have been realized *via* an optimization method using the program ORIGIN-5.0.

The kinetics of postpolymerization of dimethacrylates, namely MGPh-9: {M-O-$(CH_2CH_2O)_3$-C(O)-C_6H_4-C(O)-$(OCH_2CH_2)_3$-O-M}; OCM-2: {M-O-(CH_2CH_2O)-C(O)-$(CH_2$-$CH_2O)_2$-C(O)-O-(CH_2CH_2O)-O-M}; TGM-3: {M-O-$(CH_2$-CH_2-O-$)_3$-M-} and DMEG: {M-O-CH_2-CH_2-O-M-}, where M is CH_2=C(CH_3)-C(O)-, have been studied in the temperature range 5–50°C.

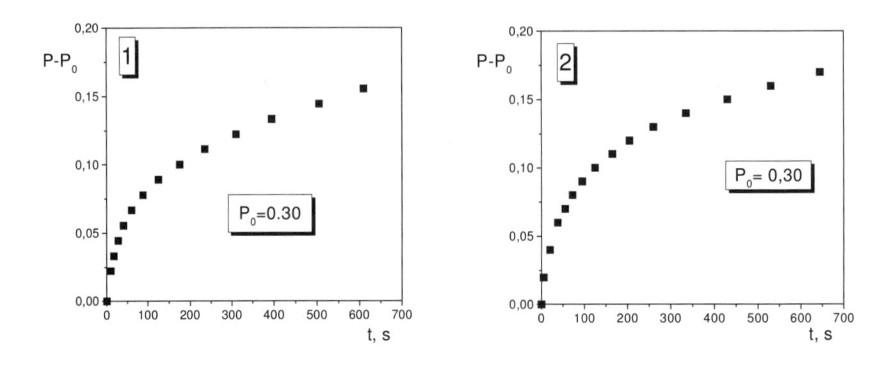

Figure 7.1. Typical kinetic curves for DMEG postpolymerization at 5 (1) and 15°C (2).

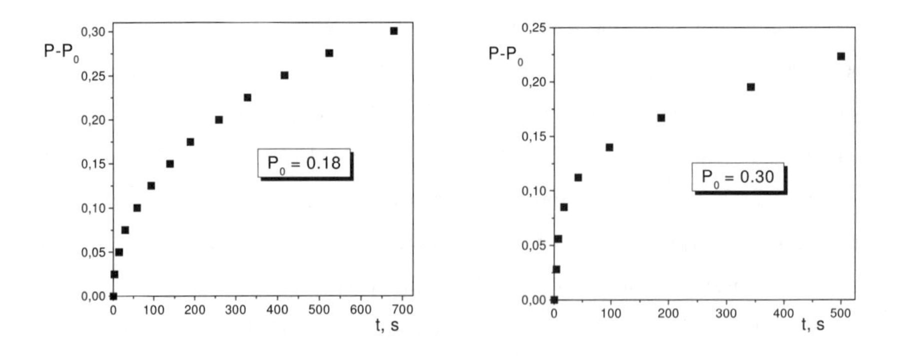

Figure 7.2. Typical kinetic curves for BDMC postpolymerization at 20°C and different stages of conversion at the start of the dark period: $P_0 = 0.18$ (1) and $P_0 = 0.30$ (2).

Typical experimental kinetic curves of the postpolymerization of dimethacrylates at different temperatures are presented in Figures 7.1–7.5 as dependence of conversion increment $P–P_0$ on the duration (time t) of the dark period.

In order to discover the nature of monomolecular chain termination, the influence of the nature and the concentration of plastifying additives on the rate of dimethacrylate OCM-2 postpolymerization have been investigated.

Kinetics of postpolymerization have also been investigated at 25°C, UV-illumination intensity of 37.4 W/m^2 and concentration of the photoinitiator 1% (by mass).

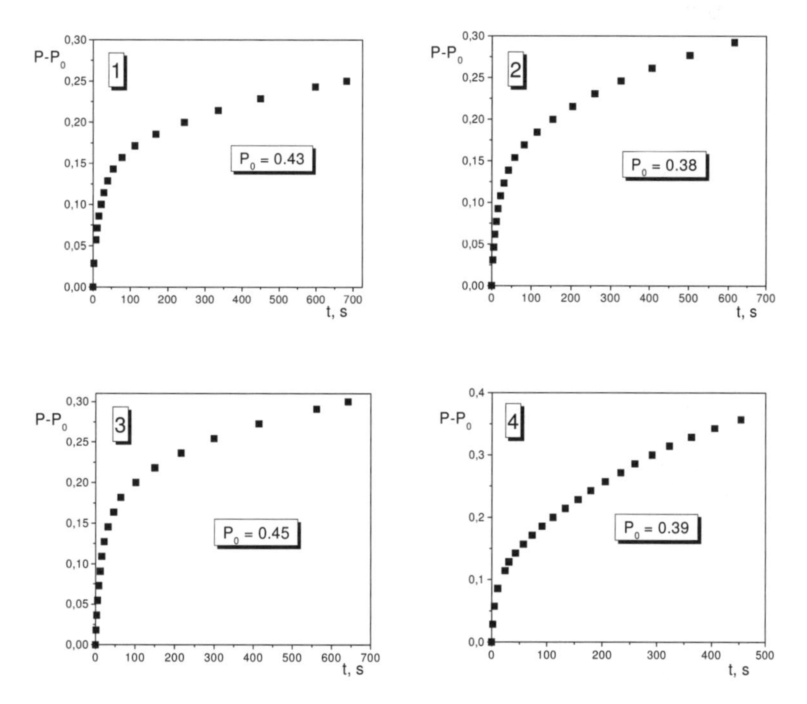

Figure 7.3. Typical kinetic curves for TGM-3 postpolymerization at 5 (1), 10 (2), 15 (3) and 20°C (4).

Typical kinetic curves for OCM-2 postpolymerization in the presence of plastifying additives are presented in Figures 7.6–7.9 as dependencies of conversion increment $P–P_0$ on the duration (time t) of the dark period.

The following reagents have been used as plastifying additives:

- oligoperoxide (OP), having the following formula

```
~[CH2–CH]n1–[CH2–CH]n2–[CH2–CH]n3~
      |            |            |
      O           C≡C      O=C–O–C4H9
      |            |
  O=C–CH3      C(CH3)2–O:O–C(CH3)3
```

at concentrations of 1.0, 2.5 and 5.0% (by mass), and also its complexes with Cu^{2+} ions ($OP(Cu^{2+})$; 1.5%, by mass) and Cr^{3+} ($OP(Cr^{3+})$; 1.0%, by mass);

- glycidylmethacrylate, 5.0% (by mass)

$$CH_2{=}C(CH_3){-}C(O){-}O{-}CH_2{-}\underset{\diagdown\ \diagup \atop O}{CH{-}CH_2}$$

and

- dioctylphtalate (DOPh)

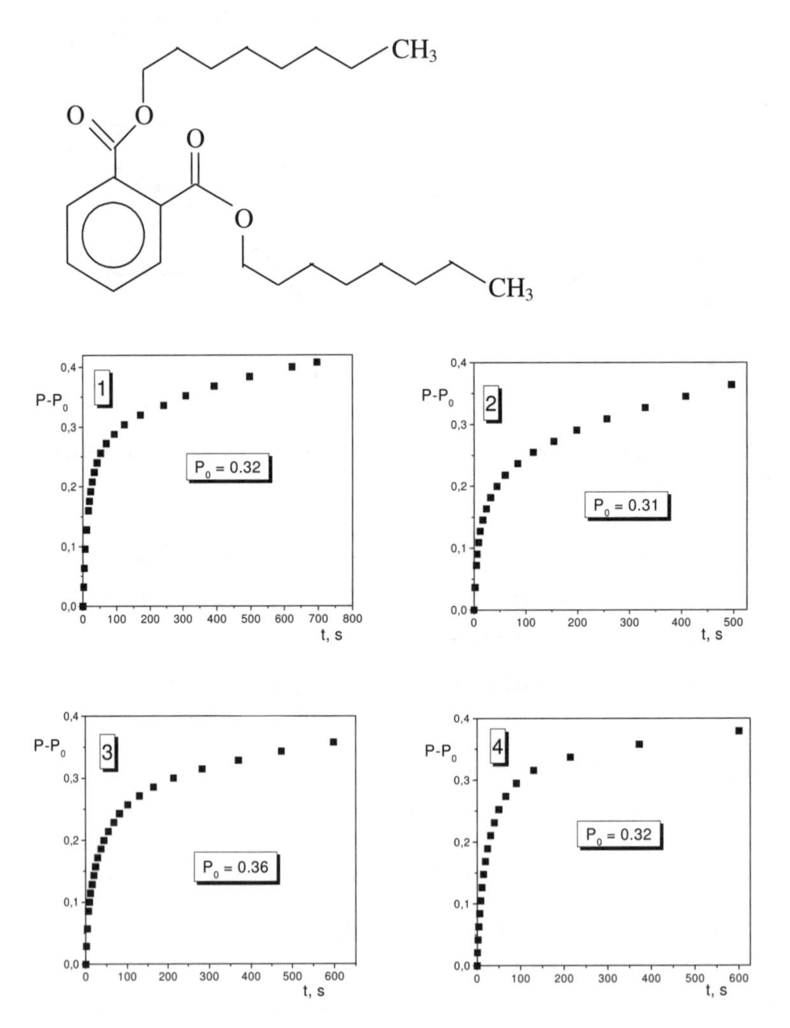

Figure 7.4. Typical kinetic curves for OCM-2 postpolymerization at 5 (1), 10 (2), 15 (3) and 20°C (4).

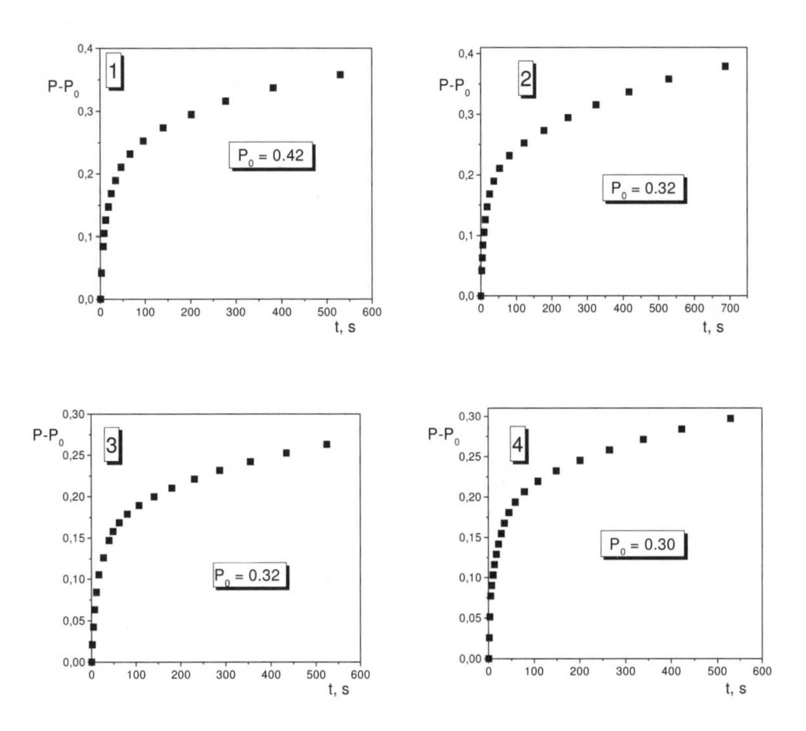

Figure 7.5. Typical kinetic curves for MGPh-9 postpolymerization at 10 (1), 15 (2), 20 (3), 25 (4), 30 (5), 35 (6), 40 (7) and 50°C (8).

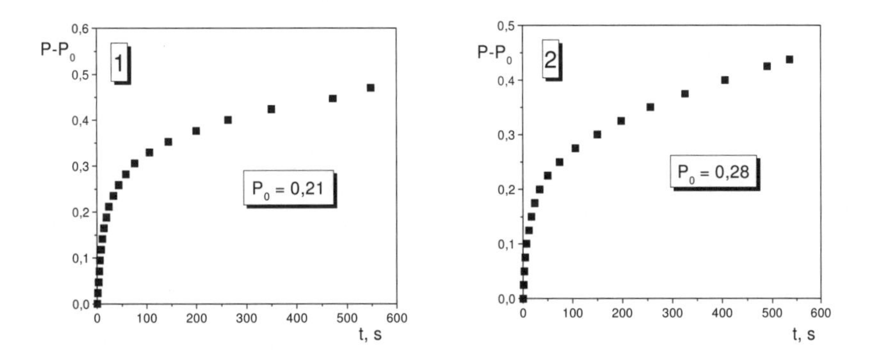

Figure 7.6. Typical kinetic curves for OCM-2 postpolymerization in the presence of the plastifying additive DOPh at 2.5 (1) and 5.0% (by mass) (2).

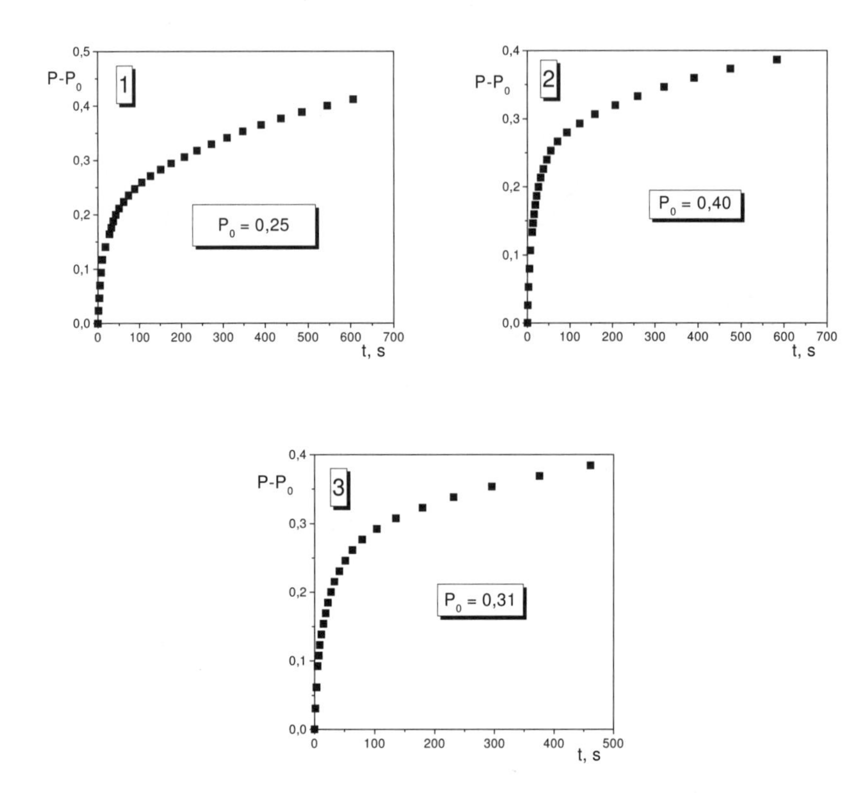

Figure 7.7. Typical kinetic curves for OCM-2 postpolymerization in the presence of the plastifying agent OP at 2.5 (1), 2.5 (2) and 5.0% (by mass) (3).

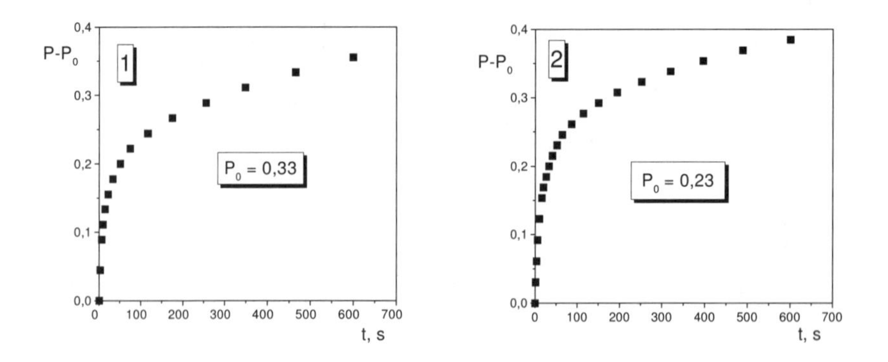

Figure 7.8. Typical kinetic curves for OCM-2 postpolymerization at different concentrations and different plastifying agents: 1.5% (by mass) OP(Cu^{2+}) (1) and 1.0% (by mass) OP(Cr^{3+}) (2).

Note that the experimental error in the construction of an individual kinetic curve of a postpolymerization is sufficiently small, as observed from the location of the points in the Figures 7.1–7.9. At the same time, kinetic curves scattering, performed under the same experiemtnal conditions, greatly exceeds the error of an individual kinetic curve. Therefore, 4–15 kinetic curves were obtained for each experimental condition. This increased the reliability of obtained estimations.

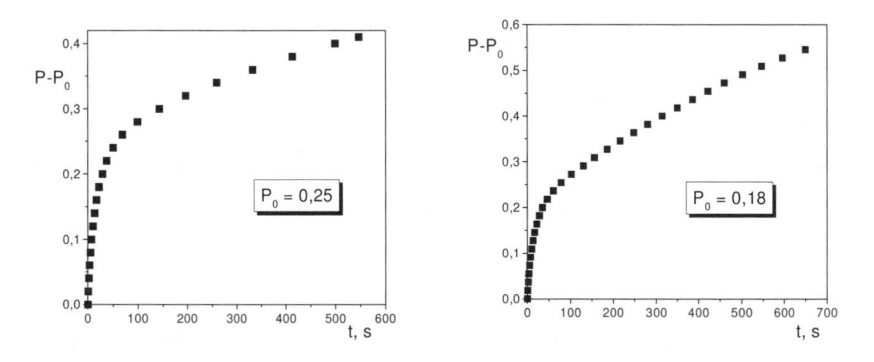

Figure 7.9. Typical kinetic curves for OCM-2 postpolymerization in the presence of glycidyl methacrylate (5%, by mass) at different conversions at the start of the dark period.

As we can see from Figures 7.1–7.9, there are two sections on the kinetic curve. The first is rapid and short, and the second is slow and long, with a relaxation time of 100 s. This proves the fact that the polymerization is led by radicals with different characteristic life times.

7.2. The kinetic model and the stretched exponential law

In the previous analysis of experimental data, it was observed that these data can be described using the main principles of the microheterogeneous conception. However, in the interpretation of experimental data as a whole, a range of facts excluding one another is discovered. This proves the imperfection of the presented model and, therefore, the simplification to assume the presence in the system of only two types of macroradicals is made, since it is more feasible to expect the presence of a wide spectrum of characteristic times of monomolecular chain termination. That is why the main principles for deduction of the kinetic model of non-stationary (so-called dark, since it occurs after the UV-illumination is stopped) photoinitiated polymerization have been formulated as follows [1, 2]:

(i) the observed polymerization rate is the sum of the homophaseous polymerization process taking place in the volume of liquid monomer, according to the classical kinetic scheme with bimolecular chain termination, and the rate of the heterophaseous process taking place in the interface layer at the solid polymer–liquid monomer boundary in the gel-effect regime with monomolecular chain termination;

(ii) radicals in the liquid phase are characterized by short life times and, therefore, cannot make an essential contribution to the postpolymerization process, of which the rate is determined by the rate of the heterophaseous process at the interface;

(iii) the monomolecular chain termination rate is controlled by the rate of its propagation, and the acts of chain propagation and chain termination represent two different results of active radical interaction with the functional group(s) of a monomer, causing the formation of active radicals (chain propagation), or their 'freezing", i.e., trapping that leads to formation of non-active radicals;

(iv) postpolymerization is led by radicals with different characteristic life times and, therefore, the relaxation function of the monomolecular chain termination is obeys the stretch exponential law.

Taking into account the above, the following scheme of monomolecular chain termination has been presented by us:

$$R_m + M_m \begin{cases} \xrightarrow{k_p} R_m \\ \xrightarrow{k_t} R_z \end{cases} \tag{7.1}$$

According to scheme (7.1) the acts of chain propagation and termination at the interface represent two results of the interaction of active radical R_m with a functional group of monomer M_m. Also, the first route leads to chain growing (propagation) and the second to monomolecular chain termination, i.e., monomolecular chain termination represents the act of its propagation leading to a trap.

In accordance with the scheme (7.1) the rate of chain propagation (w_p) can be written as follows:

$$w_p = k_p\,[M_m]\,[R_m], \tag{7.2}$$

and, respectively, the rate of the monomolecular chain termination (w_t)

$$w_t = k_t\,[M_m]\,[R_m] \tag{7.3}$$

here $[M_m]$ and $[R_m]$ are concentrations of monomer and active radicals at the interface, respectively.

Thus, the rate of radical's concentration changing under non-stationary process conditions can be described by the expression

$$d[R_m]/dt = v_{im} - k_t\,[M_m]\,[R_m], \tag{7.4}$$

in which v_{im} is the rate of initiation.

Since a liquid phase (sol) in the polymerization process of polyfunctional monomers represents the starting monomer, we can assume that the concentration of monomer at the interface is constant ($[M_m]$=constant) and, respectively, $k_t[M_m]$ will also have a constant value. Therefore, taking into account that the dark process proceeds without initiation ($v_{im}=0$), the concentration of radicals as a function on time can be described by equation

$$[R_m] = (v_{im0}/\beta)\exp\{-\beta t\}. \tag{7.5}$$

In this case time t is calculated from the start of dark period, v_{im0} is the rate of initiation at the moment the UV-illumination is stopped, $v_{im0}/\beta=[R_{m0}]$ is starting concentration of radicals for dark period in reactive zone and parameter

$$\beta = k_t\,[M_m] = \tau_t^{-1} \tag{7.6}$$

determines the characteristic life time τ_t of a radical, or relaxation time.

In accordance with the starting principle we assume that the process at the interface takes the main share in the kinetics of postpolymerization and, therefore, the polymerization rate can be written as follows:

$$-d[M]/dt = w_m\varphi_m = w_p\varphi_m, \tag{7.7}$$

where [M] is the molar-volumetric concentration of monomer calculated on the total volume of the system, w_m is the specific rate of polymerization at the interface,, equal to the chain propagation rate w_p and φ_m is the volumetric part of the interface layer.

After substitution of equation (7.5) in equation (7.2) the specific rate of a polymerization at an interface can be written as follows:

$$w_m \equiv w_p = v_{im0}\,(k_p/k_t)\exp\{-\beta t\} \tag{7.8}$$

The volumetric part φ_m of the interface layer is approximated *via* function [1]

$$\varphi_m \approx h\,(F_s/F_v)\,\varphi_s\,(1-\varphi_s) \approx h\,(F_s/F_v)\,p\,(1-p), \tag{7.9}$$

here φ_s is the volumetric part of the solid polymeric phase; h is the thickness of the interface layer, F_s/F_v is the ratio of fractal characteristics of the surface and volume of solid polymeric phase micrograins in liquid monomeric phase and microdrops of liquid monomeric phase in the solid polymeric matrix and p is relative conversion.

Let us express the process rate in units of relative conversion (p):

$$\frac{dp}{dt} = -\frac{d[M]/dt}{[M_0]\Gamma_0}, \tag{7.10}$$

in which $[M_0]$ is the concentration of monomer in bulk and Γ_0 is the limited achieved conversion in the solid polymeric phase.

By combining equations (7.7)–(7.10) we will obtain the expression describing the postpolymerization rate:

$$dp/dt = kp(1-p)\,v_{im0}\exp\{-\beta t\}; \tag{7.11}$$

here, k is $(k_p/k_t)\,h\,(F_s/F_v)\,[M_0]\,\Gamma_0$.

In the presented equation, the initiation rate v_{im0}, the polymerization rate dp/dt and the relative conversion p are functions both of time and also of the coordinate of layer x. However, experimentally measured parameters are their integral analogies, averaged upon the total thickness of layer ℓ. They can be written *via* transformations:

$$W = dP/dt = \ell^{-1}\int_0^{\ell}(dp/dt)dx, \qquad P = \ell^{-1}\int_0^{\ell} p dx \tag{7.12}$$

The differential initiation rate we will write as a function of layer coordinate x at the end of an illumination period based on equation (5.39) (see Chapter 5.3):

$$v_{im0} = \gamma_{\text{i}}\varepsilon c_0 J_0 e^{\tau_0 - y_0}[1+e^{-y_0}(e^{\tau_0}-1)]^{-1}. \tag{7.13}$$

here $y_0 = \varepsilon c_0 x$, $\tau_0 = \gamma_i \varepsilon J_0 t'$; c_0 is initial concentration of the photoinitiator, J_0 is the intensity of light falling on the surface of the photocomposition, ε is molar extinction coefficient of the photoinitiator, γ_i is quantum yield of photoinitiation and t' is illumination time.

By integration of expression (7.12), taking into account equations (7.11) and (7.13), and using the mean value theorem we will obtain

$$\frac{\mathrm{d}P}{\mathrm{d}t} = k\frac{P(1-P)\exp\{-\beta t\}\gamma_{\text{i}} J_0(1-\exp\{-y_0\})}{\ell(1+\exp\{-y_0\})(\exp\{\tau_0\}-1)} \tag{7.14}$$

At the beginning of the dark period (t=0), $P=P_0$ and equation (7.14) expresses the initial rate of the postpolymerization:

$$\left(\frac{\mathrm{d}P}{\mathrm{d}t}\right)_{t=0} \equiv W_0 = \frac{kP_0(1-P_0)\gamma_{\text{i}} J_0(1-\exp\{-y_0\})}{\ell(1+\exp\{-y_0\})(\exp\{\tau_0\}-1)} \tag{7.15}$$

Taking into account equation (7.15), the expression (7.14) for postpolymerization rate can be written as follows

$$\frac{\mathrm{d}P}{\mathrm{d}t} = W_0\frac{P(1-P)}{P_0(1-P_0)}\exp\{-\beta t\} \tag{7.16}$$

As we expected, equation (7.16), with single characteristic time $\tau_t = \beta^{-1}$, satisfactorily describes only the starting rapid and short section of the experimental kinetic curve. This means that monomolecular chain termination is characterized by a wide spectrum of relaxation times; therefore, the stretched exponential law has been used for a full description of the kinetic curve. In accordance with this law the function of relaxation $\psi(t)$ is written as follows:

$$\psi(t)=A\exp\{-bt\}^{\gamma},\quad 0<\gamma<1. \tag{7.17}$$

The value $0 < \gamma < 1$ gives the effect of the stretched exponential to equation (7.17).

The stretched exponential law in the form of equation (7.17) has been proposed first in 1864 by Kohlrausch [3] for the description of material creeping and has been used by Williams and Watts [4] in 1970 for the analysis of dielectric relaxation of polymers. It was disclosed from this time, that the above-mentioned law describes the different types of relaxation, mainly in the non-ordered systems, in which spatial inhomogeneity forms the hierarchy of relaxation times subordinated to the properties of the fractal set [5]. Since then it has been discovered that the above-said law describes the different types of relaxation, mainly in the non-ordered systems, in which the dimensional inhomogeneity causes the formation of relaxation times hierarchy characterized by fractal properties [6–8]. Among these properties, in Ref. [8] the electron transfer reaction from the cation of methyl violet (P) to colloidal particles of platinum (unsaturated trap T) in aqueous solution is considered. When in contact with a trap mono-cations give one electron and disappear. The unsaturation of a trap means, that the reaction $P + T \rightarrow T$ can be repeated unlimited times (in the presented case one particle of platinum can assemble up to 10^4 electrons). It is considered, that in the case when the concentrations of cations and traps are high, or the reaction proceeds under intensive stirring, then the process is described by the equation applied for a classical first-order reaction. The time-course of the concentration decrease will be characterized by the usual exponential shape. However, if the concentration of accidentally located traps is low, then there will be fields of space which are practically free from the traps. Cations in such a field can reach traps only after a long time and, consequently, decreasing their quantity *via* time will be slower. Attempts to use the stretched exponential law for the description of the rate constants of the bimolecular chain termination in the solid polymeric matrix at high temperatures (above 100°C) [9] also can be found in the literature.

There are two main approaches for the deduction of the stretched exponential law. The first is based on the fractal properties of the characteristic relaxation times spectrum. In order to give a short mathematical description of these variants the so-called forsteric model of direct transfer [5] can be applied. The above-mentioned model was the result of investigations on excitation transfer from donor to statistical defects in condensed media. A law of excitation decrease of the selected donor, which is located in the origin of coordinates, at the expense of direct energy transfer to a defect located in the junction R_i of a lattice with the defined structure is considered. The relaxation function $\psi_i(t)$ is the probability of fact, that for the same time t a donor is retained in the exciting state. This function is equal to

$$\psi_i(t)=\exp\{-tW(R_i)\}. \tag{7.18}$$

Here $W(R_i)$ is the relaxation rate, depending on the distance between the donor and defect, and determines the relaxation time $\tau(R_i)=1/W(R_i)$.

If all defects are located accidentally and the probability that this junction is occupied by a defect is equal to p, then we will obtain for the relaxation function of the whole system

$$\psi(t)=\prod_{i=1}[1-p+p\exp\{-tW(R_i)\}]. \tag{7.19}$$

In the case that $p << 1$, equation (7.19) can be written as

$$\psi(t)\cong\exp\left\{-p\sum_i[1-\exp\{-tW(R_i)\}]\right\}. \tag{7.20}$$

After introduction of the lattice junctions density $\rho(R)$, the sum can be written as integral

$$\psi(t)\cong\exp\left\{-p\int dR\rho(R)[1-\exp\{-tW(R)\}]\right\}. \tag{7.21}$$

Now, an empirical function $W(R)=aR^{-s}$ (a and s are constants) is introduced in equation (7.21) and, if the dimensional structure of system is uniform ($\rho(R)$=constant), then we will obtain the following equation for D-space:

$$\psi(t)\cong\exp\left\{(-t/\tau)^{D/s}\right\} \tag{7.22}$$

The second approach is based on the different activation energies of separate particles. However, if at the adsorption on the inhomogeneity surface this approach is physically grounded, then under consideration of processes in the liquid phase the assumption about activation energy dependence upon time looks unconvincingly. In spite of this fact, the mathematical description of problem practically does not differ from the previous one [10–12].

However, the most interesting and incomprehensible in the nature of the stretched exponential law is the fact that for more experimental regularities the index degree is near to 0.6.

After the introduction of the stretched exponential effect in equation (7.16) we obtain the final kinetic equation of the postpolymerization

$$\frac{dP}{dt} = W_0 \frac{P(1-P)}{P_0(1-P_0)} \exp\{-\beta t\}^{\gamma} \tag{7.23}$$

7.3. Dependence of the monomolecular chain termination rate constant on temperature, molecular mass of monomer and additives nature

The integral form of equation (7.23) has not a simple analytic view, and that is why integral kinetic curves $P–P_0 = f(t)$ (see Figures 7.1-7.9) were numerically differentiated for comparison of equation (7.23) with experimental data. Thus, the differential kinetic curves $dP/dt = f(t)$ were obtained; typical examples of such curves are presented in Figures 7.10–7.14 by points. On the basis of comparison of the obtained curves and equation (7.23) using the optimization method all three constants of equation (7.23) have been found, namely w_0, β and γ.

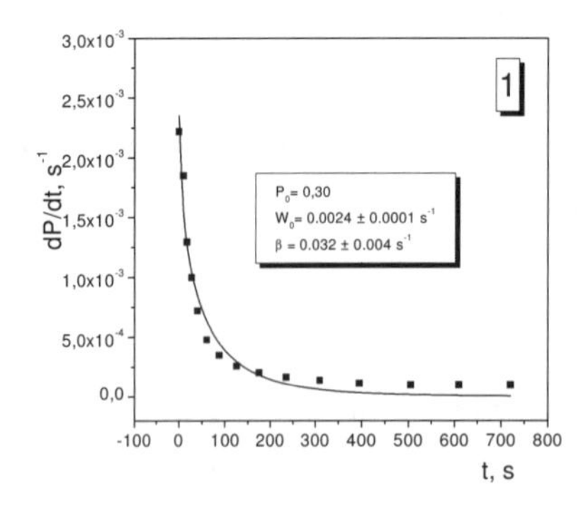

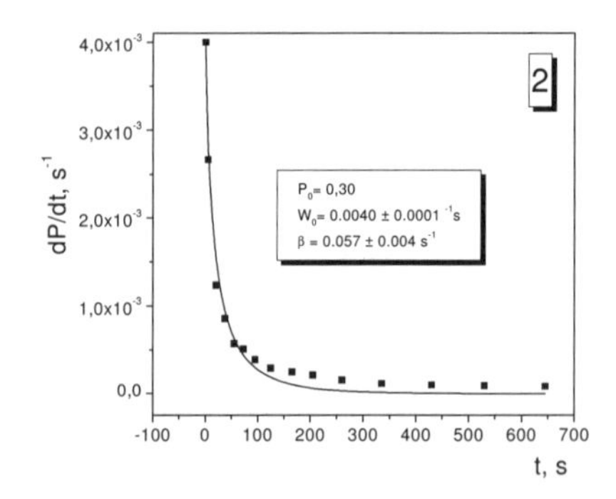

Figure 7.10. Interpretation of experimental kinetic curves of DMEG postpolymerization in the differential form (points) and compared with calculated values using equation (7.23) (lines) at 5 (1) and 15°C (2).

As was discovered, the spread in numerical values for parameter γ from 0.5 to 0.7 was less satisfactory in the selection of all parameters for optimization on the experimental series. This can be explained by the fact that the most important section of the kinetic curve for estimation of parameter γ, namely the slow and long section, is characterized by low values dP/dt and forms only a small share of the function of number squared of calculated deviations from experimental ones. Fixing the parameter γ showed that the minimal standard error in the estimation of w_0 and β is observed at the value γ=0.6. Thus, all experimental curves dP/dt = $f(t)$ were compared with equation (7.23) at fixed value γ=0.6 with optimization only on two parameters, namely w_0 and β. Examples of comparison of values calculated using equation (7.23) (lines) in accordance with experimentally determined kinetic curves (points) are presented in Figures 7.10–7.14. Note that in all cases satisfactory correspondence is found between experimental kinetic curves and kinetic curves calculated using equation (7.23).

As was mentioned earlier [13, 14] the experimental error in the construction of separate kinetic curve of postpolymerization is sufficiently small. This can be proven by the location of the points in Figures 7.10–7.14.

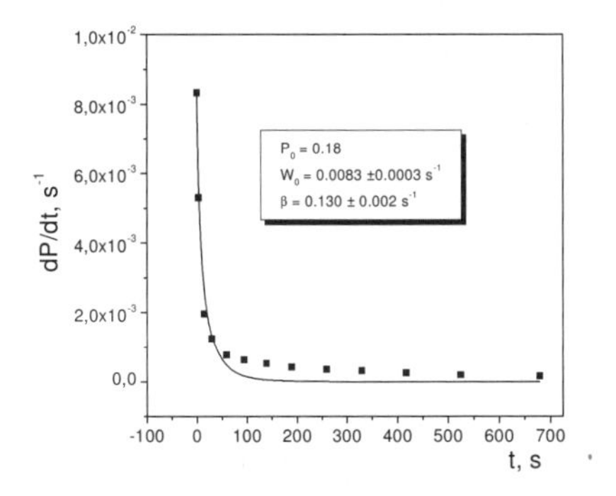

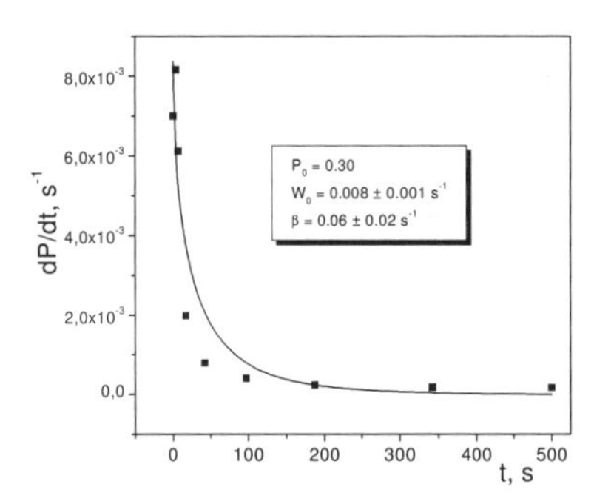

Figure 7.11. Interpretation of experimental kinetic curves for BDMC postpolymerization in the differential form (points) and compared with curves calculated using equation (7.23) (lines) at different starting conversions.

At the same time, the spread in kinetic curves under the same experimental conditions considerably exceeds the error of the individual kinetic curve. This phenomenon is well-known as a bad reproduction the kinetic measurements and, in accordance with Refs. [13, 14] represents direct proof of the microheterogeneity of the polymerization system, a fluctuated mechanism of the formation and propagation of the solid polymeric phase and interface layer at the "liquid monomer–solid polymer" boundary, and also its active role *via* postpolymerization.

Therefore, 4–15 kinetic curves have been obtained for each experimental condition. Based on these curves kinetic parameters of process have been averaged and, thus, the statistical truth of obtained estimations was improved. For example, in Table 7.1 values of the parameter β obtained from individual kinetic curves and also results of their average are given.

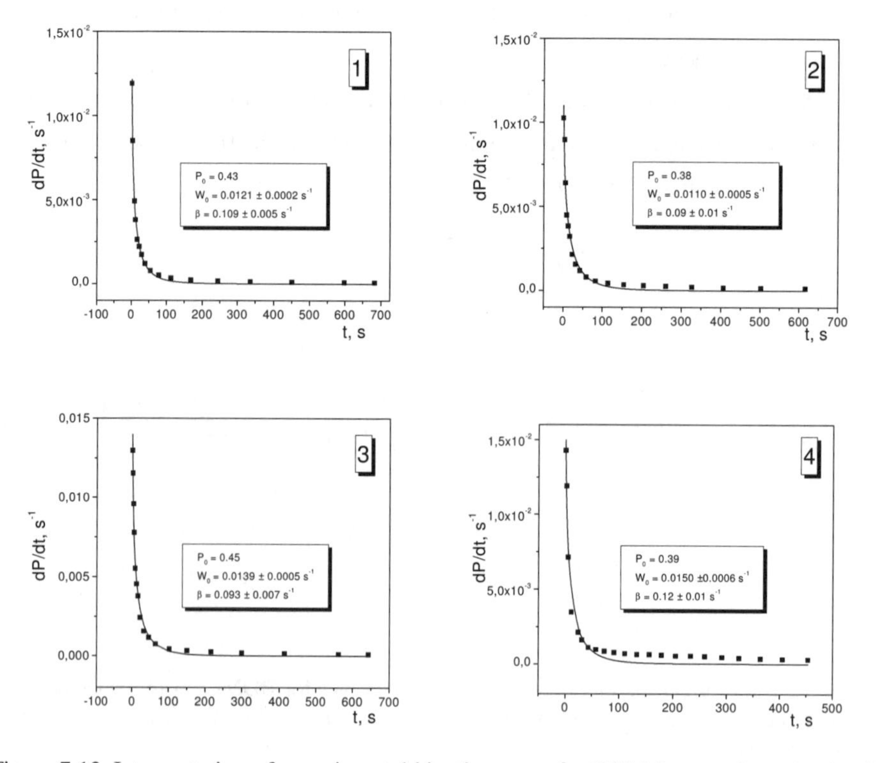

Figure 7.12. Interpretation of experimental kinetic curves for TGM-3 postpolymerization in the differential form (points) compared with curves calculated using equation (7.23) (lines) at 5 (1), 10 (2), 15 (3) and 20°C (4).

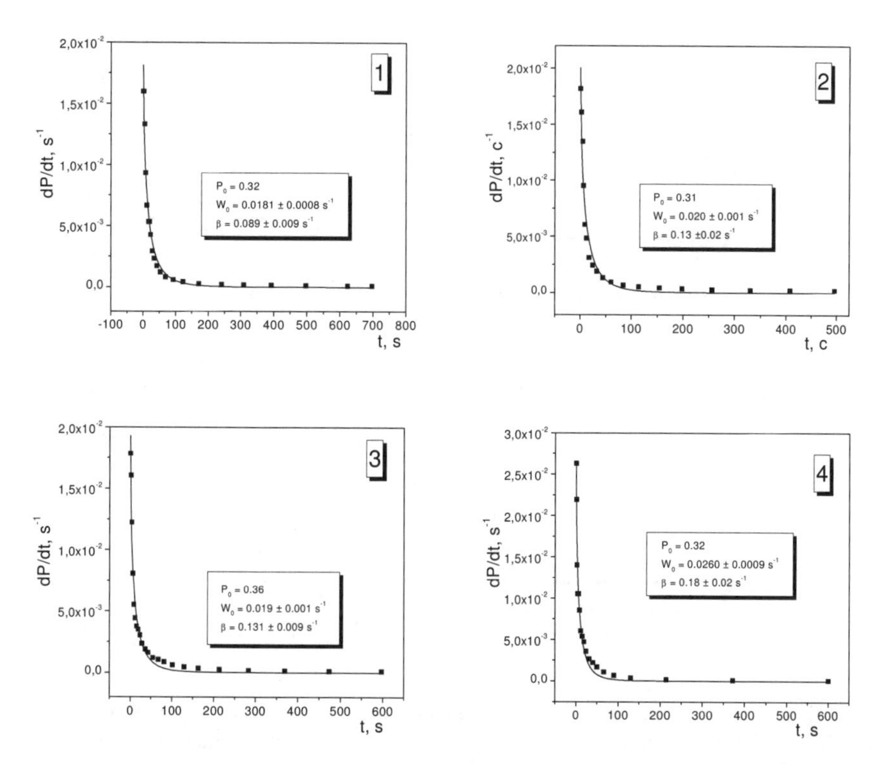

Figure 7.13. Interpretation of experimental kinetic curves for OCM-2 postpolymerization in the differential form (points) compared with curves calculated using equation (7.23) (lines) at 5 (1), 10 (2), 15 (3) and 20°C (4).

Table 7.1. Spread of parameters W_0 and β, calculated using equation (7.23) for postpolymerization of DMEG at 15°C

No.	P_0	τ (s)	$W_0\times10^3$ (s^{-1})	$\beta\times10^2$ (s^{-1})
1	0.30	650	4.0 ± 0.1	5.7 ± 0.4
2	0.29	700	3.7 ± 0.1	4.6 ± 0.3
3	0.16	650	3.4 ± 0.1	7.4 ± 0.7
4	0.22	750	4.1 ± 0.1	5.6 ± 0.5
5	0.19	700	3.6 ± 0.2	5.5 ± 0.6
6	0.21	750	3.2 ± 0.1	5.3 ± 0.6
Average $\beta\times10^2$ (s^{-1})				5.7 ± 0.9

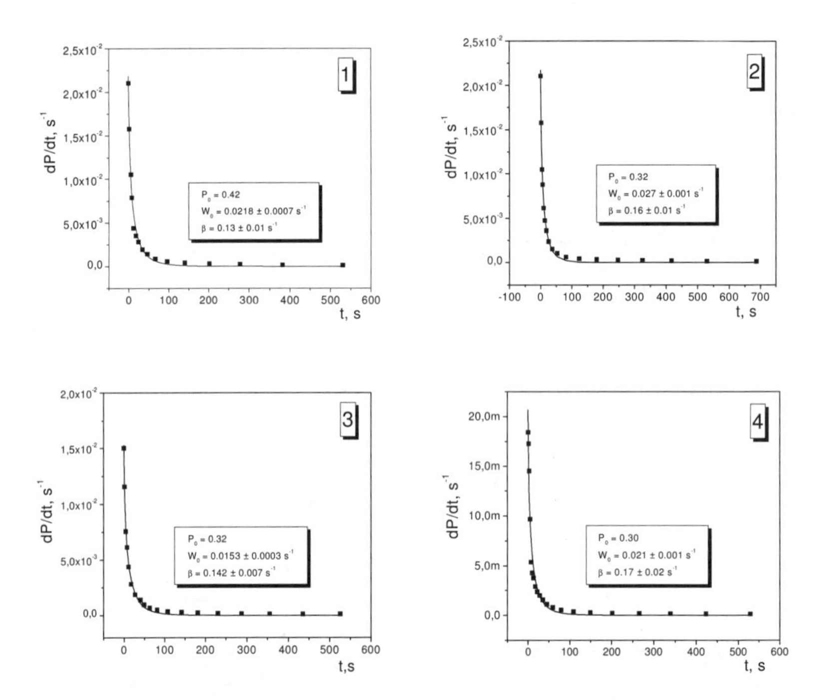

Figure 7.14. Interpretation of experimental kinetic curves for MGPh-9 postpolymerization in the differential form (points) compared with curves calculated using equation (7.23) (lines) at 10 (1), 15 (2), 20 (3) and 25°C (4).

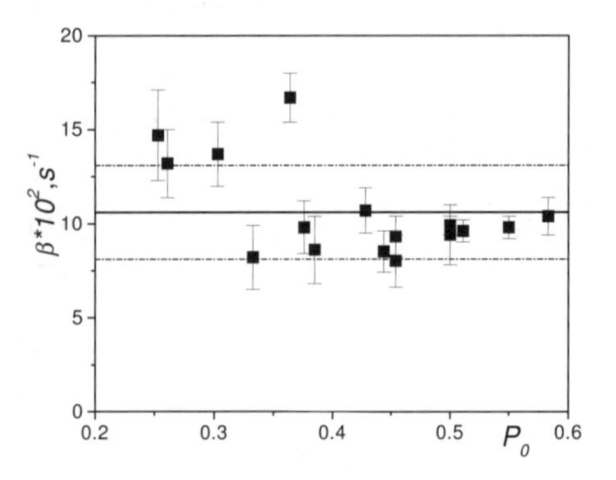

Figure 7.15. Values of β calculated using the optimization method at different starting conversions of postpolymerization for TGM-3 at 15°C.

We can see from Figure 7.15 that, in spite of the spread in values of the parameter β, these values do not depend on the starting conversion P_0 of the dark period. The same picture was observed for other experimental conditions.

7.3.1. Dependence of the monomolecular chain termination rate constant on temperature

As was mentioned above, the liquid phase (sol), *via* the polymerization process of polyfunctional monomers, represents the initial monomer. Thus, we can assume that the concentration of monomer at the interface layer will be equal to its molar-volumetric concentration in the bulk ($[M_m]=[M_0]$). On the basis of this assumption we can estimate the values of the monomolecular chain termination rate constant $k_t=\beta/[M_0]$. Average values of β and monomolecular chain termination rate constants k_t for the dimethacrylates investigated at different temperatures are presented in Tables 7.2–7.5.

Table 7.2. Average values of β and k_t values for DMEG calculated according to experimental data at different temperatures.

T (°C)	$(\beta\pm\Delta\beta)\times10^2$ (s^{-1})	$(k_t\pm\Delta k_t)\times10^5$ (m^3/mol×s)	No. of kinetic curves
5	4.6±2.3	0.87±0.43	5
15	5.7±0.9	1.08±0.17	7
25	6.4±2.8	1.21±0.53	13

We can conclude on the basis of data presented in Tables 7.2–7.5 that the parameter β and the monomolecular chain termination constants rate k_t depend on temperature. This proves the activation nature of constant k_t.

Table 7.3. Average values of β parameter and k_t values for TGM-3 calculated according to experimental data at different temperatures.

T (°C)	$(\beta\pm\Delta\beta)\times10^2$ (s^{-1})	$(k_t\pm\Delta k_t)\times10^5$ (m^3/mol×s)	No. of kinetic curves
5	10.7±3.0	2.36±0.66	5
10	10.7±2.1	2.36±0.46	7
15	10.6±2.5	2.34±0.55	16
20	12.2±0.7	2.70±0.15	5
25	14.6±4.5	3.22± 0.99	9
35	14.3±4.1	3.16±0.9	6
50	14.9±4.0	3.29±0.88	5

Table 7.4. Average values of β and k_t values for OCM-2 calculated according to experimental data at different temperatures.

T (°C)	$(\beta \pm \Delta\beta)\times10^2$ (s^{-1})	$(k_t \pm \Delta k_t)\times10^5$ (m^3/mol×s)	No. of kinetic curves
5	10.9±3.0	3.77±1.04	7
10	13.3±2.1	4.60±0.73	9
15	15.3±2.1	5.29±0.73	5
20	15.0±2.0	5.19±0.69	5
25	13.8±2.6	4.78±0.90	5
35	16.5±4.0	5.71±1.4	6
50	19.1±2.8	6.61±0.97	7

Table 7.5. Average values of β and k_t values for MGPh-9 calculated according to experimental data at different temperatures

T (°C)	$(\beta \pm \Delta\beta)\times10^2$ (s^{-1})	$(k_t \pm \Delta k_t)\times10^5$ (m^3/mol×s)	No. of kinetic curves
10	12.8±1.2	6.21±0.58	5
15	15.6±2.8	7.57±1.36	8
20	13.8±3.4	6.70±1.65	5
25	16.5±2.3	8.01±1.12	5
30	17.6±2.2	8.54±1.07	5
35	19.7±7.4	9.56±3.59	5
40	19.0±4.4	9.22±2.14	5
45	19.7±5.2	9.56±2.52	5
50	17.6±6.1	8.54±2.96	5

Interpretation of the experimental values of the β parameter obtained at different temperatures in Arrhenius coordinates is presented in Figure 7.16. The respective linear regression equations used for calculation of the activation energy of the monomolecular chain termination are also presented in Figure 7.16 and in Table 7.6 (E_A^t).

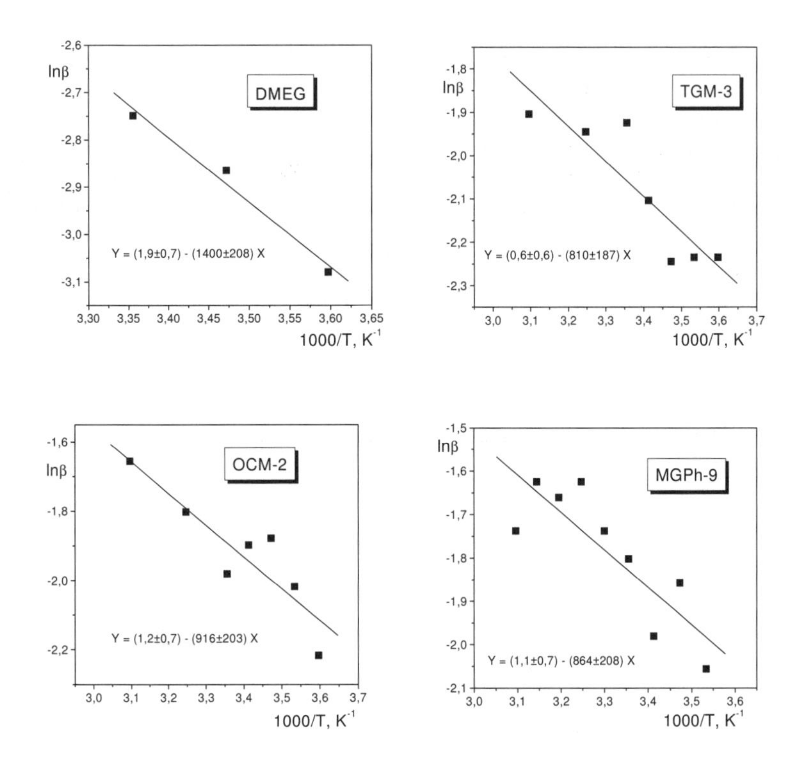

Figure 7.16. Temperature dependencies of the β parameter in the Arrhenius coordinates

As we can see from the data presented in Table 7.6, for TGM-3, OCM-2 and MGPh-9 activation energies of the monomolecular chain termination are in the range of 6.7–7.6 kJ/mol. For DMEG activation energy is higher, 11.6 kJ/mol. This once more confirms the fact that the monomolecular chain termination is an activated process with the values of activation energy the same as chain propagation activation energy. In order to prove that the obtained values of the activation energies of the monomolecular chain termination are not determined by diffusion of macroradicals and, respectively, by viscous flow activation energy, the dependence of viscosity for some dimethacrylates (DMEG, OCM-2 and MGPh-9) on temperature was investigated.

Table 7.6. Activation energies of the monomolecular chain termination (E_A^t) and viscous flow (E_A^η) *for* dimethacrylates

Dimethacrylate	$E_A^t \pm \Delta E_A^t$ (kJ/mol)	$E_A^\eta \pm \Delta E_A^\eta$ (kJ/mol)
DMEG	11.6 ± 1.7	13.6 ± 0.9
TGM-3	6.7 ± 1.6	–
OCM-2	7.6 ± 1.7	55.5 ± 2.3
MGF-9	7.2 ± 1.7	44.1 ± 1.6

The dependencies obtained in the Arrhenius coordinates and respective regression equations are represented in Figure 7.17.

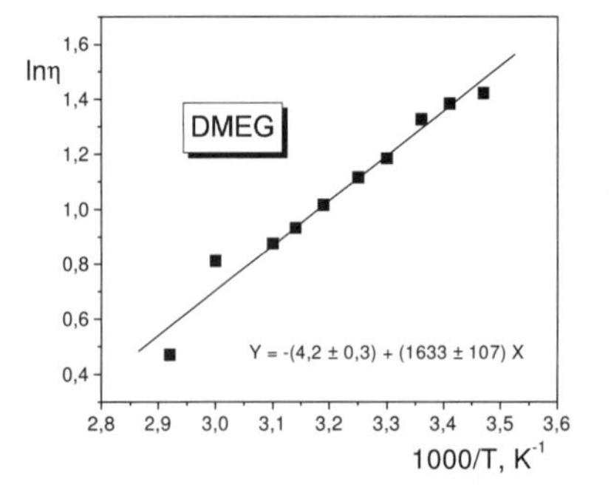

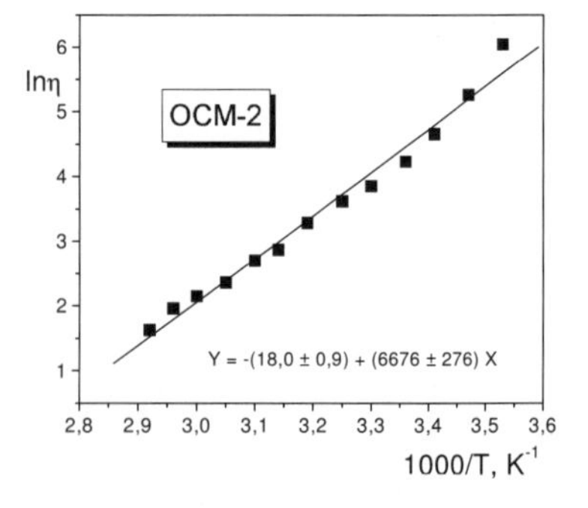

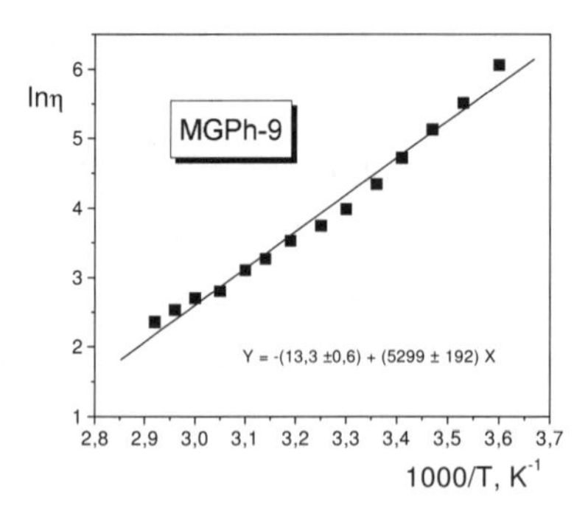

Figure 7.17. Dependence of the viscosity on temperature in the Arrhenius coordinates for DMEG, OCM-2 and MGPh-9.

On the basis of presented data we can conclude that the activation energies of the viscous flow are higher than the activation energies of the monomolecular chain termination found *via* kinetic curves of the postpolymerization. It once more confirms the assumption we made when we deduced the kinetic equation, according to which the monomolecular chain termination rate is controlled by the rate of its propagation. Thus, monomolecular chain termination cannot be considered as passive active radical 'freezing".

7.3.2. Scaling dependence of the monomolecular chain termination on the molar-volumetric concentration of the monomer in bulk

From the analysis of the data presented in Tables 7.2–7.5 we draw conclusions on the dependence of β parameter (and k_t) on the concentration of monomer in the bulk $[M_0]$. This dependence has a scaling view. Values of the β parameter have been calculated at 20°C *via* linear regression equations in order to prevent the accidental factor in the estimation of β parameter under the interpretation of the scaling dependence of β and k_t on $[M_0]$. After that values $k_t=\beta/[M_0]$ were calculated at 20° and the results of these calculations are presented in Table 7.7.

Table 7.7. Kinetic parameters, β, k_t and $k_p(\rho_0)^\xi$, of dimethacrylate postpolymerization at 20°C.

Dimethacrylate	$[M_0]$ (mol/m^3)	Average $\beta\times10^2$ (s^{-1})	Average $k_t\times10^5$ (m^3/mol×s)	$k_p\rho_0^\xi$
DMEG	5297	6.2	1.17	19.4
BDMC	4526	9.8	2.16	27.6
TGM-3	3814	12.2	3.20	30.6
OCM-2	2890	14.6	5.05	30.4
MGPh-9	2060	15.1	7.33	25.1

Note: $k_p\rho_0^\xi = \bar{k}_t\rho_m^\xi = \bar{k}_t[M_0]^\xi$, where ξ=1.67.

Interpretation of the scaling dependence of the calculated values of β on the concentration of monomer in bulk $[M_0]$ is presented in Figure 7.18. If we designate this dependence as $\beta \sim [M_0]^{1-\xi}$, then, in accordance with linear regression equation of (see Figure 7.18) we obtain $1-\xi=-0.86\pm0.27$. From this follows that the scaling dependence $k_t \sim [M_0]^{-\xi}$ and $\xi=1.86\pm 0.27$ for the monomolecular chain termination rate constant.

As was maentioed above, the scaling dependence $k_t \sim [M_0]^{-\xi}$ has been determined earlier [15], despite the fact that postpolymerization kinetics in Ref. [15] were described on the basis of kinetic scheme (7.1) with two characteristic relaxation times. This means that the scaling dependence $k_t \sim [M_0]^{-\xi}$ is determined by fundamental factors but not by choice of postpolymerization kinetics equation (exponential law with two characteristic relaxation times based on scheme (7.1) or stretched exponential law based only on scheme (7.1). However, at the same time the stretched exponential law requires the spectrum of characteristic relaxation times of the properties of fractal set [5]. Scaling of $k_t \sim [M_0]^{-\xi}$ is also requires these values to be known. Thus, we conclude that the relation between scaling form and stretched exponential law should be present. Maybe this is why $1/\xi \approx \gamma$, but, however, a special theoretical discussion is needed.

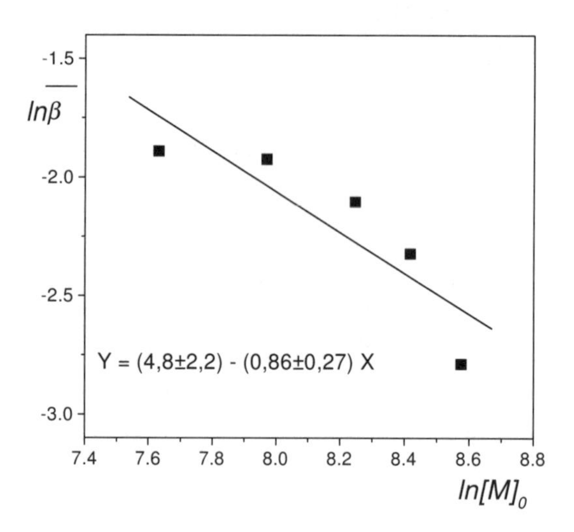

Figure 7.18. Dependence at 20ºC of calculated values of the average β parameter on molar-volumetric concentration of dimethacrylate in bulk $[M_0]$.

Note that, when comparing values of the monomolecular chain termination constants rate k_t, which have been obtained in Ref. [16] *via* a kinetic model with two characteristic relaxation times and exponential relaxation law we can see, that they are practical the same. For example, we found for BDMC 2.2 and 1.4, for OCM-2 5.0 and 3.6, and for MGPh-9 7.3 and 5.2 (all in 10^{-5} m^3/mol×s). So, both kinetic models, despite their differences, give comparable estimations of k_t, emphasizing the main characteristic of dimethacrylate postpolymerization, i.e., monomolecular chain termination according to kinetic scheme (7.1).

Also, as mentioned above, the scaling dependence $k_t \sim [M_0]^{-\xi}$ has been determined earlier [16], although in this dependence the kinetics of postpolymerization is described on the basis of kinetic schemes (7.24), (7.25) and (7.26) with two characteristic times of relaxation:

$$R_m + M_m \begin{cases} \xrightarrow{k_p} R_m \\ \xrightarrow{k_t} R_z \end{cases} \qquad (7.24)$$

$$R_z + M_m \xrightarrow{k_m} R_m' \qquad (7.25)$$

$$R_m' + M_m \begin{cases} \xrightarrow{k_p'} R_m' \\ \xrightarrow{k_t'} R_z' \end{cases} \qquad (7.26)$$

For us, this clearly means, that the scaling dependence $k_t \sim [M_0]^{-\xi}$ is not determined by choice of the kinetic equation of postpolymerization (exponential law with two characteristic relaxation times on the basis of schemes (7.24), (7.25) and (7.26) or the stretched exponential law on the basis of scheme (7.1)) but by fundamental causes. However, at this the stretched exponential law requires the spectrum of the characteristic times of the relaxation characterizing by fractal properties to be known [6]. This is also proven by the scaling form of the dependence $k_t \sim [M_0]^{-\xi}$. Evidently, between this dependence and the stretched exponential law there should be a relation.

7.3.3. The influence of plastifying agents on dimethacrylate postpolymerization rate

With the aim of providing evidence for the monomolecular chain termination nature by interpretation of kinetic scheme (7.1), we investigated the influence of nature and concentration of plastifying additives on OCM-2 postpolymerization.

Typical examples of experimental kinetic curves of postpolymerization in the plot $dP/dt=f(t)$ are denoted in Figures 7.19–7.22 as points.

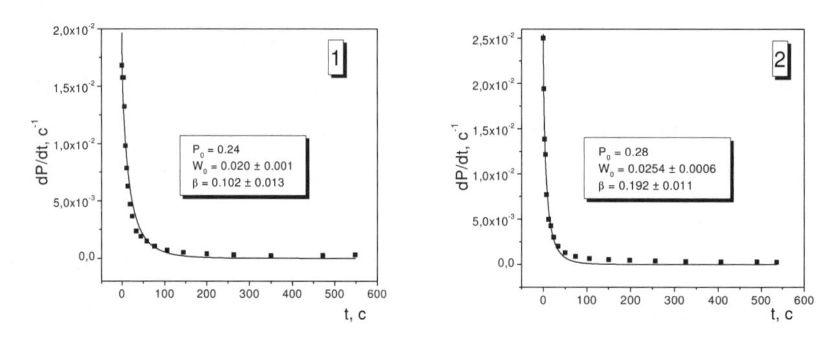

Figure 7.19. Interpretation of the experimental kinetic curves for OCM-2 postpolymerization in the differential form (see points) and their comparison with curves calculated using equation (7.23) (see lines) at DOPh (plastifying agent) concentrations of 2.5 (1) and 5.0% (by mass) (2).

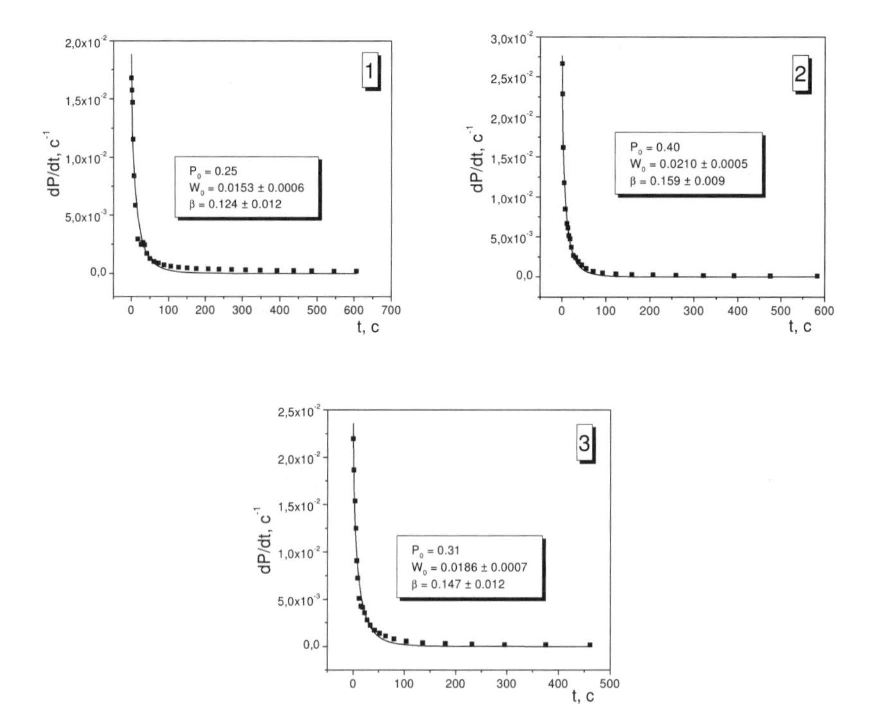

Figure 7.20. Interpretation of the experimental kinetic curves for OCM-2 postpolymerization in the differential form (see points) and comparison with curves calculated using equation (7.23) (see lines) at OP (plastifying agent) concentrations of 1.0 (1), 2.5 (2) and 5.0% (by mass) (3).

By comparison with equation (7.23) *via* the optimization method, parameters W_0 and β were found at fixed value of γ=0.6. Examples of a comparison of curves calculated using equation (7.23) (see lines) and experimental of kinetic curves (see points) is also presented in Figures 7.19–7.22. Note that in all cases a satisfactory agreement is observed between experimental curves and curves calculated using equation (7.23).

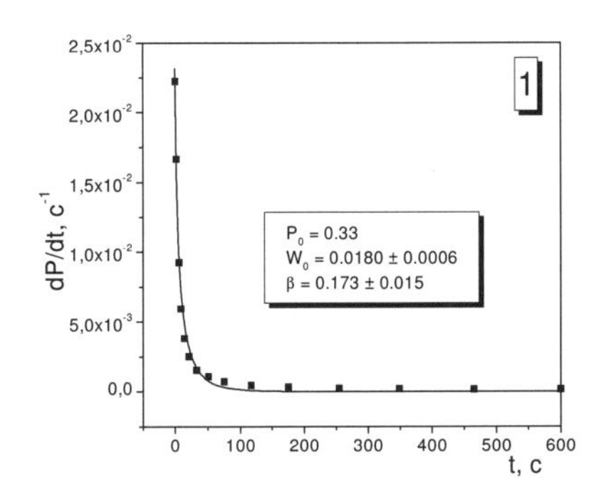

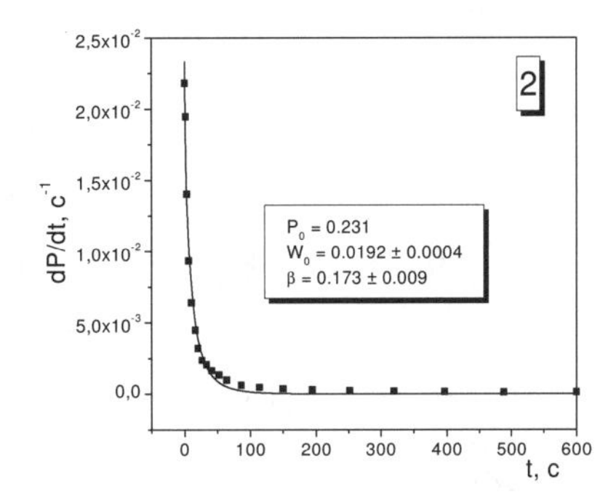

Figure 7.21. Interpretation of the experimental kinetic curves for OCM-2 postpolymerization in the differential form (see points) and their comparison with curves calculated using equation (7.23) (see lines) with different plastifying agents, namely 1.5% (by mass) OP(Cu^{2+}) (1) and 1.0% (by mass) OP(Cr^{3+}) (2).

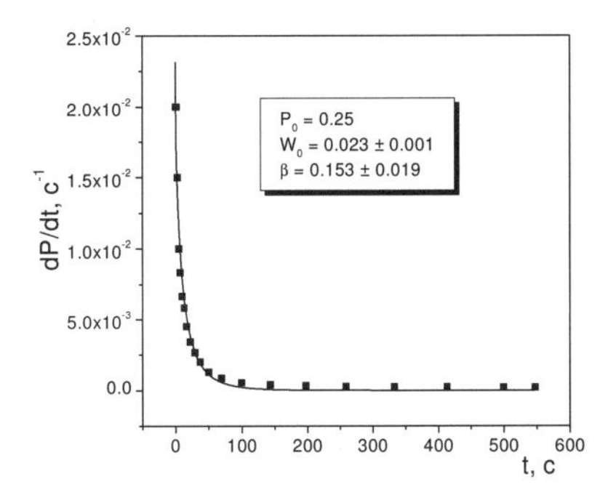

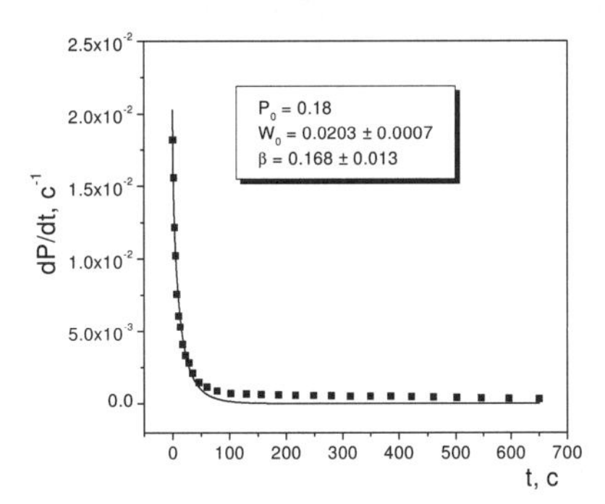

Figure 7.22. Interpretation of the experimental kinetic curves for OCM-2 postpolymerization in the differential form (see points) and comparison with curves calculated using (7.23) (see lines) with the addition of 5% (by mass) glycidyl methacrylate at different starting conversions of dark period.

For example, in Table 7.8 the values of the β parameter obtained on the basis of individual kinetic curves and also a result of their approximation are shown.

Table 7.8. Character of scattering of the parameters W_0 and β, calculated according to experimental data of kinetic equation (7.23) for the postpolymerization of OCM-2 with the additive of 2.5% (by mass) OP.

No.	P_0	τ (s)	$W_0\times10^3$ (s^{-1})	$\beta\times10^2$ (s^{-1})
1	0.06	550	7.2±0.4	14.3±1.3
2	0.27	550	16.9±0.6	13.3±0.9
3	0.24	600	14.9±0.6	11.6±1.1
4	0.40	600	21.0±0.5	15.9±0.9
5	0.30	750	21.4±0.7	15.4±1.2
6	0.21	500	15.1±0.4	15.8±1.1
Average $\beta\times10^2$ (s^{-1})				14.4±1.7

In the Figure 7.23 is shown, that despite the scattering of β parameter values, they do not depend on starting conversions P_0 of dark period. The same situation was observed also for all other plastifying additives.

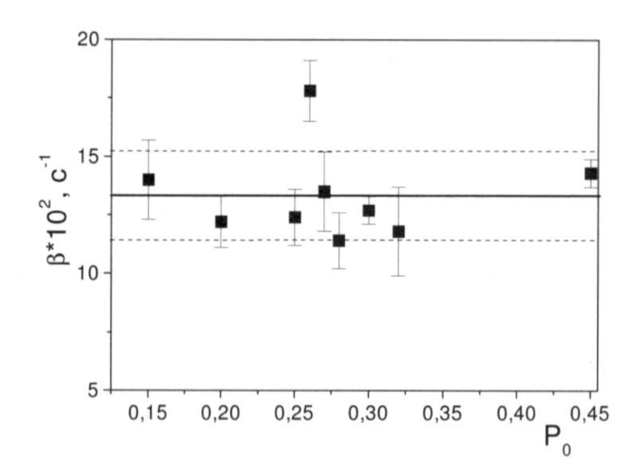

Figure 7.23. β parameter values calculated using the optimization method at different starting conversions of OCM-2 postpolymerization with 1% (by mass) OP.

Average values of the β parameter for investigated plastifying additives at different concentrations are represented in Table 7.9.

Taking into account the approximation $\overline{k}_t = \overline{\beta}/[M_0]$ (here $[M_0]$ is the concentration of monomer in bulk and is 2890 mol/m^3 for OCM-2), the values of the monomolecular chain termination rate constants for OCM-2 with the addition of plastifying agents have been calculated on the basis of data presented in Table 7.9.

Table 7.9. Average values of the β parameter calculated on the basis of experimental data according to equation (7.23) for OCM-2 postpolymerization depending on the nature and concentration of plastifying agents.

No.	Additive	ω of additive (%, by mass)	$\overline{\beta} \pm \Delta\beta \times 10^2$ (s^{-1})	No. of kinetic curves
1	None	0.0	13.8 ± 2.6	5
2	DOPh	2.5	13.9 ± 3.9	4
3	DOPh	5.0	17.2 ± 5.0	5
4	OP	1.0	13.3 ± 1.9	9
5	OP	2.5	14.4 ± 1.7	6
6	OP	5.0	14.4 ± 0.6	5
7	OP(Cu^{2+})	1.5	16.3 ± 1.4	4
8	OP(Cr^{3+})	1.0	17.9 ± 2.0	7
9	GMC	5.0	13.6 ± 3.2	5

The concentration of photoinitiator is 1% (by mass), the intensity of UV-illumination is 37.4 W/m^2.

As we can see from Table 7.10, all plastifying agents (except GMC) increase the value of the monomolecular chain termination rate constant. A complex of OP with metal ions causes a bigger increase of the monomolecular chain termination rate constant. This can probably be explained by the occurrence of a chain-transfer reaction. Such a reaction represents the induced decomposition of peroxydic bonds and is activated by the coordinate-connected cation of the metal. At this, a more reactive radical is changed by a less reactive one and gets into the conformation trap. The experimental results obtained once more confirm the assumption that monomolecular chain termination is a radical self-burial act, namely an act of interaction of the active center of a radical with the trap of the functional groups of a monomer in which it is blocked through side links of the propagation chain.

Table 7.10. Kinetic constants for OCM-2 postpolymerization in the presence of plastifying agents.

Additive	Average molecular weight (g/mol)	Concentration (mol/m^3)	$\bar{k}_t \pm \Delta k_t \times 10^5$ ($m^3/mol \times s$)
None	–	0.0	4.78±0.90
DOPh	390.56	77.0	4.81±1.35
DOPh	390.56	155.0	5.95±1.73
OP	1200	10.0	4.60±0.66
OP	1200	25.0	4.98±0.59
OP	1200	50.0	4.98±0.21
OP(Cu^{2+})	1200	15.0	5.64±0.48
OP(Cr^{3+})	1200	10.0	6.19±0.69
GMC	142.16	425.0	4.70±1.11

Thus, the probability of monomolecular chain termination first of all depends on the concentration of traps which, in turn is determined by the conformation of a propagating polymeric chain. From this point of view any additive that does not contain a functional group able to induce polymerization, displacing the monomer from the conformation volume of propagating polymeric chain, increases the traps concentration. In other words, the probability that the active center of a radical will be blocked from the functional groups of a monomer is increased. The stronger this effect, the longer the chain, namely its conformation volume.

Glycidyl methacrylate, possessing a functional group able to induce polymerization, cannot increase the quantity of traps in the conformation volume of a polymeric chain. That is why it practically does not influence the monomolecular chain termination rate constant.

7.4. Experimental regularities of monomethacrylate postpolymerization

Stationary kinetics of mono- and bifunctional monomers polymerizations show great differences, but both of them are described satisfactory using the conception of different reactive zones. They also have in common that in the interface layer at the liquid monomer–solid polymer boundary in 3-D polymerization and in the polymer–monomeric phase in linear polymerization the same mechanism of monomolecular chain termination controlled by the rate of chain propagation is postulated. Therefore, we studied the kinetics of monomethacrylate postpolymerization to obtain estimations of the monomolecular chain termination rate constant in the polymer–monomeric phase [17].

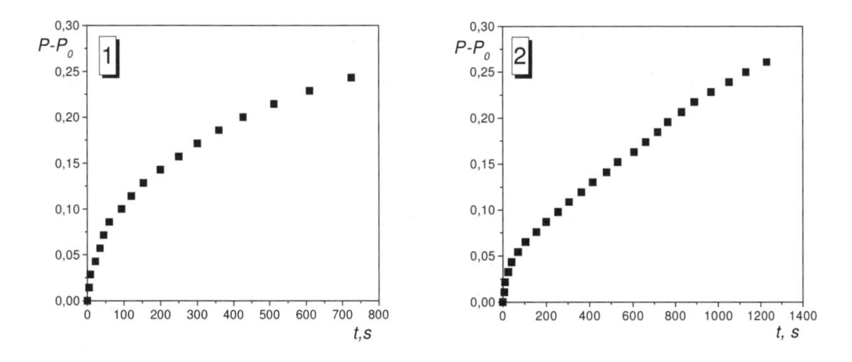

Figure 7.24. Typical kinetic curves for glycidyl methacrylate postpolymerization at c_0=3.0% (by mass), E_0=37.4 W/m^2 *and* T=20°C for the interval $P_0 \approx$ 0.5–0.6. P_0=0.58 (1), P_0=0.53 (2).

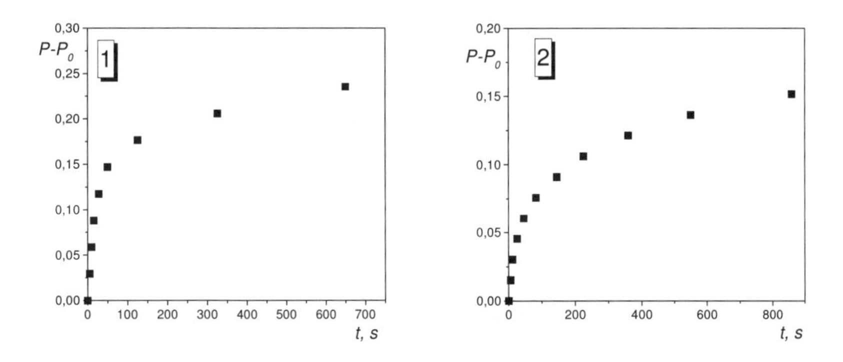

Figure 7.25. Typical kinetic curves for glycidyl methacrylate postpolymerization at c_0=3.0% (by mass), E_0=37.4 W/m^2 and T=20°C for the interval $P_0 \approx$ 0.6– 0.7. P_0=0.706 (1), P_0=0.69 (2).

Postpolymerization kinetics have been studied, for example, in photocompositions based on glycidyl methacrylate (at 0.5 and 3.0 % (by mass) of the photoinitiator) and iso-butylmethacrylate (at 3.0% (by mass) of the photoinitiator). Investigations were carried out in a wide range of conversions at the start of the dark period ($P_0 \geq 0.5$) as follows: the UV-illumination was stopped at some stage of the polymerization, the initial conversion of polymerization P_0 for dark period was changed *via* the change of the process durability under the light regime, contraction of the layer of polymeric composition was fixed during the dark period and after that the light was switched on again, polymerization was allowed to finish and the limited contraction of the layer of polymeric composition was determined.

Starting kinetic curves were obtained form the plot of the conversion increment $\Delta P=P-P_0$ for dark period at various reaction times.

All data of kinetic curves obtained from the plot of the conversion increment $\Delta P=P-P_0$ *versus* time of the dark period were divided into three groups, depending upon the value of the starting conversion on intervals: $P_0 \approx 0.5$–0.6 (Figure 7.24), $P_0 \approx 0.6$–0.7 (Figure 7.25) and $P_0 \geq 0.7$ (Figure 7.26).

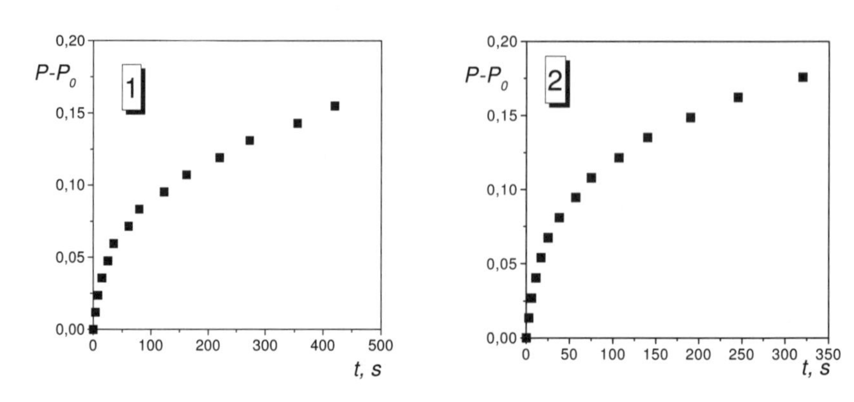

Figure 7.26. Typical kinetic curves for the glycidyl methacrylate postpolymerization at c_0=3.0% (by mass), E_0=37.4 W/m^2 and T=20°C for the interval $P_0 \geq 0.7$. P_0=0.785 (1), P_0=0.74 (2).

The first series of kinetic curves (Figure 7.24) is characterized by three sections: the first is rapid and short (the characteristic time is $\tau \approx 10^2$ s), the second is practically linear with an almost constant rate of the process (the characteristic time is $\tau \approx 6 \times 10^2$ s) and in the third section the polymerization rate gradually and slowly decreases (the characteristic time is $\tau \approx 10^3$ s).

The second series of kinetic curves (Figure 7.25) represents the intermediate link between the first and the third sections and is characterized only by a weakly expressed short linear section.

The third series of kinetic curves (Figure 7.26) is characterized by only two characteristic sections: the first is rapid and short (the characteristic time is $\tau \approx 10^2$ s) and the second is slow and long (the characteristic time is $\tau \approx 10^3$ s).

Similar characteristics are observed for kinetic curves for glycidyl methacrylate postpolymerization at an initial concentration of photoinitiator c_0=0.5% (by mass) (Figure 7.27) and iso-butylmethacrylate at an initial concentration of photoinitiator c_0=3.0% (by mass) (see Figure 7.28).

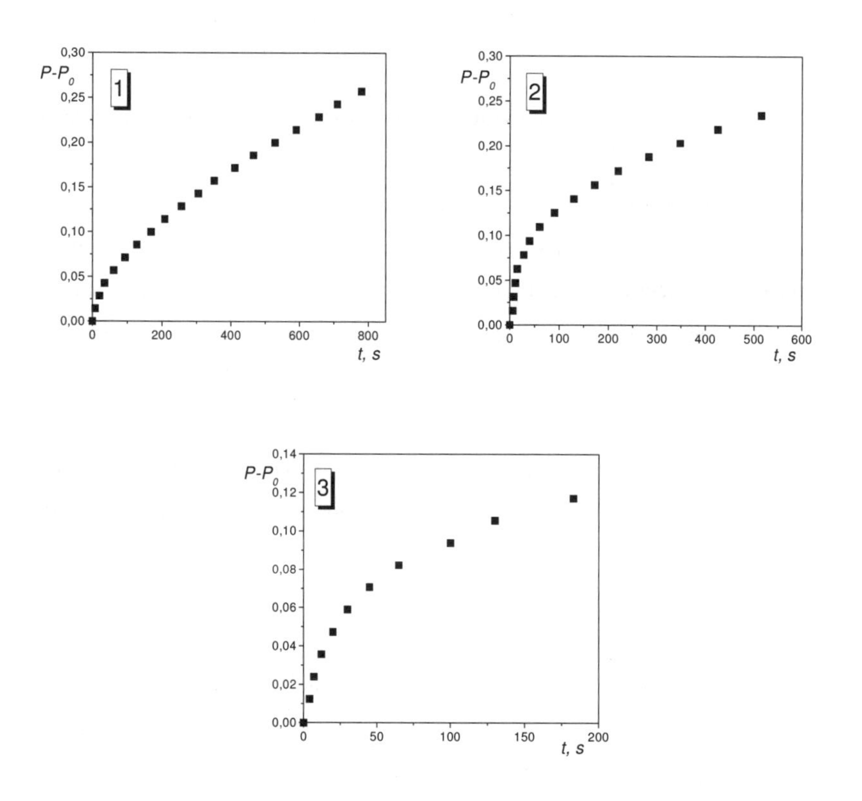

Figure 7.27. Typical kinetic curves of the glycidylmethacrylate postpolymerization at c_0=0.5% (by mass), E_0=37.4 W/m^2 and T=20°C for $P_0 \approx 0.5$–0.6 (P_0=0.557, 1), $P_0 \approx 0.6$–0.7 (P_0=0.64, 2) and for $P_0 \geq 0.7$ (P_0 = 0.848, 3).

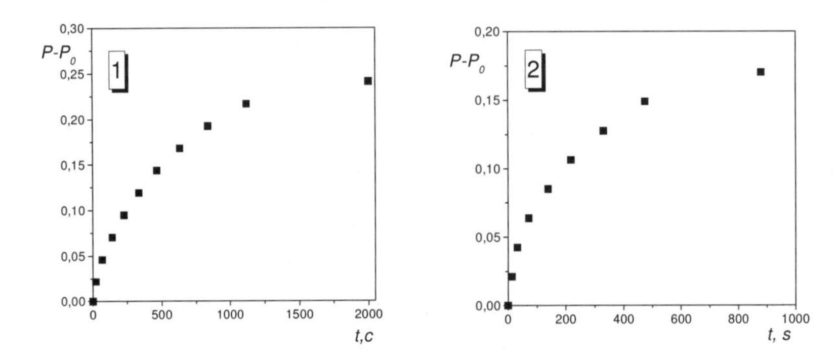

Figure 7.28. Typical kinetic curves of iso-butylmethacrylate postpolymerization at c_0=3.0% (by mass), E_0=37.4 W/m^2 and T=20°C for $P_0 \approx 0.6$–0.7 (P_0=0.70, 1) and $P_0 \geq 0.7$ ($P_{0=}$0.76, 2).

7.5. Kinetic model of postpolymerization in the polymer–monomeric phase

The types of postpolymerization kinetic curves shown can be qualitatively explained using the conception of the three reactive zones proposed earlier. Up to the moment when the polymerizing system is in the liquid monomer–polymeric phase (MPPh), namely up to the moment of conversion $P_v^0 \approx 0.5$, only a weak post-effect is observed. MPPh is characterized by low concentration of free radicals with a short life time. A visible post-effect is observed in the autoacceleration stage $P_0 > P_v^0$, that is, at the beginning of polymer–monomeric phase elimination and formation of the interface layer at the phase division boundary between MPPh and the polymer–monomeric phase PMPh, which are new reactive zones. In such reactive zones the translational and segmental mobilities of the macroradicals are sharply decreased and the life times are sharply increased. This explains the essential post-effect.

Apart from the change in concentration of active radicals also the volume of reactive zones (or their volumetric parts φ_v, φ_{vs} and φ_s) changes during the postpolymerization process. The volumetric part of MPPh (φ_v) is decreased, the volumetric part of PMPh (φ_s) is increased and the volumetric part of the interface layer, taking into account the assumption that $\varphi_{vs} \approx \varphi_s(1-\varphi_s)$, is increased at the beginning, reaches a maximum and then decreases.

Thus, at starting conversions of the dark period $P_0 \geq 0.5$ a complete description of the postpolymerization kinetics needs to take into account the proceeding of the process in two reactive zones. This fact complicates the kinetic models describing postpolymerization kinetics and makes it very difficult to compare these models with experimental data.

Therefore, we restrict ourselves to the analysis of the postpolymerization kinetics at the starting conversions of the dark period $P_0 \geq 0.7$. In this case, according to the previous analysis, we can assume with sufficient exactness $\varphi_s \approx 1$ that the postpolymerization process is concentrated manly in PMPh and is described by the last term in the equation

$$\frac{\mathrm{d}P}{\mathrm{d}\tau} = (P_s^0 - P_v^0)\frac{\mathrm{d}\varphi_s}{\mathrm{d}\tau} + \int_0^\tau \frac{w_s}{[\mathrm{M}_0]}\exp\{-\frac{w_s}{[\mathrm{M}_s^0]}(\tau - t)\}\frac{\mathrm{d}\varphi_s}{\mathrm{d}t}\mathrm{d}t \qquad (7.27)$$

According to starting positions of the kinetic model proposed by us earlier (Chapter 6) the polymerization process in PMPh is characterized by two properties. The first is the fact that the control on the chain termination rate is passed to the chain propagation rate due to a great loss of translational and segmental mobility of the macroradicals. This means that the act of the chain propagation and the act of the chain termination are two different consequences of the interaction of the active radical R_s with a functional group of the monomer, leading again to formation of active radical (chain propagation) or to 'frozen" (or, so-called self-burial) non-active radical R_z (monomolecular chain termination) (scheme (7.28)):

$$R_s + M \begin{cases} \xrightarrow{k_{ps}} R_s \\ \xrightarrow{k_{ts}} R_z \end{cases} \qquad (7.28)$$

The second is the fact that a new polymeric phase, polymer with a limited conversion near to 1, cannot separate as an independent phase due to the great viscosity of PMPh. Therefore, the polymerization within the micrograins can be considered as the glass process leading to reduction of the reactive volume of PMPh, i.e., the fraction β of the non-glassed share. This process is propagated non-uniformly on the radius of the micrograin, corresponding to different individual layers separation times. Thus, the sum rate of the polymerization in PMPh is characterized by an integral and in the case of stationary kinetics is represented by an integral in the right side of equation (7.27). However, in the analysis of the postpolymerization kinetics in PMPh the integral character of the process is taken into account only in the last stage of the deduction of the kinetic scheme.

In accordance with scheme (7.28) the specific rate of the polymerization W_s and also the rate of monomolecular chain termination dR/dt are determined as follows:

$$W_s = k_p\,[M_s^0]\,[R] \tag{7.29}$$

$$dR/dt = -k_t\,[M_s^0]\,[R], \tag{7.30}$$

where $[M_s^0]$ is a concentration of monomer in PMPh, which is in equilibrium at the moment of its separation from the MPPh.

It follows from equations (7.29) and (7.30) in the non-stationary process:

$$[R]=[R_0]\,\exp\{-\beta\, t\} \tag{7.31}$$

$$W_s = k_p\,[M_s^0]\,[R_0]\,\exp\{-\beta\, t\}, \tag{7.32}$$

where $[R_0]$ is the start concentration of active radicals in PMPh for the dark period, parameter $\beta = k_t\,[M_s^0] \equiv \tau_t^{-1}$ determines the characteristic macroradical life time.

However, since macroradicals are characterized by a wide range of relaxation times, we use the stretched exponential law and instead of (7.31) we obtain

$$[R]=[R_0]\,\exp\{-\beta t\}^{\gamma} \tag{7.33}$$

If the fraction ϕ is a part of the non-glassed share of the PMPh, then the polymerization rate will be equal to

$$d[M_s]/dt = -W_s\phi \tag{7.34}$$

It follows from the condition of the material balance that

$$d[M_s]/dt = [M_s^0]\, d\phi/dt \tag{7.35}$$

That is, when taking into account equation (7.32), we obtain

$$d\phi/\phi = -k_p\,[R_0]\,\exp\{-\beta t\}\,dt \tag{7.36}$$

Taking into account the restriction that $\phi=\phi^0$ at the beginning of dark period of time (that is at t=0) we obtain the following expression after integration of equation (7.36):

$$\phi = \phi^0 \exp\left\{-k_\mathrm{p}[\mathrm{R}_0]\int_0^t \exp\{-\beta t\}^\gamma \mathrm{d}t\right\} \tag{7.37}$$

Assuming that $\varphi_s \cong 1$, the main contribution into the total kinetics is made by the postpolymerization process in PMPh and that is why we adopt that $\mathrm{d}P/\mathrm{d}t=-\mathrm{d}[\mathrm{M_s}]/[\mathrm{M_0}]\,\mathrm{d}t$; instead of equation (7.34), taking into account equations (7.32) and (7.37) we obtain the following expression

$$\frac{\mathrm{d}P}{\mathrm{d}t} = \frac{k_\mathrm{p}[\mathrm{M_s^0}][\mathrm{R}_0]\exp\{-\beta t\}^\gamma}{[\mathrm{M}_0]}\phi^0 \exp\left\{-k_\mathrm{p}[\mathrm{R}_0]\int_0^t \exp\{-\beta t\}^\gamma \mathrm{d}t\right\} \tag{7.38}$$

According to the starting principle that $[\mathrm{M_s}^0]/[\mathrm{M_0}]=1-P_\mathrm{s}^0$ we can rewrite equation (7.38) as follows

$$\frac{\mathrm{d}P}{\mathrm{d}t} = \phi^0(1-P_\mathrm{s}^0)k_\mathrm{p}[\mathrm{R}_0]\exp\{-\beta t\}^\gamma \exp\left\{-k_\mathrm{p}[\mathrm{R}_0]\int_0^t \exp\{-\beta t\}^\gamma \mathrm{d}t\right\} \tag{7.39}$$

Note that the integral in the right part of equation (7.39) does not have a simple analytical solution; however, it can be written in a row. As a result, the optimization equation thus obtained can be written as follows

$$\frac{\mathrm{d}P}{\mathrm{d}t} = a\cdot b\cdot \exp(-\beta t)^\gamma \cdot \exp(-b\cdot t\cdot c). \tag{7.40}$$

Here multipliers a, b and c are

$$a=\phi^0(1-P_\mathrm{s}^0) \tag{7.41}$$

$$b=k_\mathrm{p}[\mathrm{R_0}] \tag{7.42}$$

$$c = \frac{1}{t}\int_0^t \exp\{-\beta t\}^\gamma \mathrm{d}t = 1 - \frac{(\beta t)^\gamma}{(\gamma+1)} \pm \ldots \pm \frac{(\beta t)^{n\gamma}}{n!(n\gamma+1)} \tag{7.43}$$

Equations (7.40)–(7.43) thus obtained describe the non-stationary kinetics of methacrylate polymerization (postpolymerization) in the polymer–monomeric phase, taking into account the wide spectrum of characteristic times of the postpolymerization inside the micrograins in the polymer–monomeric phase.

7.6. Comparison of the model with experimental data and estimation of the monomolecular chain termination rate constant

The integral form of equation (7.40) has not a simple analytic look, and that is why integral kinetic curves $P–P_0 = f(t)$ (Figures 7.26–7.28) were numerically differenced for comparison of equation (7.40) with experimental data. After this differential kinetic curves $dP/dt=f(t)$ were obtained; typical examples of such curves are denoted in Figures 7.29–7.31 as points. On the basis of comparison of these curves and equation (7.40) using the optimization method all four parameters of equation (7.40), a, b, β and γ, were determined,

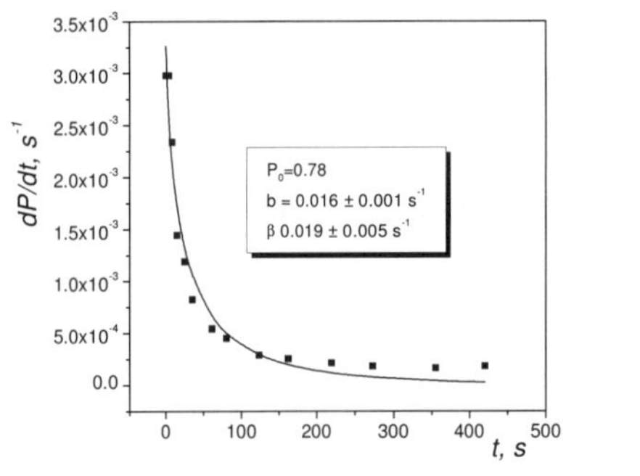

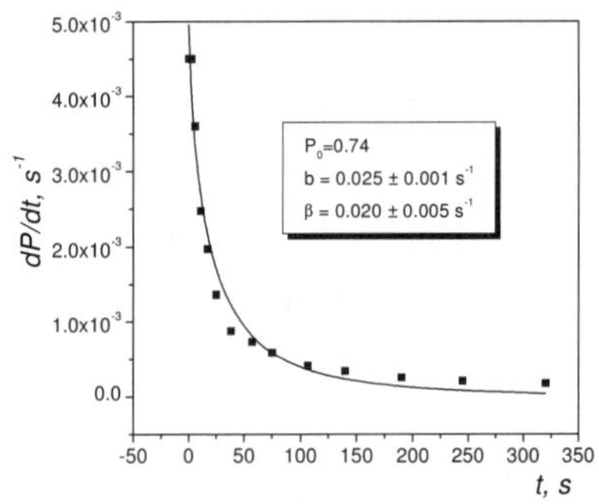

Figure 7.29. Experimental kinetic curves (points) and curves calculated using equation (7.40) (lines) for glycidyl methacrylate postpolymerization at c_0=3.0% (by mass), E_0=37.4 W/m^2 and T=20°C.

The treatment of the experimental data in accordance with the kinetic curves of the postpolymerization by equation (7.40) was carried out using the optimization method. It has been determined that optimization on all four parameters a, b, β and γ gives unsatisfactory results. Therefore, in accordance with the data on stationary kinetics, we used $P_S^0 \approx 0.8$ and $\phi^0 \approx 1.0$ and, after that, the parameter a=0.2 was fixed for all kinetic curves. The optimization on three parameters b, β and γ at a fixed value of parameter a=0.2 showed that the value of the parameter γ varies between 0.5 and 0.7. Thus, we decided to fix the value of the parameter γ at 0.6 and accepted it as the "standard" most often used in practice for the description of relaxation processes by the stretched exponential law.

Comparison of kinetic curvesm calculated using equation (7.40) at fixed values a=0.2 and γ=0.6, and parameters b and β selected by optimization, with the experimental kinetic curves for the polymerization of glycidyl methacrylate is presented in Figures 7.29 and 7.30.

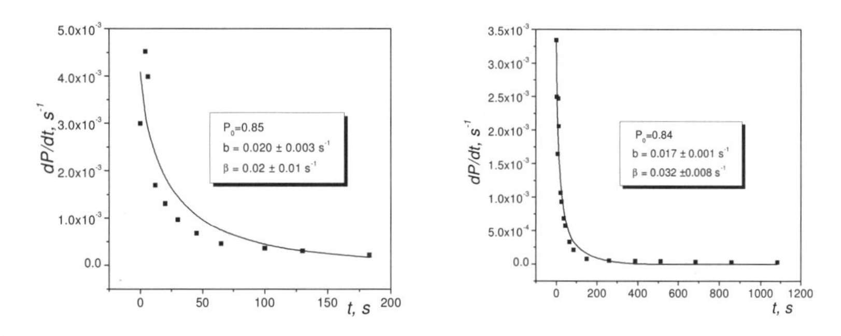

Figure 7.30. Experimental kinetic curves (points) and curves calculated using equation (7.40) (lines) for glycidyl methacrylate postpolymerization at c_0=0.5% (by mass), E_0=37.4 W/m^2 and T=20°C.

Figure 7.31 presents the comparison of kinetic curves calculated using equation (7.40) (see lines) at a=0.2 and γ=0.6 with the experimental kinetic curves for the polymerization of iso-butylmethacrylate.

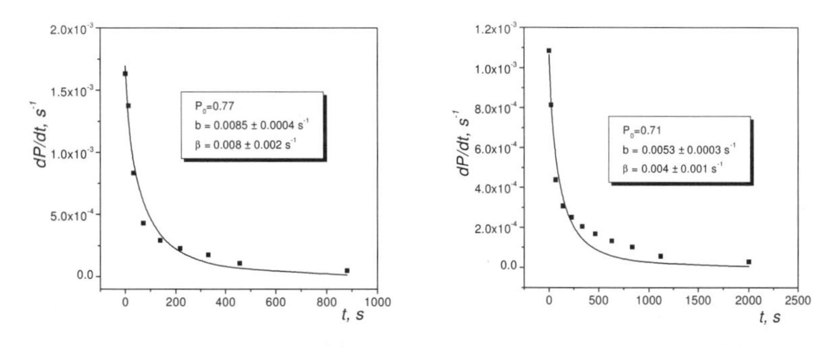

Figure 7.31. *Experimental (points) and calculated via the eq.* (7.40) (*see lines*) *kinetic curves of the iso-butylmethacrylate postpolymerization at* c_0=3.0% by mass, E_0=37.4 *W/m*^{2} *and* T=20 ^0C

As we can see from the comparison, in all cases the calculated kinetic curves are in good agreement with the experimental ones at all time intervals. This means that the proposed kinetic model and the kinetic equation (7.40) are true for the postpolymerization process of methacrylates in the polymer–monomeric phase, taking into account monomolecular chain termination.

Kinetic parameters *b* and *β* for glycidyl methacrylate postpolymerization, calculated using equation (7.40), at two concentrations of the photoinitiator are represented in Tables 7.11 and 7.12.

Table 7.11. Parameters for glycidyl methacrylate postpolymerization calculated using equation (7.40) at c_0=3.0% (by mass), E_0=37.4 W/m^2 and T=20°C.

No. of kinetic curve	Starting conversion of the dark period (P_0)	Polymerization time (s)	$b=k_p$ [R_0] (s^{-1})	$\beta=k_t$ [M_S^0] (s^{-1})
1	0.70	900	0.014±0.001	0.015±0.008
2	0.75	350	0.035±0.002	0.026±0.009
3	0.78	550	0.016±0.001	0.019±0.005
4	0.74	350	0.025±0.001	0.020±0.005
5	0.72	550	0.0165±0.0005	0.034±0.004

Table 7.12. Parameters for glycidyl methacrylate postpolymerization calculated using equation (7.40) at c_0=0.5% (by mass), E_0=37.4 W/m^2 and T=20°C.

No. of kinetic curve	Starting conversion of the dark period (P_0)	Polymerization time (s)	$b=k_p$ [R_0] (s^{-1})	$\beta=k_t$ [M_S^0] (s^{-1})
1	0.85	200	0.020±0.003	0.02±0.01
2	0.76	600	0.0064±0.0004	0.002±0.001
3	0.75	900	0.0061±0.0003	0.004±0.001
4	0.70	1000	0.0073±0.0004	0.007±0.002
5	0.76	800	0.0100±0.0007	0.008±0.003
6	0.84	1100	0.017±0.001	0.032±0.008

The kinetic parameters *b* and *β* for iso-butylmethacrylate postpolymerization, calculated using equation (7.40), are presented in Table 7.13.

The data in Tables 7.11–7.13 once more confirm the process's fluctuative sensitivity, which first of all is connected with the microheterogeneity of the polymerization system.

Table 7.13. Parameters for iso-butylmethacrylate postpolymerization calculated using equation (7.40) at c_0=3.0% (by mass), E_0=37.4 W/m^2.

No. of kinetic curve	Starting conversion of the dark period (P_0)	Polymerization time (s)	$b=k_p$ [R$_0$] (s^{-1})	$\beta=k_t$ [M$_S$^{0}] (s^{-1})
1	0.77	900	0.0085±0.0004	0.008±0.002
2	0.82	1100	0.0029±0.0002	0.006±0.001
3	0.75	500	0.0065±0.0004	0.008±0.002
4	0.71	2000	0.0053±0.0003	0.004±0.001

Taking into consideration the average value of the parameter $b=k_t$ [M$_S$^{0}]=k_t [M$_0$] (1–P_S^{0}), monomolecular chain termination rate constants k_t have been calculated *via* the postpolymerization process in the polymer–monomeric phase of glycidylmethacrylate (c_0=3.0% and c_0=0.5% (by mass() and also of iso-butylmethacrylate (c_0=3.0% (by mass), E_0=37.4 W/m^2). Results are presented in Table 7.14.

Table 7.14. Average values of parameters for methacrylate postpolymerization calculated according to equation (7.40).

Monomer+c_o photoinitiator	$b=k_p$ [R$_0$] (s^{-1})	$\beta=k_t$ [M$_S$^{0}] (s^{-1})	b/β	k_t×10^5 (m^3/mol×s)
Glycidyl methacrylate+3.0%	0.021±0.008	0.023±0.007	0.8	1.5±0.5
Glycidyl methacrylate+0.5%	0.013±0.006	0.016±0.012	0.9	1.1±0.8
Iso-butyl methacrylate+3.0%	0.006±0.002	0.006±0.002	1.0	0.48±0.16

Average estimations of the parameters $b=k_p$ [R$_0$] and $\beta=k_t$ [M$_S$^{0}] allow to reach a very important conclusion: values of the parameters b and β for glycidylmethacrylate decreased visibly as the photoinitiator concentration decreased from 3.0 to 0.5% (by mass). Parameters b and β for iso-butylmethacrylate decreased more than for glycidyl methacrylate. At the same time, the ratio $b/\beta=k_p$ [R$_0$] /k_t[M$_S$^{0}], within the range of error, is practically the same and is near to 1. If we take the ratio k_p [R$_0$] /k_t [M$_S$^{0}] ≈ 1 then we can write

$$\frac{k_p[R_0][M_s^0]}{k_t[R_0][M_s^0]} \times \frac{[R_0]}{[M_s^0]} \approx 1$$

or

$$V_p / V_t \approx [M_S^0] / [R_0], V_t \approx V_p [R_0] / [M_S^0] \quad (7.44)$$

This expression means, that the chain termination rate $V_t = k_t [R_0][M_S^0]$ is controlled or determined by the chain propagation rate, namely $V_p = k_p [R_0][M_S^0]$. This fact is in good agreement with kinetic scheme (7.28).

REFERENCES

[1] Medvedevskikh Yu., Kytsya A., Bazylyak L., Bratus., Turovskij A., Zaikov G. *Monomolecular chain termination in kinetics of dimethacrylates postpolymerization /In: Chemical reactions in liquid and solid phase: Kinetics & Thermodynamics*, Nova Science Publishers, New York, NY, **2003**, p. 117.

[2] Medvedevskikh Yu., Kytsya A., Bazylyak L., Bratus., Turovskij A., Zaikov G. *Kinetika i kataliz*, **2004**, 45 (4), p. 1-8.

[3] Montroll E. W., Bendler J. T. *J. Stat. Phys.,* **1984**, v. 34, p. 129.

[4] Williams G., Watts D. C. *Trans. Faraday Soc.,* **1970**, v. 66, p. 80.

[5] Shlezinger M., Klafter J. *Abstracts of the International Symposium on Fractals in Physics*, Moscow, Myr, **1988**, p. 553.

[6] Blumen A., Klafter J., Tsumophen G. *Book of Abstracts of the Int. Symp. on Fractals in Physics,* Moscow, Myr, **1988**, p. 561-574.

[7] Sokh D. *Reneval theory*, Chapman & Hall, London, **1962**, p. 135.

[8] Djordjevič Z. *Book of Abstracts of the Int. Symp. on Fractals in Physics,* Moscow, Myr, **1988**, p. 581-585.

[9] Oliva C., Selli E., Di Blas S. and Termignone G. J. *Chem. Soc. Perkin Trans.*, **1995**, 2, p. 2133.

[10] Plonka A. *Annu. Rep. Prog. Chem., Sect. C.*, **2001**, 97, p. 91-147.

[11] Plonka A. *Annu. Rep. Prog. Chem., Sect. C.*, **1998**, 94, p. 89-175.

[12] Plonka A. *Prog. Reaction Kinetics and Mechanism*, **2000**, 25, p. 109-218.

[13] Medvedevskikh Ju., Zaglad'ko E., Turovskij A., Zaikov G. *Int. J. Polym. Mater.,* **1999**, 43, p. 157.

[14] Zaglad'ko E., Medvedevskikh Ju., Turovskij A., Zaikov G. E. *Int. J. Polym. Mater.,* **1998**, 39, p. 227.

[15] Medvedevskikh Yu., Zaglad'ko E., Bratus' A., Turovskyj A. *Dokl. NASU,* **2000**, 10, p. 148.

[16] Medvedevskikh Yu., Zaglad'ko E., Turovskyj A., Zaikov G. R*uss. Polym. News*, **1999**, 4 (3), p. 33.

[17] Bratus' A. *Kinetic model of the bulk linear photoinitiated polymerization till the high conversions*, Ph. D *Thesis in chemical sciences*, Lviv, **2003**, 120 p.

Chapter 8. Statistics of self-avoiding random walks and the stretched exponential law

From the mathematical point of view, the stretched exponential law for the relaxation function requires a set of the characteristic times for the fractal properties, as mentioned in the previous chapter. That is why the statistics of self-avoiding random walks has been studied on the basis of a stretched exponential law for the monomolecular chain termination. These statistics determine exactly these fractal properties of the propagating polymeric chain. Let us demonstrate these statistics in accordance with Ref. [1].

8.1. Statistics of self-avoiding random walks

8.1.1. The Flory method and Pietronero conception

One of the characteristics of long flexible polymer chains is that their thermodynamic and dynamic properties in solution are to a great extent determined not by the nature of the links but by their quantity, i.e., by conformation of macromolecules. The basis of their description is a model of free-jointed Kuhn chains and statistics of random walks (RW) of its growing end [2]. In terms of these statistics, the distance distribution between the chain ends is described by a Gaussian function with two features: a distribution mode that corresponds to zero distance between chain ends, and root-mean-square radius R_o of polymer coil and chain length N correlate as $R_o \sim N^{\nu}$, where $\nu=1/2$. Experimental data give values close to 0.6.

Taking into account that the effects of short-range ordering, e.g., of fixed valence angles and hindrance of internal rotation [3, 4], do not change the shape of the distribution function and the correlation $R_o \sim N^{1/2}$ but increase R_o, that can be interpreted in terms of increase of a statistical length of an equivalent Kuhn chain segment [5].

It has been noticed several times [3, 4] that the configuration of soft polymer chains differs from the RW trajectory in one important aspect: it must not intersect. This limitation, known as long-range ordering effect or excluded volume effect, requires new statistics, i.e., statistics of self-avoiding walks (SAW). The attempts made so far [3] have not succeeded in solving this problem completely.

Therefore, the authors of Refs. [6, 7] find the success of the Flory method (after de Gennes [8]) to be especially unexpected. In this method, the free energy F of polymer chain conformation in kT units is presented by

$$F=\upsilon N^2/R^d + dR^2/2a^2N, \tag{8.1}$$

where $\upsilon=a^d(1-2\chi)$ is the excluded or effective volume of monomer links, a is its statistical length, χ is the Flory–Huggins parameter, R is the distance between chain ends and d is the Euclidean space dimension.

The augend in (8.1) represents the pairwise repulsion energy of chain links in the self-consistent field approximation and the addend represents the polymer chain elastic energy determined *via* entropy (F =$-TS$) of its conformation in RW statistics.

Minimization of F by R allows to find the equilibrium or the most probable radius R_f of Flory coil:

$$R_f=a_f N^\nu, \tag{8.2}$$

$$\nu=3/(d + 2), \tag{8.3}$$

where $a_f=a(1-2\chi)^{1/(d+2)}$.

Formula (8.3) appears to be universal: it is in good accordance with the results of physical experiments and computer simulation of SAWs on a lattice [6–9]. At this time, formula (8.3) gives the unaffected value $\nu=1$ in a case of $d=1$; values $\nu=0.6$ and 0.75 for a space of $d=3$ and 2 dimensions, respectively, differ from the ones obtained by computer simulation of SAWs on a lattice only within the limits of 1% error.

The Flory conception in expression (8.1) cannot give a full explanation for these results. Therefore. Pietronero [6] has proposed another conception of the Flory method, in accordance with which both components of expression (8.1) possess an entropic nature, but conformation entropy is determined by SAW statistics.

The density distribution of distance between the chain ends corresponding to SAWs statistics has been presented by Pietronero as multiplication of two functions: $w(R) \sim S(R) * G(R)$, where one is the Gaussian function $G(R) \sim \exp\{-dR^2/2a^2N\}$ and determines the probability that the trajectory of end-chain RW *via* N steps ends in a spherical layer R, R + dR. The survival probability function $S(R)$ selects these trajectories of RW which are not self-avoided. Such function has been determined starting from the following point of view (only a little change has been done by us in the interpretation [6, 7]). Let us assume, that $\rho=\upsilon N/R^d$ is the share of the excluded volume in the conformative volume of the polymeric chain. If one propability of striking the chain end into the excluded volume upon some step of walk is proportional to ρ, then the probability that it does not happen is proportional to $1-\rho$. On all trajectories consisting of N steps the probability of the chain end striking into the excluded volume at any time will be equal to $S(R) \sim (1-\rho)^N \approx \exp\{-N\rho\}$, since $\rho << 1$. By summation of the $S(R)$ and $G(R)$ in $w(R)$ we obtain

$$w(R) \sim \exp\{-\upsilon N^2/R^d - dR^2/2a^2N\}, \qquad (8.4)$$

that is equivalent to expression (8.1), if the transformation $w(R)$ into chain conformation $S \sim k\ln w(R)$ and after that into free energy $F = -TS$ would occur.

Pietronero' s conception provides an explanatiorfor the good correlation of the Flory method formula (8.3) with the results of SAW computer simulation on the lattice and gives a heuristic landmark for the construction of strict SAW statistics.

8.1.2. Statistics of SAWs on a *d*-dimensional lattice

In accordance with the standard procedure [2] (see also [3]) of Gaussian distribution in the deduction of RW statistics, we will take into account that the one-dimensional chain-end walk has a step equal to root-mean-square chain link length projection on the walking direction, and the number of steps for each direction is equal to the length of chain N. So, each step of walk is actually d-dimensional, and transition to d-dimensional distribution function is unnecessary.

Let us incorporate the numbers n_i of chain end steps in the i direction of the d-dimensional lattice with cell size equal to the statistical length of the chain link:

$$\Sigma_i n_i = N,\ i=1,d. \qquad (8.5)$$

The quantity of realization variants of random steps in i directions is equal to $n_i!/n_i^+!n_i^-!$, where the numbers of steps in positive (n_i^+) and negative (n_i^-) directions of wandering are correlated by $n_i^+ + n_i^- = n_i$. Considering that the probability of wandering in the positive or negative direction chosen is assumed to be the same and equal to 1/2, the probability $\omega(n_i)$ that at a given n_i, n_i^+ positive and n_i^- negative steps will be made is determined by the Bernoulli distribution:

$$\omega(n_i)=(1/2)n_i\, n_i!/n_i^+!n_i^-!. \tag{8.6}$$

Then, incorporating the quantity of effective steps in i directions of wandering $s_i=n_i^+-n_i^-$, we obtain $n_i^+=(n_i+s_i)/2$, $n_i^-=(n_i-s_i)/2$. Then (8.6) can look as follows:

$$\omega(n_i)=(1/2)\, n_i!/((n_i+s_i)/2)!\, ((n_i-s_i)/2)!. \tag{8.7}$$

For d-dimensional wandering we have

$$\omega(n)=(1/2)^N \textstyle\prod_i n_i!/((n_i+s_i)/2)!\, ((n_i-s_i)/2)!. \tag{8.8}$$

One can see that the change of sign at s_i does not change the value $\omega(n)$. Thus, $\omega(n)$ represents the probability that the trajectory of random wandering after n_i steps in i directions will end in one of 2^d cells $M_p(s)$, coordinates of which are determined by vectors $s=(s_i)$, $i=1,d$ having a distinction in signs of their components s_i only. These cells or states of chain end are equiprobable.

The condition of self-avoidance of a RW trajectory on d-dimensional lattice demands the step not to fall twice in the same cell. From the point of view of chain link distribution over cells it means that every cell cannot contain more than one chain link. Chain links are inseparable. They cannot be torn off one from another and placed in cells in random order. Consequently, the numbering of chain links corresponding to wandering steps is their significant distinction. Therefore, the quantity of different variants of N distinctive chain links placement in Z identical cells under the condition that one cell cannot contain more than one chain link is equal to $Z!/(Z-N)!$.

Considering the identity of cells, the *a priori* probability that a given cell will be filled is equal to $1/Z$, and that it will not be filled is $(1-(1/Z))$. Subsequently, the probability $\omega(z)$ that N given cells will be filled and $Z-N$ cells will be empty, considering both the above-mentioned conditions of placement of N distinctive links in Z identical cells and the quantity of its realization variants, will be determined by the following expression

$$\omega(z)=Z!(1/Z)^N(1-(1/Z))^{Z-N}/(Z-N)!. \tag{8.9}$$

The probability $\omega(s)$ of a simultaneous event, meaning that the RW trajectory is also the SAW trajectory and, at given Z, N and n_i, with its last step it will fall into one of 2^d equiprobable cells $M_p(s)$, will be equal to

$$\omega(s)=\omega(z)\ \omega(n). \qquad (8.10)$$

Let us find the asymptotic (8.10), assuming $Z >> 1$, $N >> 1$ and $n_i >> 1$ under the condition $s_i << n_i$, $N << Z$. Using the approximated Stirling formula $\ln x! \approx x \ln x - x + \ln(2\pi)^{1/2}$ for all $x >> 1$ and expansion $\ln(1-(1/Z)) \approx -1/Z$, $\ln(1-(N/Z)) \approx -N/Z$, $\ln(1 \pm s_i/n_i) \approx \pm s_i/n_i - (s_i/n_i)^2/2$, and assuming also that $N(N-1) \approx N^2$, we obtain

$$\omega(s) \approx \beta \exp\{-N^2/Z - 1/2\ \Sigma_i s_i^2/n_i\}, \qquad (8.11)$$

where $\beta^{-1}=e(2\pi)^{d/2}$.

With increasing quantity of effective steps s_i the chain end moves away from its origin, increasing the conformational volume where those SAW trajectories are localized at that end in one of 2^d equiprobable cells $M_p(s)$. That is why the cell quantity Z, allowed for a SAW trajectory, is not a fixed parameter of distribution (8.11), but a function of vector $s=(s_i)$: $Z=Z(s)$. This function choice can be made based on different geometrical estimations (see Figure 8.1).

For instance, if one considers that the conformational volume of SAW trajectories is localized in d-dimensional rectangle with vertexes $M_p(s)$, then $Z=2^d\ \Pi_i |s_i|$. If one assumes that the conformational volume is localized in a sphere with radius $R_s=(\Sigma_i s_i^2)^{1/2}$, then the expression $Z=(\Sigma_i s_i^2)^{d/2}$ can be used for the definition. If one assumes that fiducial cells $M_p(s)$ belong to an ellipsoid surface, on which semi-axes, in accordance with the canonical equation $\Sigma_i s_i^2/b_i^2=1$ must be equal to $b_i= d^{d/2} |s_i|$, then $Z=d^{d/2}\ \Pi_i |s_i|$. Analysis has shown that this definition is the most suitable (as will be confirmed below); that is why we assume

$$Z=d^{d/2}\ \Pi_i |s_i|. \qquad (8.12)$$

The first link of chain makes the first step, that is why $1 \leq |s_i| < n_i$. However, taking into account that $\omega(s)$ is sharply decreased at the end $|s_i|$ to n_i, we can assume without great error, that $|s_i|$ is determined in the section $[1, n_i]$. Putting together (8.11) and (8.12), we have

$$\omega(s)=\beta\exp\{-N^2/d^{d/2}\,\Pi_i\,/s_i/-1/2\,\Sigma_i s_i^2/n_i\},\ 1\le s_i\le n_i. \qquad (8.13)$$

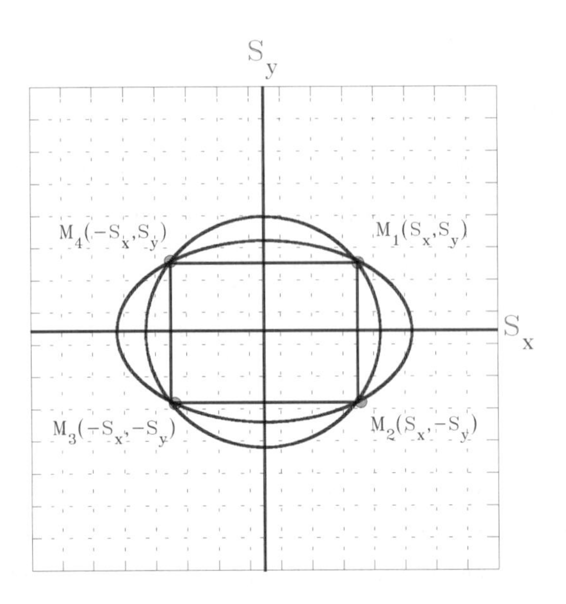

Figure 8.1. Geometrical variants of determination of SAW trajectories conformational volume at given coordinates of equiprobable states $M_p(s)$ of the chain end for two-dimensional lattice.

The function $\omega(s)$ determines the probability of a complicated event, meaning that the RW trajectory is also a SAW trajectory at the same time and with it last step it falls into one of 2^d equiprobable cells $M_p(s)$: one can also say that it realizes the state $M_p(s)$. This implies that it is numerically equal to the part of those SAW trajectories from the whole quantity of RW trajectories $(2d)^N$, which realize the state $M_p(s)$. The quantity $L(s)$ of such trajectories define the thermodynamical probability of realization of the state $M_p(s)$

$$L(s)=(2d)^N\omega(s). \qquad (8.14)$$

Summing $L(s)$ over the whole set Ω of chain end states, we find the total quantity L of SAW trajectories

$$L=(2d)^N\,\Sigma\omega(s). \qquad (8.15)$$

Designating

$$c(s)=\beta\Sigma \exp\{-N^2/d^{d/2}\Pi_i/ s_i/ -1/2 \Sigma_i s_i^2/n_i\},\qquad (8.16)$$

then function

$$w(s)=(1/c(s)) \exp\{-N^2/d^{d/2} \Pi_i/ s_i / -1/2\Sigma_i s_i^2/n_i\}\qquad (8.17)$$

is normalized to unity and defines the mathematical probability of chain-end distribution over the states $M_p(s)$ of d-dimensional lattice. It is equal to the relation of quantity $L(s)$ of those SAW trajectories, which realize the state $M_p(s)$, to the total quantity of SAW trajectories L: $w(s)=L(s)/L$.

In turn, the ratio $L/(2d)^N$ equals the part of the total number of SAW trajectories from the total number of RW trajectories and is a function $g(N)$ of SAW trajectories fatigue, in accordance to terminology of Refs. [6, 7], and quantitatively it can be determined *via* the normalizing constant of distribution (8.17):

$$g(N)=L/(2d)^N=\beta\, c(s)\qquad (8.18)$$

8.1.3. Statistics of SAWs in continuous space

Expressions (8.13)–(8.18) represent the SAW statistics at d-dimensional lattice. In a metric space, incorporating the variable of shifting the x_i-semi-axis of conformational ellipsoid, with the states $M_p(s)$ belonging to its surface

$$x_i=a / s_i / d^{1/2}\qquad (8.19)$$

and parameter σ_i is the standard deviation of the Gaussian part of the distribution

$$\sigma_i^2=a^2\, n_i\, d.\qquad (8.20)$$

In accordance with (8.5), the coupling is laid on the value σ_I

$$\Sigma_i\sigma_i^2=a^2\, N\, d.\qquad (8.21)$$

Since $s_i^2/n_i=x_i^2/\sigma_i^2$, $d^{d/2}\Pi_i/ s_i/ =\Pi_i x_i/a^d$, distribution (8.13) can be rewritten as follows:

$$w(x)=(1/c(x)) \exp\{-a^d\, N^2/\Pi_i x_i-1/2\Sigma_i x_i^2/\sigma_i^2\},\ a \le x_i \le an_i.\qquad (8.22)$$

If $x_{i\min}=a << x_{i\max}=an_i$, distribution (8.22) can be considered as a continuous one with a normalizing constant

$$c(x)=\int \exp\{-a^dN^2/\Pi_i x_i - 1/2\,\Sigma_i x_i^2/\sigma_i^2\}dx,\ a \le x_i \le an_i \tag{8.23}$$

where $dx=\Pi_i dx_i$ and the integral is d-aliquot. In this case: $c(x)=a^d d^{d/2} c(s)$.

Substitution of (8.19) induces an essential distinction between $w(x)$ and $w(s)$: $w(x)$ determines the probability $w(x)dx$ that the SAW trajectory at given parameters N and σ_i will end in an elementary volume dx, laying on an ellipsoid surface with semi-axes x_i, $i=1,d$. In the other case, all the surface of ellipsoid is a geometrical place of points or chain end states with the same corresponding distribution density $w(x)$.

The maximum of $w(x)$ at given N and σ_i corresponds to the most probable or equilibrium state of a polymer chain. The semi-axes x_i^o of the equilibrium conformational ellipsoid can be found from the condition

$$\partial \ln w(x)/\partial x_i = a^d N^2/x_i \Pi_i x_i - x_i/\sigma_i^2 = 0, \text{ when } x_i = x_i^o. \tag{8.24}$$

Solving the system of algebraic equations (8.24), we obtain

$$x_i^o = \sigma_i (a^d N^2/\Pi_i \sigma_i)^{1/(d+2)}. \tag{8.25}$$

We will continue to consider the situation when all the directions of chain steps are equiprobable, i.e.,

$$n_i = N/d, \tag{8.26}$$

$$\sigma_i^2 = \sigma_o^2 = a^2 N. \tag{8.27}$$

Substitution of (8.27) into (8.25) makes the semi-axes of equilibrium ellipsoid equal, and equal to the radius of the Flory ball

$$x_i^o = R_f = aN^\nu, \tag{8.28}$$

where ν is defined by (8.3).

So, expression (8.28) represents the partial case of equation (8.25) and is correct only for random walks equiprobable in directions of d-dimensional space. If this condition is not met, then equilibrium conformation of polymeric chain has an ellipsoid shape with semi-axes x_i^o determined by (8.25). Such conformations can be formed at deformation of Flory polymeric coils under the influence of external forces, for instance, in a solution, that leads to the double-ray refraction effect [3] in the saturated adsorption layer of polymeric molecules with functional groups, where forces of compression in absorption space lead to conformation of stretched out along the normality to ellipsoid surface rotation [10].

8.1.4. Properties of distribution in SAW statistics

To avoid the utilization of parameter a, i.e., the length of chain link in calculations, we will use a dimensionless variable

$$\lambda_i = x_i / R_f. \quad (8.29)$$

Moreover, we only consider the distribution corresponding to condition (8.26) of RW equiprobability over the axes of d-dimensional space. In this case the expressions (8.22) and (8.23) can be rewritten as follows

$$w(\lambda) = (1/c(\lambda)) \exp\{-(R_f/\sigma_o)^2 (1/\Pi_i \lambda_i + 1/2 \sum_i \lambda_i^2)\}, \quad (8.30)$$

$$c(\lambda) = I_0, \quad (8.31)$$

$$I_0 = \int \exp\{-(R_f/\sigma_o)^2 (1/\Pi_i \lambda_i + 1/2 \sum_i \lambda_i^2)\} d\lambda, \quad (8.32)$$

where $d\lambda = \Pi_i d\lambda_i$, $\lambda_{min} = ad^{1/2}/R_f$ and $\lambda_{max} = aN/R_f d^{1/2}$. Normalizing constants of distributions (8.22) and (8.30) are correlated with $c(\lambda) = c(x)/R_f^d$; therefore, we have a coupling

$$c(s) = (R_f/a)^d c(\lambda)/d^{d/2} = c(x)/a^d d^{d/2}. \quad (8.33)$$

In accordance to (8.27) and (8.28)

$$(R_f/\sigma_o)^2 = N^{(4-d)/(d+2)}. \quad (8.34)$$

Quantitative estimation of I_0 (and of the following ones) is performed using the Romberg algorithm in Mathcad software with accuracy parameter Tol=10^{-4}. I_0 values are listed in Table 8.1.

Values λ_i^{o}=1, i=1,d, relating to maximal distribution density, correspond to the equilibrium state of the polymer coil

$$w(\lambda^o) = (1/c(\lambda)) \exp\{-(R_f/\sigma_o)^2 (d+2)/2\} \quad (8.35)$$

The distribution density decreases at any deviation of values λ_i at λ^o=1: it this case not only at $\lambda_i \to \lambda_{max}$, but also at $\lambda_i \to \lambda_{min}$ it tends to zero. Consequently, the probability of chain end placement close to its origin is negligibly small. The latter makes a radical distinction between SAW statistics and Gaussian RW statistics.

Table 8.1. Characteristics of distribution density $w(\lambda)$

$N^{1/5}$	I_0	$I_0(L)$	$M(\lambda)$	$M(\lambda^2)$	λ_0	$D(\lambda)$	$\sigma(\lambda)$
2	7.72×10^{-3}	8.39×10^{-3}	1.139	1.487	1.219	0.191	0.437
4.0	1.91×10^{-5}	2.00×10^{-5}	1.073	1.246	1.116	0.094	0.307
6.0	7.11×10^{-8}	7.33×10^{-8}	1.048	1.163	1.078	0.065	0.255
8.0	3.13×10^{-10}	3.21×10^{-10}	1.035	1.125	1.060	0.053	0.230
10.0	1.52×10^{-12}	1.55×10^{-12}	1.026	1.099	1.048	0.045	0.217
12.0	7.79×10^{-15}	7.90×10^{-15}	1.024	1.083	1.041	0.034	0.185

Figure 8.2 represents the distribution density behavior for several variants of function $w(\lambda)$ for a three-dimensional space d=3 at $(R_f/\sigma_o)^2=N^{1/5}$, equal to 3 and 4, and change of one variable λ_z at fixed values of $\lambda_x=\lambda_y=1$, $\lambda_x=\lambda_y=0.8$ and $\lambda_x=\lambda_y=1.25$. The two latter variants clearly show what will happen at the deformation of a polymer coil: at compression along the x and y axes, the polymer chain stretches along the z axis, transforming to an oblong ellipsoid of revolution; in contrast, at compression along the z axis, polymer chain stretches along the x and y axes, transforming to oblate ellipsoid of revolution.

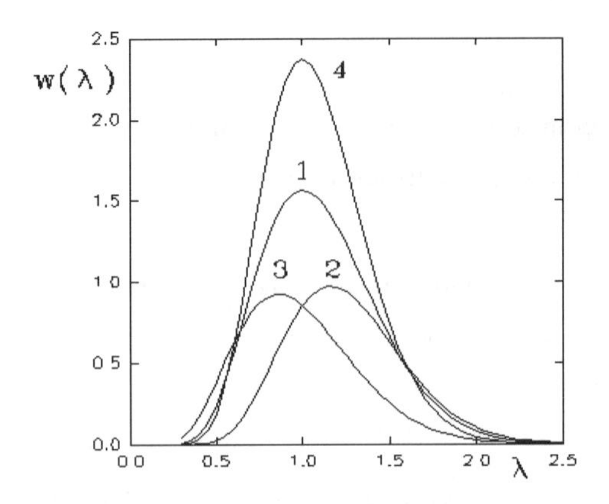

Figure 8.2. Behavior of distribution density $w(\lambda)$ at the variation of $\lambda=\lambda_z$ and fixed values of λ_x and λ_y for d=3. $N^{1/5}$=3 (1,2,3), $N^{1/5}$=4 (4), $\lambda_x=\lambda_y$=1 (1,4), $\lambda_x=\lambda_y$=0.8 (2), $\lambda_x=\lambda_y$=1.25 (8.3).

Starting moments of distribution $M(\lambda^k)$ (k=1,2) (8.30) are calculated (as mentioned before) using the correlation

$$M(\lambda^k)=\int \lambda_i^k\, w(\lambda)\mathrm{d}\lambda. \qquad (8.36)$$

For a Flory coil, values $M(\lambda^k)$ do not depend on the index i. Root-mean-square value $\lambda_o=(M(\lambda^2))^{1/2}$, dispersion $D(\lambda)=M(\lambda^2)-(M(\lambda))^2$ and standard deviation $\sigma(\lambda)=(D(\lambda))^{1/2}$ of distribution (8.30) are calculated *via* the values $M(\lambda)$ and $M(\lambda^2)$. A part of the data calculated is presented in Table 8.1.

Since $M(x)=R_fM(\lambda)$, $x_0=R_f\lambda_0$, where $M(x)=M(R)$ is a square radius and $x_0=R_0$ is the root-mean-square radius of a Flory coil, the data presented Table 8.1 show that $M(R)$, $R_0 > R_f$. This can be explained by the asymmetry of functions $w(\lambda)$ and $w(x)$, that relatively have the most probable values $\lambda^0=1$ and $x^0=R_f$. However, with increasing chain length $M(R)$ and R_0 are asymptotically approach R_f; thus, at $N > 10^4$ we can assume, that $M(R) \cong R_0 \cong R_f$. However, at $N < 10^4$ the difference between R_f and R_0 is significant and can be reflected in the experimental estimation v if it is based on the determination of the root-mean-square radius of the Flory coil.

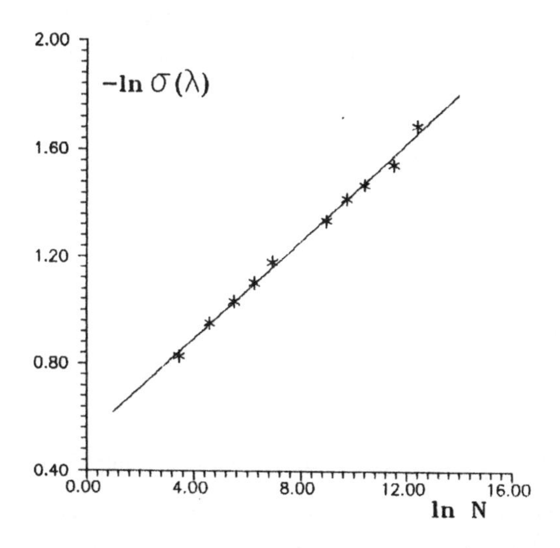

Figure 8.3. Dependence of the standard deviation $\sigma(\lambda)$ density distribution $\omega(\lambda)$ on chain length N on the basis of integration results calculated using (8.32) and (8.36) for d=3. Linear regression equation: $\ln(\sigma)=-0.528-0.091\ln N$.

As we can see from Table 8.1, dispersion $D(\lambda)$ and standard deviation $\sigma(\lambda)$ are decreased with the increasing chain length, thus confirming the self-organization of polymeric chain. This effect is relative, since dispersion $D(x)=R_f^2D(\lambda)$ and standard deviation $\sigma(x)=R_f\sigma(\lambda)$ distribution (8.22) are increased with the increasing N. Estimation of the dependence of $\sigma(\lambda)$ on N (Figure 8.3) allows to obtain the expressions $\sigma(\lambda) \approx e^{-1/2}N^{-0.1}$ and $\sigma(x) \approx e^{-1/2}\sigma_0$, respectively. Thus, although the absence of a self-avoiding random walk trajectory increases the polymeric coil radius at the expense of v increasing from 0.5 to 0.6, the standard deviation $\sigma(x)$ in SAW statistics is less than σ_0 in RW. It once more confirms the fact that limitation on self-avoiding RW trajectories leads to significant self-organization of polymeric chain, leading to a decrease in its conformation entropy.

8.1.5. The fatigue function

The fatigue function $g(N)$ is defined by (8.18) and expresses the part of the total quantity of SAW trajectories from the total quantity of RW trajectories. The latter quantity $(2d)^N$ is strictly defined for the lattice only, i.e. for the discrete space of wanderings. The total quantity of RW trajectories for the continuous space is unknown; however, the fatigue function can still be defined by (8.18) *via* the normalizing constant $c(x)$. Considering coupling (8.33), we can express the fatigue function *via* the normalizing constant $c(\lambda)$:

$$g(N)=\beta(R_f/a)^d\, d^{-d/2}\, c(\lambda). \tag{8.37}$$

For calculations one can use the quantitative values $c(\lambda)=I_0$ presented in Table 8.1. However, in the case of $d < 4$ space, an approximated analytical expression I_0 can be formulated, using the Laplace method [11]. With this aim we have rewritten (8.32) as a d-multipled Laplace integral

$$I_0=\int \exp\{(R_f/\sigma_0)^2 f(\lambda)\}\mathrm{d}\lambda, \tag{8.38}$$

where $f(\lambda)=-1/\Pi_i\lambda_i-1/2\ \Sigma_i\lambda_i^2$. In accordance with the Laplace method, the space of the point, in which $f(\lambda)$ is maximum, i.e., in this case $\lambda_i^0=1$, $i=1, d$, is the main part of the asymptote of integral (8.38). This assumption has been proven to be correct, as parameter $(R_f/\sigma_0)^2$ has a higher value. Therefore, condition $(R_f/\sigma_0)^2=N^{(4-d)/(d+2)} >> 1$ is required, which is met only at $d < 4$, taking into account that the main augend of Laplace' s asymptote can be written as follows [11]:

$$I_0\,(L)=(2\pi)^{d/2}(R_f/\sigma_0)^{-d}\exp\{(R_f/\sigma_0)^2 f(\lambda^0)\}/|\det(f''_{ij}(\lambda^0))|^{1/2}. \quad (8.39)$$

Here, $f(\lambda^0)=-(d+2)/2$, $(f''_{ij}(\lambda))$ is a matrix of secondary derivatives $f(\lambda)$ in point $\lambda_i^0=1$, $i=1,d$. Secondary derivatives are as follows: $f''_{ii}(\lambda)=-2/\lambda_i^2\Pi_i\lambda_i-1$ and $f''_{ij}(\lambda)=-1/\lambda_i\lambda_j\Pi_i\lambda_i$ at $i \neq j$. At $\lambda_i^0=1$, $i=1,d$ all diagonal derivatives $f''_{ii}=-3$ and all non-diagonal one's $f''_{i,j}=-1$. Thus, for instance, in a space of $d=3$ we obtain $|\det f''_{ij}(\lambda^0)|^{1/2}=2\sqrt{5}$, and

$$I_0(L)=(2/5)^{1/2}\pi^{3/2}N^{-3/10}\exp\{-5/2N^{1/5}\} \quad (8.40)$$

When $I_0(L)$ values in Table 8.1 are compared with I_0 values, we can see that they are similar, and for theoretical analysis we can use the approximation $c(\lambda) \approx I_0(L)$. Then, the fatigue function for $d=3$ can be written as follows:

$$g(N) \approx g_o N^{3/2}\exp\{-5/2\,N^{1/5}\} \quad (8.41)$$

where $g_o=\ (e2\sqrt{5}\ \ 3^{3/2})^{-1}$. Function (8.41) has a maximum at $N^{1/5}=3$. Its graphical representation is given in Figure 8.4.

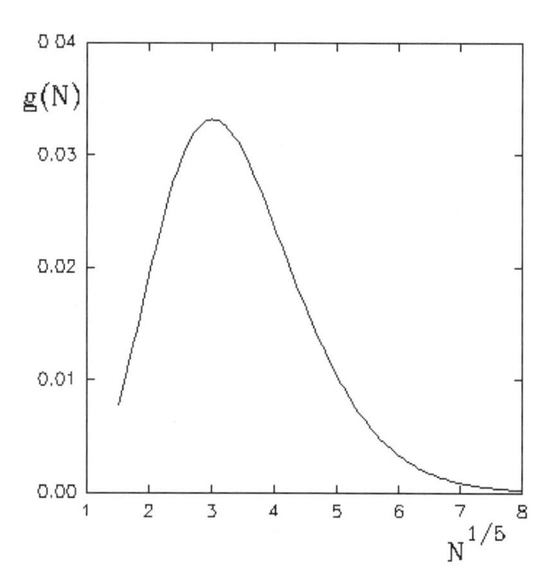

Figure 8.4. Dependence of fatigue function $g(N)$ (8.41) on chain length N.

8.1.6. Conformation of the polymeric chain

The configuration of the polymeric chain as one of the possible trajectories of SAW can be an unforeseen factor in the process. However, its conformation, i.e., its statistically average configuration, has a simple structure. In accordance with RW statistics, the conformation of polymeric chain can be a coil form, in which the volumic density of links is maximal in the centre and diffusionally decreases with increasing distance from the coil centre. SAW statistics hint at another structure of the conformative space of the polymeric chain.

Let us select a relative value, namely the density of chain links per cell $N/Z=(a/R_f)^d N/\Pi_i\lambda_I$, as a measure of chain links distribution in its conformative space. Then, the density of chain links distribution per cell in the conformative space of the chain will be determined by expression (8.42)

$$\rho_i(\lambda)=(a/R_f)^d N/\Pi_i\lambda_I\, w(\lambda) \qquad (8.42)$$

By integration of (8.42) we obtain the expression for average density:

$$<\rho(\lambda)>=(a/R_f)^d N/<\lambda^3> \qquad (8.43)$$

From the previous analysis follows, that we can assume $<\lambda^3> \approx (\lambda^0)^3=1$, so $<\rho(\lambda)> \approx N^{2(1-d)/(d+2)}$, the average settling of cells, decreases with increasing chain length. For a $d=3$ space its value decreases in accordance with the law $<\rho(\lambda)> \approx N^{-4/5}$.

With the aim of supervision we use the radial density $\rho(\lambda)$ which is determined from the $\rho_i(\lambda)$ under the condition that $\lambda_i=\lambda$, $i=1,d$. Then,

$$\rho(\lambda)=(a/R_f)^d N/\lambda^d c(\lambda)\exp\{-(R_f/\sigma_0)^2(1/\lambda^d + 3/2\lambda^2)\} \qquad (8.44)$$

Charts of functions $\rho(\lambda)$ and $\rho(\lambda)/<\rho(\lambda)>$ are presented in Figure 8.5a and 8.5b. As we can see, the main share of chain links is concentrated not in the nucleus of the conformative chain space, but in a spherical shell of thickness R_f (from $\lambda \approx 0.5$ to $\lambda \approx 1.5$) diffusely blurred both outside and inside the polymeric coil. Statistically, the nucleus of the polymeric coil is empty.

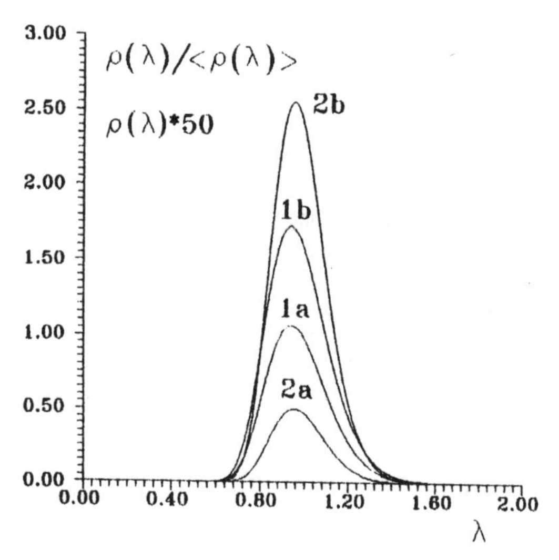

Figure 8.5. Radial density distribution $\rho(\lambda)$ (a) and its ratio to average density $\rho(\lambda)/<\rho(\lambda)>$ (b) of the number of links per cell of the conformative space of the polymeric chain. 1, $N^{1/5}=3$; 2, $N^{1/5}=4$.

Densities $\rho(\lambda)$ and $<\rho(\lambda)>$ decrease (see Figure 8.5a) with increasing N, but the ratio $\rho(\lambda)_{max}/<\rho(\lambda)>$ is increased (see Figure 8.5b). The, with increasing chain length the shell begins to form the most important factor determining the structure of the conformative space of the polymeric chain. The effect of self-organization in SAW statistics is also shown in this case.

8.2. The stretched exponential law for monomolecular chain termination

In accordance with the kinetic scheme (7.1), monomolecular chain termination represents the act of chain propagation as a result of which the active radical R_m transits into a non-active R_z radical. This state can be presented as physically blocking the active center of radical by its own and someone else's links and, therefore, such a radical is un-accessible for interaction with monomer molecules. From this follows, that self-burial of the active radical represents the act of chain propagation leading to the conformation trap. *Via* such understanding of monomolecular termination the kinetic scheme (7.1) can be represented as [12]:

$$\mathrm{R_m + M \xrightarrow{k_p} R_m}$$

$$\mathrm{R_m + L \xrightarrow{k_p} R_z} \qquad (8.45)$$

where L is the trap.

8.2.1. Deduction of the stretched exponential law

We can see from this scheme that the chain termination is controlled by its propagation rate, but the latest act of chain propagation leads to self-burial of the active radical in the trap.

According to (7.1) and (8.45) the rate of chain propagation for one chain is equal to $k_p\,[M_m]=k_p\,\rho_m$ and the termination rate can be written in two forms: $k_t\,[M_m]=k_t\,\rho_m$ or $k_p\,[L]=k_p\,\rho_L$, thus

$$\beta=k_t\,\rho_m=k_p\,\rho_L \qquad (8.46)$$

Molar-volumetric concentrations of monomer $[M_m]$ and traps [L] are denoted here as corresponding densities ρ_m and ρ_L for convenience.

The characteristic chain propagation time τ_p and chain termination time τ_t can be written as

$$\tau_p=(k_p\,\rho_m)^{-1},\ \tau_t=(k_t\,\rho_m)^{-1}=(k_p\,\rho_L)^{-1} \qquad (8.47)$$

The growing macroradical represents the fractal and the most probable distance between the beginning and the end of the chain (the so-called radius of the Flory ball r_f) connected with its length N by scaling (8.28):

$$r_f \sim N^{\nu_f} = N^{1/d_f}\,, \qquad (8.48)$$

where the Flory index ν_f is determined by the expression

$$\nu_f=3/(d+2), \qquad (8.49)$$

where d is the dimension of Euclidean space. At this, $d_f=(\nu_f^{-1})$ is the dimension of fractal of chain.

Both expressions (8.48) and (8.49) follow from strict statistics of self-avoiding random walks [1].

Let us adopt, that the set of traps is also a fractal and, as follows from this:

$$r_{\mathrm{f}} \sim L^{1/d_{\mathrm{L}}} \tag{8.50}$$

Here L is the number of traps in the sphere of radius r_{f}, and d_{L} is the dimension of fractal of traps.

The relation between the number of traps in the limits of the Flory ball and length of growing macroradical follows from the comparison of equations (8.48) and (8.50).

$$L \sim N^{d_{\mathrm{L}}/d_{\mathrm{f}}} \tag{8.51}$$

Let us introduce the density of traps as $\rho_{\mathrm{L}}=L/r_{\mathrm{f}}^{d}$. Then,

$$\rho_{\mathrm{L}} \sim N^{(d_{\mathrm{L}}-d)/d_{\mathrm{f}}} \tag{8.52}$$

From equation (8.47), for characteristic times follows $\tau_{\mathrm{t}}=\tau_{\mathrm{p}}\ \rho_{\mathrm{m}}/\rho_{\mathrm{L}}$ or, taking into account equation (8.52) we obtain

$$\tau_{\mathrm{t}} \sim \tau_{\mathrm{p}}\rho_{\mathrm{m}}N^{(d-d_{\mathrm{L}})/d_{\mathrm{f}}} \tag{8.53}$$

Since ρ_{L}, according to equation (8.52), represents the average concentration of traps in the medium of the Flory ball, expression (8.53) also determines the spectrum of average characteristic times of monomolecular chain termination for macroradicals of different length N.

Let us introduce the differential radial density of traps $\rho_{\mathrm{L},diff}$ in the spherical layer $r_{\mathrm{f}}+\mathrm{d}r_{\mathrm{f}}$ as ratio of traps number increment dL to increment of volume $\mathrm{d}V=\mathrm{d}r_{\mathrm{f}}^{d}$ of the Flory ball:

$$\rho_{\mathrm{L},dif}=dL/\mathrm{d}r_{\mathrm{f}}^{d} \tag{8.54}$$

At sufficiently great $N >> 1$ it can be taken as continuous variable. Then, from equation (8.48) and equation (8.51) follows $\mathrm{d}r_{\mathrm{f}}^{d} \sim N^{(d-d_{\mathrm{f}})/d_{\mathrm{f}}}\mathrm{d}N$, $\mathrm{d}L \sim N^{(d_{\mathrm{L}}-d_{\mathrm{f}})/d_{\mathrm{f}}}\mathrm{d}N$, and we have

$$\rho_{\mathrm{L,dif}} \sim N^{(d_{\mathrm{L}}-d)/d_{\mathrm{f}}} \qquad (8.55)$$

Expressions (8.52) and (8.55) reflect the main property of fractal structures, namely scale invariance or self-similitude.

Therefore, in accordance with equation (8.55), differential characteristic times of monomolecular chain termination for macroradicals of length N are subordinated to the same dependence according to equation (8.53). This means that spectra of average and differential characteristic times of monomolecular chain termination can be characterized not only by differences in scales, but also by a range of similarities in functional dependences from N according to equation (8.55). Therefore, we cannot discriminate between average and differential values τ_{t}.

There is another important property of the self-similitude: if the chain with length N=m forms a fractal with radius $r_{\mathrm{f}m} \sim m^{1/d_{\mathrm{f}}}$ and the chain N=$m + n$ is the fractal of the radius $r_{\mathrm{f}(m+n)} \sim (m+n)^{1/d_{\mathrm{f}}}$, then the newly constructed section of the chain with the length N=n represents also the fractal with the most probable distance between the beginning N=m and the end of the chain N=$m + n$, depending only on the length N=n of the constructed chain $r_{\mathrm{f}n} \sim n^{1/d_{\mathrm{f}}}$.

The starting state of macroradicals is different and unknown at the moment t=0 of the dark period of time of the postpolymerization. However, the property of scale invariance or self-similitude of fractal structures allows to consider the parameter N in previous expressions not only as general length of a radical at time t, but also as its increment or constructed length of the chain at time t of the dark period, respectively, to its starting state at t=0.

These reflections allow to connect the true chain length of the macroradical with the time of its propagation:

$$t=\tau_{\mathrm{p}} N \qquad (8.56)$$

Substituting equation (8.56) into (8.52), we obtain

$$\tau_{\mathrm{t}} \sim \tau_{\mathrm{p}} \rho_{\mathrm{m}} (t/\tau_{\mathrm{p}})^{(d-d_{\mathrm{L}})/d_{\mathrm{f}}} \qquad (8.57)$$

If we rewrite equation (8.57) in the form

$$\tau_t \sim t^{1/d_t} \tag{8.58}$$

then it can be affirmed that a set of characteristic times τ_t of the monomolecular chain termination represents the fractal, the scale invariant of which is time, and the dimension of a quantity is equal to

$$d_t = d_f/(d - d_L) \tag{8.59}$$

Taking into account the self-similitude, the fractal set of characteristic times τ_t holds true in the form of equation (8.57) for all macroradicals, independent of their starting state.

Therefore, the relaxation function for macroradical *i*, taking into account its starting state, can be written as follows:

$$\varphi_i(t) = g_i \exp\{-t/\tau_t\} \tag{8.60}$$

By substituting into equation (8.60) the expression for τ_t, according to equation (8.57) we obtain:

$$\varphi_i(t) \sim g_i \exp\{-t/\tau_p\, \rho_m{}^{\xi}\}^{1/\xi}, \tag{8.61}$$

where

$$\xi = d_f/(d_f + d_L - d) \tag{8.62}$$

Let us introduce in the index of the exponential curve (see equation 8.61) the constant value ρ_0, with the aim of eliminating the "~" sign. This constant value should be characterized with a dimension that is antipodal to ρ_m:

$$\varphi_i(t) = g_i \exp\{-t/\tau_p\, (\rho_m/\rho_0)^{\xi}\}^{1/\xi} \tag{8.63}$$

Taking the summation with respect to all *i* radicals, finally we will obtained the stretched exponential law in the form

$$\psi(t) = g_0 \exp\{-t/\tau_p\, (\rho_m/\rho_0)^{\xi}\}^{1/\xi} \tag{8.64}$$

where $g_0 = \sum_i g_i$ takes into account general starting state of macroradicals at $t=0$.

Using the replacement $\tau_p=(k_p\ \rho_m)^{-1}$ we obtained the second form of the stretched exponential law

$$\psi(t)=g_0 \exp\{-k_p\ \rho_m\ (\rho_m/\rho_0)^{-\xi}\ t\}^{1/\xi} \qquad (8.65)$$

As a result of comparison of equation (7.23) with equation (8.65) we found

$$\gamma=1/\xi=(d_f + d_L-d)/d_f \qquad (8.66)$$

$$\beta=k_p\ \rho_m\ (\rho_m/\rho_0)^{-\xi} \qquad (8.67)$$

Correspondently, using equation (8.46) we obtained

$$k_t=k_p\ (\rho_m/\rho_0)^{-\xi} \qquad (8.68)$$

$$\rho_L=\rho_m\ (\rho_m/\rho_0)^{-\xi} \qquad (8.69)$$

$$\overline{N}=(\rho_m/\rho_0)^{\xi}, \qquad (8.70)$$

where $\overline{N}$ is the average length of constructed chains in the dark period relative to starting states of macroradicals at t=0.

8.2.2. Conclusions

In summary, the stretched exponential law and scaling form $\beta \sim \rho_m^{1-\xi}$ and $k_t \sim \rho_m^{-\xi}$ are interconnected and this completely agrees with experimental data. At this time, if we use the experimental value ξ=1,86, then value γ should be equal to 0.54; this value is near to the value γ=0.6. However, scaling dependencies $\beta \sim \rho_m^{1-\xi}$ and $k_t \sim \rho_m^{-\xi}$ for the presented range of dimethacrylates can be appreciably distorted in the case that constants of chain propagation k_p and constant ρ_0 also are a function of the dimethacrylate nature in equations (8.67) and (8.68). Therefore, the most truthful value is γ=0.6, according to which ξ=1.67; this value is within the standard deviation of the experimental value ξ= 1.86.

$k_p\ \rho_0^{\xi}$ values, presented in Table 7.7, have been calculated on the basis of data of Table 7.7, adopting that the value ξ=1.67 is the most truthful. As we can see, values $k_p\ \rho_0^{\xi}$ are similar for all dimethacrylates, except DMEG; this allows to assume similar chain propagation constants k_p and constant ρ_0 for these dimethacrylates. The significantly lower values $k_p\ \rho_0^{\xi}$ for DMEG are in good agreement with the values found for DMEG activation energy of chain termination and chain propagation $E \cong 11$ kJ/mol in comparison with others dimethacrylates, for which $E \cong 7$ kJ/mol. Thus, DMEG is characterized by a lower k_p value in comparison with other dimethacrylates. Studying the stationary kinetics of polymerization of the presented dimethacrylates qualitatively confirms this conclusion.

Let us assume that $k_p\ \rho_0^{\xi} \cong 30$ and $k_p \cong 0.1$ m^3/mol×s [13] for a rough estimate of constant ρ_0. Then we will obtain $\rho_0 \cong 300$ mol/m^3. The physical sense of this constant is a consequence of expression (8.69): at the same time condition $\rho_L=\rho_0$ holds at $\rho_m=\rho_0$; thus, ρ_0 is a density or molar-volumetric concentration of traps in the reactive zone. Expression (8.70) agrees with this fact: probabilities of chain propagation and chain termination are equal and the average chain length is equal to 1 in this case. However, it is unknown why the value ρ_0 is determined and if this value remains constant in the investigated range of dimethacrylates.

If we again adopt that $1/\xi=\gamma=0.6$, then the value $\xi=1.67$ accurately coincides with the dimensions of the fractal of the chain $d_f=(d + 2)/3=5/3$ and the value $\gamma=\nu_f$ coincides with the Flory index. In this case the dimensions of fractal of traps d_L and characteristic times d_t, according to equations (8.62) and (8.59), is equal to $d_L=2.33$ and $d_t=2.5$. At this point, it should be noted that the property of fractality for characteristic times set τ_t of macroradical relaxation is completely determined by the fractality of traps, i.e., by dimension d_L. In the case that the set of traps is not a fractal (for example, at the uniform distribution of traps in the reactive zone when the dimension of the fractal of traps formally coincides with the dimension of the Euclidean space $d_L=d$), then, in accordance with equation (8.59) the set of characteristic times τ_t would be characterized by a fractal dimension equal to infinity and τ_t would not depend on chain length and time of chain propagation, and would be presented by a constant value $\tau_t \sim \tau_p\ \rho_m$. Under this variant we have $\xi=1$, $\gamma=1$ and stretched exponential law transforms into simple exponential law.

On the other hand, if traps would be created by only one chain at the expense of self-avoidance effects, then we can expect that the dimension of fractals of such "own" traps coincides with the dimension of the fractal of the chain: $d_L=d_f$. Under these circumstances, according to equation (8.62) $\xi=5$ and $\gamma=0.2$ and we would obtain the law of greatly stretched exponential. Evidently, such a variant would be true for a single separate chain.

Thus, the ratio $d_f < d_L < d$ indicates that in the reactive zone of interface layer there are both 'strange" traps created by external for presented radical polymeric chains and own traps, but those which are created by propagation of the chain of the macroradical under investigation in the medium of other polymeric chains. This means that the dimension of fractal of traps for macroradicals results from the interaction of the fractal of the chain with the dimension d_f with the fractal of 'strange" traps with the dimension $d=3$; this leads to ratio $d_f < d_L < d$.

If we would adopt as a reference point $\xi=1/\gamma=1.67=d_f$, then, according to equation (8.62), we would obtain

$$d_L = \frac{(d_f + d)}{2} = \frac{2d+1}{3} \tag{8.71}$$

The formula (8.71) impresses by its universality; it gives the value $d_L=7/3$ at $d=3$ which agrees with experimental results and gives the physically expected value $d_L=1$ for 1-D space ($d=1$). Then, we have $\xi=1$, $\gamma=1$ at $d=1$ and for the 1-D space the stretched exponential law transforms into the simple exponential law. This is fully according to expectation, since the monomolecular chain of macroradical in the 1-D space cannot create its own traps and the value $\xi=1$ certifies the availability of only 'strange" traps with the fractal dimension $d_L=d=1$, which is equal to the dimension of the Euclidean space.

However, despite all the above, in order to prove the universality of formula (8.71) a more scrupulous analysis of interaction between the fractal of the chain and fractal of the 'strange" traps is needed.

REFERENCES

[1] Medvedevskikh Yu. *Condensed Matter Physics*, **2001**, 4 (2, 26), p. 209

[2] Kuhn W.; *Koll. Zs.,* **1934**, 68, 2–20.

[3] Volkenshtein M.V. *Configurational Statistics of Polymer chains.* Moscow-Leningrad, USSR Academy of Sciences, **1959,** 466 p. (in Russian).

[4] Flory P. *Statistical Mechanics of Chain Molecules.* Moscow, Mir, **1971**, 440 p. (in Russian).

[5] Kuhn W.; *Koll. Zs,* **1936**, 76. 258–272.

[6] Pietronero L.; *Phys. Rev. Lett.*, **1985**, 55 (19), 2025–2027.

[7] Peliti L. *Fractals Physics*.–In: *Proc. of the sixth Intern. Symp. on Fractals in Physics. ICTP, Trieste*, Italy, July 9–12, **1985,** 91 p.

[8] De Gennes P. G. *Scaling Concepts in Polymer Physics*. Ithaca, Cornell Univ. Press, **1979,** 300 p.

[9] Gould H., Tobochnik J. *An Introduction to Computer Simulation Methods. Applications to Physical Systems*. Part 2, Addison-Wesley, Reading, MA, **1988** (Russian translation: Moscow, Mir), **1990**, 270 p.

[10] Jones R. A. L., Norton L. J., Scull K. R.; *Macromolecules*, **1992**, 25 (9), 2359–2369.

[11] Fedoryuk M. V. *Saddle–Point Technique*. Moscow, Nauka, **1977**, 254 p. (in Russian).

[12] Kytsya A. *Monomolecular chain termination in the postpolymerization kinetic of bifunctional monomers*, Ph. D *Thesis in chemical sciences*, Lviv, **2004**, 125 p.

[13] *Encyclopedia of polymers,* Edited by V. A. Kargin**,** Moscow, Myr, **1972**, v. 2, p. 185.

Appendix

A new method for surfaces gluing based on the effect of the postpolymerization

Glues based on synthetic polymers are applied in different fields of modern technology. At this time, an application of glue in most cases decreases the weight of the constructions, keeps the products surface smooth and creates hermetic joints. Furthermore, when metals are glued together there is no need to prepare apertures for bolts and the risk of corrosion is decreased [1].

During the last years a new application of synthetic glues has been developed, the so-called production of "non-woven" materials, in which the structure is formed by gluing of individual fibers. This method is used in the production of the clothes and preparation of electro- and heat-isolation materials by a simple and cheap technology. In mass production, a preparation technique using glue is simpler and economically advantageous compared to other possible techniques.

These days special attention is given to glue compositions on the basis of epoxy oligomers. They are characterized by a complex of unique properties, including high adhesion to most materials, chemical stability and good electro-isolation properties [2]. However, such compositions are also characterized by a range of drawbacks, the most important of which is high curing temperature (50–100°C). In most fields of technology such a temperature is too high. This problem can be solved by applying photopolymerized compositions [3–5].

There are a number of methods and recipes for gluing transparent materials, including glass and mirrors [6, 7], optical discs [8–11] and optical cables [12]. However, these methods are not effective for gluing non-transparent materials (including metals).

The higher adhesive ability of glycidyl methacrylate when glued to a metal and glass surface has been shown in the study of photoinitiated polymerization kinetics. This result initiated a search for new glue compositions, recipes and techniques. As a possible technique gluing of non-transparent surfaces based on the postpolymerization effect (so-called dark process of the polymerization) has been proposed [13, 14].

Investigations of the postpolymerization effect of dimethacrylates and monomethacrylates show, that the ΔP conversion increment *via* the postpolymerization process can achieve high values, namely $\Delta P \approx 0.4$ for dimethacrylates and $\Delta P \approx 0.25$ for monomethacrylates. It was the basis for the application of postpolymerization as a practically new method of gluing different surfaces, including the non-transparent UV-illuminated ones.

Despite the fact that the conversion increment ΔP *via* dimethacrylate postpolymerization can achieve high values in comparison to monomethacrylate postpolymerization, dimethacrylates remain in the liquid monomeric phase up to the high conversion stage by 3-D polymerization. The polymerization system in such a case consists of a polymer with practically limited conversion and a liquid monomeric phase, in which the solubility of a polymer is not high. That is why two films which are polymerized under UV-illumination on two different surfaces at high conversion degrees of the light period show a low adhesion. At the same time, at low conversions of the light period they form a glue layer of low strength, despite the postpolymerization effect, as confirmed by experimental results.

Results of investigations the stationary and non-stationary kinetics of monomers with one functional group showed that at conversions $P \geq 0.5$ two polymeric phases are coexisted, namely: first is a monomer-polymeric solution with conversion $P \cong 0.5$ saturated by a polymer, and the second one is solid solution of a monomer in the polymer with the conversion $P \cong 0.8$. That is why two films which are polymerized under UV-illumination on different surfaces, even at high general conversion, are easily conjuncted when compressed and after that the layer formed *via* the postpolymerization process becomes stronger.

Glycidyl methacrylate was the basis of a photocomposition; IRGACURE-651 was used as photoinitiator at different concentrations.

The gluing technique was as follows. A photocomposition layer 1 (see Figure A1), 0.5 mm thick, was washed between metal plates 2 and 3 (15 mm gap between the plates), previously treated with an abrasive material and o-phosphoric acid and after that washed with water and acetone. Metallic surfaces thus prepared were UV-illuminated at E_0=37.4 W/m^2 in the aqueous phase for a period of time which has been chosen empirically such that the starting conversion of the dark period was equal to $P \approx 0.5$–0.7. After the UV-illumination was stopped, metallic plates 2 and 3 were conjuncted in the area of a photocomposition layer equal to 1.5 sm × 1.5 sm=2.25 sm^2 for 10–15 minutes.

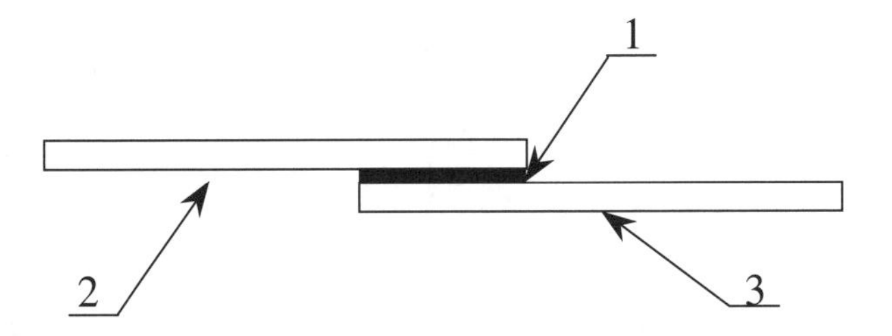

Figure A1. Single conjunction for shift strength testing. 1 is a layer of a photocomposition; 2 and 3 are metallic plates.

Strength characteristics of the adhesion conjunctions were determined in accordance with the American standard ASTM D1002. Part of the data obtained (average of 5–6 investigations) is represented in Table A1 for three variants.

Variant 1. GMC+1.5% by mass of the photoinitiator. Plates were prepared by abrasive treatment of their surface and purification with acetone.

Variant 2. GMC+1.5 % by mass of the photoinitiator+10 % by mass Al_2O_3 (powder). Plates were prepared as for Variant 1.

Variant 3.GMC+1.5 % by mass of the photoinitiator. Plates were prepared by abrasive treatment of their surface, treatment with a mordant by o-phosphoric acid and after that purification with acetone.

The results show that the characteristics of the glues used are the same as the most of industrial glues of this type and, in comparison to some of them, the glues used here are characterized by a range of advantages. In order to compare we present the data on the shift strength for the best among this type of acrylic glue (cool curing), including the glues 'Cyanakryl", based on cyanacrylate, and also 'Istmen -910", based on methyl-2-cyanacrylate, polymerization of which proceeds easily at room temperature and without heating.

Table A1. Results of investigations on glues as described above

Variant	Gluing under argon				Gluing under water			
	$\overline{F}_N$ (kN)	$\overline{\sigma}$ (MPa)	$\overline{\delta}$ (%)	σ_{max} (MPa)	$\overline{F}_N$ (kN)	$\overline{\sigma}$ (MPa)	$\overline{\delta}$ (%)	σ_{max} (MPa)
1	1.70	7.56	2.4	9.54	0.82	3.20	1.3	3.96
2	1.67	7.43	4.6	10.30	0.85	3.78	5.7	4.53
3	1.87	8.31	6.7	10.32	1.12	4.98	5.8	5.95

$\overline{F}_N$, average explosive force; $\overline{\sigma}$, average shift strength; $\overline{\delta}$, average relative deformation; σ_{max}, maximal achieved shift strength.

Maximal strength characteristics of the adhesive conjunction have been obtained at the treatment of metallic surfaces with *o*-phosphoric acid following by washing with acetone (see Variant 3). A possible explanation for this fact could be that *o*-phosphoric acid is an initiator of glycidylmethacrylate polymerization. Therefore, adsorbed *o*-phosphoric acid retained on the surface of metal after washing by acetone initiates the additional polymerization process on the surface with the formation of a strong nanolayer of polymer.

❖ Addition of aluminum powder was not effective; this probably is related to the increasing optical density of the photopolymerization composition layer as a results of the lower rate the polymerization process near the metallic surfaces as compared to the polymerization in the volume. As a result, the monolayer near a surface is weaker and the explosion proceeds in the place of a "metal –polymer" contact.

❖ As a rule, strength characteristics of the adhesive conjunctions obtained under water were significantly lower than those obtained under argon. This means that for gluing under water, the development of new, probably more hydrophobic, starting photocompositions is needed.

Generally, the proposed method for the manufacture of adhesive conjunctions based on the postpolymerization effect gave good results. Their strength characteristics are comparable to those of other well-known glued conjunctions. The method proposed here can be applied in different fields of industry after the development of restricted technological parameters and approaches.

REFERENCES

[1] *Adhesiya. Kleji, cements, pripoji.* Collection of papers edited by N. Debrojn and R. Guvinka; M.: Izd. inostr. lit., **1954.** p. 14.

[2] *Epoxide oligomers and gluing compositions.* Edited by Grekov A.; K.: Naukova dumka, Moscow, **1990.** 200 p.

[3] US Patent 6294239. Ultraviolet-curable adhesive composition (**2001**).

[4]US Patent 6284185. Ultraviolet-curable adhesive composition for bonding opaque substrates (**2001**).

[5] US Patent 5645973. Process for adjusting the sensitivity to radiation of photopolymerizable compositions (**1997**).

[6] US Patent 6216759. Method and apparatus for gluing frames, mirrors, glass sheets and the like to panels, in particular furniture panels (**2001**).

[7] US Patent 6315626. Adhesively bonded pressure-resistant glass bodies (**2001**).

[8] US Patent 6326414. Ultraviolet-curable adhesive for bonding optical disks (**2001**).

[9] US Patent 6334927. Method and apparatus for gluing together disc elements (**2002**).

[10] US Patent 6179031. Process for the adhesion of a flat plastic substrate in the form of a circular disk to a like second substrate for a digital video disc and apparatus for implementation of the process (**2001**).

[11] US Patent 6335382. Ultraviolet-curable adhesive for bonding optical disks (**2002**).

[12] US Patent 5278358. Under-sea repeter access cable (**1994**).

[13] US Patent 55871 A. Yu. Medvedevskikh, A. Zaichenko, A. Kytsya, A. Bratus, N. Mitina, L. Bazylyak, W. Brostow, The method of adhesive conjunction of the materials surfaces (I) // (**2003**).

[14] US Patent 55872 A. Yu. Medvedevskikh, A. Zaichenko, A. Kytsya, A. Bratus, N. Mitina, L. Bazylyak, W. Brostow, The method of adhesive conjunction of the materials surfaces (II) // (**2003**).

SUBJECT INDEX

E

F

G

H

I

J

K

L

M

T

U

V

W

X

Y

Z